L.C. Piccinini
G. Stampacchia
G. Vidossich

Applied Mathematical Sciences 39

Ordinary Differential Equations in R^n

Problems and Methods

D1155030

Springer-Verlag
New York Berlin Heidelberg Tokyo

Applied Mathematical Sciences

EDITORS

Fritz John
Courant Institute of Mathematical Sciences
New York University
New York, NY 10012

J.E. Marsden
Department of Mathematics
University of California
Berkeley, CA 94720

Lawrence Sirovich
Division of Applied Mathematics
Brown University
Providence, RI 02912

ADVISORS

H. Cabannes University of Paris-VI

M. Ghil New York University

J.K. Hale Brown University

J. Keller Stanford University

J.P. LaSalle Brown University

G.B. Whitham California Inst. of Technology

EDITORIAL STATEMENT

The mathematization of all sciences, the fading of traditional scientific boundaries, the impact of computer technology, the growing importance of mathematical-computer modelling and the necessity of scientific planning all create the need both in education and research for books that are introductory to and abreast of these developments.

The purpose of this series is to provide such books, suitable for the user of mathematics, the mathematician interested in applications, and the student scientist. In particular, this series will provide an outlet for material less formally presented and more anticipatory of needs than finished texts or monographs, yet of immediate interest because of the novelty of its treatment of an application or of mathematics being applied or lying close to applications.

The aim of the series is, through rapid publication in an attractive but inexpensive format, to make material of current interest widely accessible. This implies the absence of excessive generality and abstraction, and unrealistic idealization, but with quality of exposition as a goal.

Many of the books will originate out of and will stimulate the development of new undergraduate and graduate courses in the applications of mathematics. Some of the books will present introductions to new areas of research, new applications and act as signposts for new directions in the mathematical sciences. This series will often serve as an intermediate stage of the publication of material which, through exposure here, will be further developed and refined. These will appear in conventional format and in hard cover.

MANUSCRIPTS

The Editors welcome all inquiries regarding the submission of manuscripts for the series. Final preparation of all manuscripts will take place in the editorial offices of the series in the Division of Applied Mathematics, Brown University, Providence, Rhode Island.

SPRINGER-VERLAG NEW YORK INC., 175 Fifth Avenue, New York, N. Y. 10010

Printed in U.S.A.

5-23

Applied Mathematical Sciences | Volume 39

Applied Mathematical Sciences

1. John: **Partial Differential Equations**, 4th ed. (cloth)
2. Sirovich: **Techniques of Asymptotic Analysis.**
3. Hale: **Theory of Functional Differential Equations**, 2nd ed. (cloth)
4. Percus: **Combinatorial Methods.**
5. von Mises/Friedrichs: **Fluid Dynamics.**
6. Freiberger/Grenander: **A Short Course in Computational Probability and Statistics.**
7. Pipkin: **Lectures on Viscoelasticity Theory.**
8. Giacaglia: **Perturbation Methods in Non-Linear Systems.**
9. Friedrichs: **Spectral Theory of Operators in Hilbert Space.**
10. Stroud: **Numerical Quadrature and Solution of Ordinary Differential Equations.**
11. Wolovich: **Linear Multivariable Systems.**
12. Berkovitz: **Optimal Control Theory.**
13. Bluman/Cole: **Similarity Methods for Differential Equations.**
14. Yoshizawa: **Stability Theory and the Existence of Periodic Solutions and Almost Periodic Solutions.**
15. Braun: **Differential Equations and Their Applications**, 3rd ed. (cloth)
16. Lefschetz: **Applications of Algebraic Topology.**
17. Collatz/Wetterling: **Optimization Problems.**
18. Grenander: **Pattern Synthesis: Lectures in Pattern Theory, Vol I.**
19. Marsden/McCracken: **The Hopf Bifurcation and its Applications.**
20. Driver: **Ordinary and Delay Differential Equations.**
21. Courant/Friedrichs: **Supersonic Flow and Shock Waves.** (cloth)
22. Rouche/Habets/Laloy: **Stability Theory by Liapunov's Direct Method.**
23. Lamperti: **Stochastic Processes: A Survey of the Mathematical Theory.**
24. Grenander: **Pattern Analysis: Lectures in Pattern Theory, Vol. II.**
25. Davies: **Integral Transforms and Their Applications.**
26. Kushner/Clark: **Stochastic Approximation Methods for Constrained and Unconstrained Systems.**
27. de Boor: **A Practical Guide to Splines.**
28. Keilson: **Markov Chain Models—Rarity and Exponentiality.**
29. de Veubeke: **A Course in Elasticity.**
30. Sniatycki: **Geometric Quantization and Quantum Mechanics.**
31. Reid: **Sturmian Theory for Ordinary Differential Equations.**
32. Meis/Markowitz: **Numerical Solution of Partial Differential Equations.**
33. Grenander: **Regular Structures: Lectures in Pattern Theory, Vol. III.**
34. Kevorkian/Cole: **Perturbation Methods in Applied Mathematics.** (cloth)
35. Carr: **Applications of Centre Manifold Theory.**

(continued)

L.C. Piccinini
G. Stampacchia
G. Vidossich

Ordinary Differential Equations in $\mathbf{R}^n$

Problems and Methods

Translated by A. LoBello

With 38 Illustrations

Springer-Verlag
New York Berlin Heidelberg Tokyo

L.C. Piccinini
Istituto di Matematica
Informatica e Sistemistica
Università di Udine
33100 Udine
Italy

G. Vidossich
Scuola Internazionale Superiore
di Studi Avanzati
Strada Costiera 11
34014 Trieste
Italy

G. Stampacchia
Formerly of
Scuola Normale Superiore
56100 Pisa
Italy

Translator
A. LoBello
Allegheny College
Meadville, PA 16335
U.S.A.

AMS Subject Classification: 34-01

Library of Congress Cataloging in Publication Data
Piccinini, Livio Clemente.
Ordinary differential equations in R^n.
(Applied mathematical sciences; v. 39)
Translation of: Equazioni differenziali
ordinarie in R^n.
Includes bibliographical references and index.
1. Differential equations. I. Stampacchia,
Guido. II. Vidossich, Giovanni. III. Title.
IV. Series: Applied mathematical sciences
(Springer-Verlag New York Inc.) v. 39.
QA1.A647.vol 39 [QA372] 510s [515.3′52] 82-5713
AACR2

Original Italian edition: *Equazioni Differenziali Ordinarie in R^n (problemi e metodi),* © Liguori editore Srl 1978.

Printed and bound by R.R. Donnelley & Sons, Harrisonburg, Virginia.
Printed in the United States of America.

9 8 7 6 5 4 3 2 1

ISBN 0-387-90723-8 Springer-Verlag New York Berlin Heidelberg Tokyo
ISBN 3-540-90723-8 Springer-Verlag Berlin Heidelberg New York Tokyo

Preface

During the fifties, one of the authors, G. Stampacchia, had prepared some lecture notes on ordinary differential equations for a course in advanced analysis. These remained for a long time unused because he was no longer very interested in the study of such equations. We now see, though, that numerous applications to biology, chemistry, economics, and medicine have recently been added to the traditional ones in mechanics; also, there has been in these last years a reemergence of interest in nonlinear analysis, of which the theory of ordinary differential euqations is one of the principal sources of methods and problems. Hence the idea to write a book. Our text, based on the old notes and experience gained in many courses, seminars, and conferences, both in Italy and abroad, aims to give a simple and rapid introduction to the various themes, problems, and methods of the theory of ordinary differential equations.

The book has been conceived in such a way so that even the reader who has merely had a first course in calculus may be able to study it and to obtain a panoramic vision of the theory. We have tried to avoid abstract formalism, preferring instead a discursive style, which should make the book accessible to engineers and physicists without specific preparation in modern mathematics. For students of mathematics, it provides motivation for the subject of more advanced analysis courses.

At the end of almost all the sections we have proposed exercises with the intent either of illustrating the results by examining them from other points of view or of giving possible applications or complementary material. The biographical notes were conceived as a guide for further studies and are not meant to establish priorities.

G.S., L.C.P., G.V.

Dedication

This book was born of an idea of Guido Stampacchia, fruit of his renewed interest in ordinary differential equations. His untimely death did not, however, permit him to see the work through to press.

To Guido Stampacchia, who was our teacher, we affectionately dedicate this work.

L.C.P., G.V.

Contents

		Page
CHAPTER I	EXISTENCE AND UNIQUENESS FOR THE INITIAL VALUE PROBLEM UNDER THE HYPOTHESIS OF LIPSCHITZ	1
	1. General Results	1
	1.1 Definitions	1
	1.2 Geometrical Interpretation	3
	1.3 Functions Satisfying a Lipschitz Condition	4
	1.4 Existence Theorem	5
	1.5 Uniqueness Theorem	10
	1.6 Continuous Dependence on Initial Conditions and Parameters	12
	1.7 Interval of Definition and Extension of Solutions	13
	1.8 Gronwall's Lemma	17
	1.9 Application of Gronwall's Lemma to the Cauchy Problem	19
	2. Qualitative Properties of Solutions	23
	2.1 Differentiability of Solutions	23
	2.2 Analyticity of the Solutions	24
	3. Solutions as Functions of the Initial Data	29
	3.1 Differentiability with Respect to the Parameter	30
	3.2 Differentiability with Respect to the Initial Point	32
	3.3 Higher Order Differentiability and Analyticity	39
	3.4 Remark about a More General Point of View	40
	4. Systems of Equations as Particular Transformations Between Function Spaces	40
	4.1 Review of Metric Spaces	41
	4.2 Review of Banach Spaces	47
	4.3 The Cauchy Problem and Fixed Points of Certain Transformations in Banach Spaces	52
	5. Exercises	56
	5.1 Variables Separable Equations	56
	5.2 Equations Reducible to Separable Equations	57
	5.3 Linear Equations of the First Order	59
	5.4 Linear Equations of Order Higher than the First with Constant Coefficients	59
	5.5 Euler Equations	61
	5.6 Envelopes and Differential Equations	61
	5.7 Various Exercises	63
	5.8 Selected Exercises	64
	6. Bibliographical Notes	75

		Page
CHAPTER II	LINEAR SYSTEMS	79
	1. Elements of Linear Algebra	79
	1.1 Matrices and Eigenvalues	79
	1.2 Linear Operators Between Banach Spaces	84
	1.3 Canonical Form of Matrices	88
	1.4 Spectrum and Eigenvalues of a Linear Operator	93
	1.5 Limits of Operators	94
	2. Linear Systems of Ordinary Differential Equations	97
	2.1 Formal Solution of Linear Systems	97
	2.2 Fundamental Systems of Solutions and Adjoint Systems	99
	2.3 Nonhomogeneous Systems	102
	3. Operational Calculus	103
	3.1 Analytic Functions of Operators	103
	3.2 Linear Systems with Constant Coefficients	110
	4. Linear Finite Differences Equations	118
	4.1 Homogeneous Linear Finite Differences Equations	118
	4.2 Nonhomogeneous Linear Finite Differences Equations	124
	5. Examples	124
	6. Bibliography	131
CHAPTER III	EXISTENCE AND UNIQUENESS FOR THE CAUCHY PROBLEM UNDER THE CONDITION OF CONTINUITY	132
	1. Existence Theorem	133
	1.1 Characterization of Compact Sets of Continuous Functions: Ascoli's Theorem	133
	1.2 Local Existence	140
	1.3 Global Existence	146
	2. The Peano Phenomenon	155
	2.1 Approximation of all Solutions to a Given Cauchy Problem	156
	2.2 Maximal and Minimal Solutions. The Peano Phenomenon	159
	2.3 The Peano Phenomenon for Systems	162
	2.4 Maximal Solutions, Differential Inequalities, and Global Existence	168
	3. Questions of Uniqueness	171
	3.1 Continuous Dependence	171
	3.2 Uniqueness Theorems	178
	3.3 How Many Differential Equations Have the Uniqueness Property?	183
	4. Elements of G-Convergence	187
	4.1 Introduction	187
	4.2 G-Convergence for Equations Satisfying the Lipschitz Condition	187
	4.3 Homogenization	189
	4.4 G-Compactness	196
	4.5 G-Convergence and the Peano Phenomenon	198
	5. Bibliographical Notes	200

		Page
CHAPTER IV	BOUNDARY VALUE PROBLEMS	207
	1. Continuous Mappings on Euclidean Spaces	207
	1.1 The Topological Degree	208
	1.2 The Theorems of Brouwer and Miranda	213
	2. Geometric Boundary Value Problems	217
	2.1 The Boundary Value Problems of Picard and Nicoletti	217
	2.2 A Geometrical Formulation of the Boundary Value Problem	224
	2.3 Some Applications of the Geometric Formulation	233
	3. Sturm-Liouville Problems: Eigenvalues and Existence and Uniqueness Theorems	236
	3.1 Eigenvalues and Eigenfunctions	236
	3.2 Prüfer's Change of Variables	239
	3.3 Existence and Properties of the Eigenvalues	247
	3.4 Applications to Questions of Uniqueness for Problems Involving Nonlinear Equations	251
	3.5 Application to the Existence of Solutions for Problems Involving Nonlinear Equations	254
	3.6 Further Properties of Eigenvalues and Eigenfunctions	261
	4. Periodic Solutions	265
	4.1 The Case of First Order Equations	265
	4.2 The Case of Second Order Equations	268
	4.3 The Case of Systems	272
	4.4 On the Structure of Periodic Solutions	279
	5. Functional Boundary Value Problems	284
	5.1 Linear Functional Problems	286
	5.2 Nonlinear Functional Problems	296
	6. Bibliographical Notes	299
CHAPTER V	QUESTIONS OF STABILITY	311
	1. Stability of the Solutions of Linear Systems	312
	1.1 Definition of Stability	312
	1.2 Stability for Autonomous Linear Systems	314
	1.3 Autonomous Linear Systems of the Second Order	317
	1.4 Certain Stability Problems for Nonautonomous Linear Systems	325
	2. Some Methods for the Determination of the Stability of Nonlinear Systems	332
	2.1 Definitions	332
	2.2 Liapunov's Method	337
	2.3 The Fixed Point Method: Asymptotic Equivalence	346
	2.4 Olech's Method	352
	2.5 The Method of the Logarithmic Norm	358
	2.6 Invariant Sets	361
	3. Some Applications	365
	3.1 Problems in Biology and Chemistry	365
	3.2 Problems in Automatic Control Theory	370

		Page
CHAPTER V	(cont.)	
	4. The Method of Runge and Kutta	373
	4.1 The Fourth Order Runge-Kutta Algorithm	374
	4.2 Practical Use of the Runge-Kutta Method	376
	5. Bibliographical Notes	378
INDEX		382

Chapter I
Existence and Uniqueness for the Initial Value Problem Under the Hypothesis of Lipschitz

In this chapter, we shall study initial value problems in their various aspects: uniqueness, existence, domain of definition of the solutions and qualitative properties of the solutions. We shall assume that the functions that appear on the right side of the equations belong to a special class of continuous functions, the Lipschitz functions. This restriction allows us to treat the questions at hand with a remarkable ease.

In Chapter III below we shall take up the same questions again but with more generality with regard to the functions on the right.

1. GENERAL RESULTS

In this first section we concern ourselves with the existence and uniqueness of the solutions that pass through a given point and study how these vary as a function of that given point.

All the proofs are based on elementary arguments. For a different point of view on the same questions, see n. 4.

1.1. Definitions

An *ordinary differential equation* is a relation among an independent variable x, an unknown function $y(x)$ of that variable, and certain of its derivatives. The most general form which an ordinary differential equation may assume is therefore

$$F(x,y,y',\dots,y^{(n)}) = 0. \tag{1.1}$$

We have here an equation involving functions . The objects that satisfy it are themselves functions, called *solutions or integrals* of the differential equation. The highest order of the derivative that appears in (1.1) is called the *order* of the differential equation.

More generally, one may consider m equations among an independent variable x, m unknown functions of the variable x, and their derivatives up to a certain order. One then obtains a system of ordinary differential equations, which may be represented in the form:

$$F_i(x,y_1,y_1',\ldots,y_1^{(\nu_1)};\ldots;y_m,y_m',\ldots,y_m^{(\nu_m)}) = 0 \qquad (i = 1,\ldots,m). \tag{1.2}$$

The number $\nu_1 + \nu_2 + \cdots + \nu_m$ is then called the *order* of the system, and any m-tuple of functions that satisfies (1.2) is called a *solution* or *integral* of the system.

A differential equation is said to be of *normal type* if one can write (1.1) in the form:

$$y^{(n)} = f(x,y,y',\ldots,y^{(n-1)}) \qquad (1.1)'$$

where f is a real valued function of n variables. More generally, a system is said to be of normal type if (1.2) can be put in the form:

$$y_i^{(\nu_i)} = f_i(x,y_1,y_1',\ldots,y_1^{(\nu_1-1)};\ldots;y_m,y_m',\ldots,y_m^{(\nu_m-1)}) \qquad (i = 1,2,\ldots,m). \qquad (1.2)'$$

We shall study below only equations and systems of equations that are of normal type.

A differential equation of order n can always be reduced to a system of n equations of the first order (that is, to a system of order 1) by putting

$$y(x) = y_1(x),\ y'(x) = y_2(x),\ldots,y^{(n-1)}(x) = y_n(x).$$

This means that by transforming a function $y(x)$ satisfying (1.1)' we obtain a solution of the system

$$y_1' = y_2;\ y_2' = y_3;\ldots;y_{n-1}' = y_n;\ y_n' = f(x,y_1,y_2,\ldots,y_n)$$

and, vice versa, a solution $y_1(x),y_2(x),\ldots,y_n(x)$ of this system furnishes a solution, with its derivatives, of equation (1.1), simply by setting $y=y_1$. More generally, every system of type (1.2)' is reducible to a system of $\nu_1 + \nu_2 + \cdots + \nu_m$ equations of the first order. It is, indeed, enough to put

$$\begin{array}{lll} y_1 = \phi_{11} & y_2 = \phi_{21} & y_m = \phi_{m1} \\ y_1' = \phi_{12} & y_2' = \phi_{22} & y_m' = \phi_{m2} \\ \cdots\cdots & \cdots\cdots & \cdots\cdots \\ y_1^{(\nu_1-1)} = \phi_{1\nu_1} & y_2^{(\nu_2-1)} = \phi_{2\nu_2} & y_n^{(\nu_m-1)} = \phi_{m\nu_m}. \end{array}$$

By virtue of these remarks, we may, without loss of generality, limit our study to systems of differential equations of the form

$$y_i' = f_i(x,y_1,y_2,\ldots,y_n) \qquad (i = 1,2,\ldots,n). \tag{1.3}$$

1.2. Geometrical Interpretation

We now see, by way of an example, how an ordinary differential equation may arise.

Given a family of curves with equation

$$\phi(x,y,c_1,c_2,\ldots,c_n) = 0, \tag{1.4}$$

we obtain by differentiation with respect to x (thinking of y as a function of x, and of $c_1,c_2,\ldots,c_n$ as constants)

$$\phi_x(x,y,c_1,c_2,\ldots,c_n) + \phi_y(x,y,c_1,\ldots,c_n)y' = 0$$

$$\phi_{xx} + 2\phi_{xy}y' + \phi_{yy}y'^2 + \phi_{yx}y'' = 0$$

$$\cdots$$

If we eliminate the constants $c_1,c_2,\ldots,c_n$ from the first n equations, then the equation of the family of curves becomes a relation of type (1.1). Without stopping here to list the hypotheses on the function $\phi(x,y,c_1,\ldots,c_n)$ that make the preceding process valid, we shall note that we are led to a differential equation of order n by eliminating the n arbitrary constants in the equation of the family of curves (1.4).

These considerations show that, in general, the solutions of a differential equation of order n are infinite in number and raise the question whether every differential equation of order n may be obtained by the indicated procedure if we begin with a suitable family of curves of type (1.4). This was the fundamental question in the classical theory of ordinary differential equations: Given a differential equation of order n, find a relation of type (1.4) from which, by elimination of the constants, one may obtain the given equation. The relation $\phi(x,y,c_1,\ldots,c_n) = 0$ is called the *general integral* of the given differential equation. At the present time, interest in this problem has notably lessened and interest in other problems has replaced it; one tries to find an integral of the differential equation that satisfies further conditions. Among these, the classical one is to find a solution of equation (1.1) that satisfies the initial conditions:

$$y(x_0) = y_0;\ y'(x_0) = y_1^{(0)};\ldots;y^{(n-1)}(x_0) = y_{n-1}^{(0)}.$$

This is called the initial value problem or *Cauchy problem*. We adopt this point of view and propose to study first of all, and in depth, the initial value problem restricted to equations and systems of normal type. We shall see below that this allows us to answer, in a certain sense, even the question posed concerning the existence of the general integral and to make precise the meaning to be given to this question.

If equation (1.1)' can be reduced to a system of equations of the first order (1.3), the initial conditions may then be expressed by

$$y_1(x_0) = y_1^{(0)};\ y_2(x_0) = y_2^{(0)};\ldots;y_n(x_0) = y_n^{(0)}. \tag{1.5}$$

In general, we shall study the initial value problem for the system of ordinary differential equations of normal type (1.3); we require that the solution satisfy the conditions (1.5).

1.3. Functions Satisfying a Lipschitz Condition

We shall first of all make the hypothesis that the functions $f_i(x,y_1,y_2,\ldots,y_n)$ appearing on the right side of the system are continuous in a closed set I of the space R^{n+1} of the variables $x,y_1,y_2,\ldots,y_n$. As a result, the differences

$$f_i(x,y_1,\ldots,y_{s-1},\bar{\bar{y}}_s,y_{s+1},\ldots,y_n) - f_i(x,y_1,\ldots,y_{s-1},\bar{y}_s,y_{s+1},\ldots,y_n), \tag{1.6}$$

where $(x,y_1,\ldots,\bar{\bar{y}}_s,\ldots,y_n)$, $(x,y_1,\ldots,\bar{y}_s,\ldots,y_n)$ are two points in I, tend to 0 as $|\bar{y}_s - \bar{\bar{y}}_s| \to 0$.

In this chapter, we shall need to impose on the functions $f_i(x,y_1,y_2,\ldots,y_n)$ a more restrictive condition than the previous one; we shall suppose that the quantities (1.6) are infinitesimal of order not less than the first, that is, that there are constants A_s such that

$$|f_i(x,y_1,\ldots,y_{s-1},\bar{\bar{y}}_s,y_{s+1},\ldots,y_n) - f_i(x,y_1,\ldots,y_{s-1},\bar{y}_s,y_{s+1},\ldots,y_n)| \le A_s|\bar{\bar{y}}_s-\bar{y}_s|. \tag{1.7}$$

This expresses the fact that the ratio of increments with respect to the variable y_s calculated at the point $(x,y_1,\ldots,\bar{y}_s,\ldots,y_n)$ is bounded. A function that is differentiable with respect to y_s at the given point satisfies (1.7) there. This condition is therefore less restrictive than differentiability but more so than continuity. If the conditions (1.7) are valid with the same constants A_s no matter what the points $(x,y_1,\ldots,\bar{\bar{y}}_s,\ldots,y_n)$, $(x,y_1,\ldots,\bar{y}_s,\ldots,y_n)$ may be in some set E, we say that the function $f_i(x,y_1,\ldots,y_n)$ *satisfies a Lipschitz condition* or that the function is a *Lipschitz function* in the set E with respect to the variable y_s. One may immediately verify that a function $f_i(x,y_1,\ldots,y_n)$ which satisfies a Lipschitz condition with respect to each variable y_s does so with respect to all the y_i if the set E is a rectangular domain; that is, it satisfies an inequality of the form

$$|f_i(x,\bar{\bar{y}}_1,\bar{\bar{y}}_2,\ldots,\bar{\bar{y}}_n) - f_i(x,\bar{y}_1,\bar{y}_2,\ldots,\bar{y}_n)| \leq A \sum_{s=1}^{n} |\bar{\bar{y}}_s - \bar{y}_s|, \tag{1.7}'$$

where A is a constant independent of the points $(x,\bar{y}_1,\ldots,\bar{y}_n)$, $(x,\bar{\bar{y}}_1,\ldots,\bar{\bar{y}}_n)$ of the rectangular domain E.

Functions with continuous first derivatives in a closed rectangular domain satisfy a Lipschitz condition, for it follows from the Mean Value Theorem that

$$|f_i(x,y_1,\ldots,\bar{\bar{y}}_s,\ldots,y_n) - f_i(x,y_1,\ldots,\bar{y}_s,\ldots,y_n)| = \left|\frac{\partial f_i}{\partial y_s}(x,y_1,\ldots,\tilde{y}_s,\ldots,y_n)\right|\cdot|\bar{\bar{y}}_s-\bar{y}_s|,$$

where $\tilde{y}_s$ is some number in the interval $]\bar{y}_s,\bar{\bar{y}}_s[$. If A is the maximum of the absolute value of $\partial f_i/\partial y$ in the rectangular domain E, (1.7)' follows. See the exercises of Sections 4.1 and 4.2 for the general properties of functions satisfying a Lipschitz condition.

1.4. Existence Theorem

In order to avoid typographical complications, we shall from now on occasionally put, so long as there is no possibility of confusion,

$$f_i(x,y)$$

in place of

$$f_i(x,y_1,\ldots,y_n),$$

letting y take the place of $y_1,\ldots,y_n$.

Let $f_i(x,y)$ be n functions continuous in I that satisfy a Lipschitz condition with respect to y_s in every rectangular domain R contained in I. If $(x_0, y_1^{(0)}, \ldots, y_n^{(0)})$ is a fixed point in I, let R be a rectangular domain contained in I defined by the conditions

$$\begin{cases} x_0 - a \le x \le x_0 + a \\ y_i^{(0)} - b \le y_i \le y_i^{(0)} + b. \end{cases} \tag{1.8}$$

The Lipschitz condition expressed by (1.7)' will therefore be satisfied in R for a suitable constant A. We are now ready to prove the following:

Existence Theorem: If the functions $f_i(x,y)$ $(i = 1,2,\ldots,n)$ are continuous and satisfy a Lipschitz condition with respect to the variables $y_1, y_2, \ldots, y_n$ in the rectangular domain R defined by (1.8), then there exist an interval $]x_0-\delta, x_0+\delta[$ and n functions $y_1(x), y_2(x), \ldots, y_n(x)$ which are differentiable and satisfy

$$y_i'(x) = f_i(x, y_1(x), \ldots, y_n(x)) \qquad (i = 1,2,\ldots,n) \tag{1.9}$$

and

$$y_i(x_0) = y_i^{(0)} \qquad (i = 1,2,\ldots,n). \tag{1.10}$$

One may take $\delta = \min\{a, \frac{b}{M}\}$, where M is such that in R:

$$|f_i(x,y)| \le M \qquad (i = 1,2,\ldots,n).$$

We begin by observing that if functions $y_1(x), y_2(x), \ldots, y_n(x)$ with the desired properties exist, they also satisfy the following relations

$$y_i(x) = y_i^{(0)} + \int_{x_0}^{x} f_i(t, y(t))\,dt, \tag{1.11}$$

as can be seen by integrating (1.9) and taking note of (1.10).

Vice versa, if there exists a system $y_1(x), \ldots, y_n(x)$ of continuous functions satisfying (1.11), one has $y_i(x_0) = y_i^{(0)}$. These functions are differentiable and satisfy (1.9).

Equation (1.11) constitutes a system of integral equations of the type of Volterra; such a system is, as we have just seen, equivalent to the Cauchy problem relative to the system (1.9) under the initial conditions (1.10).

Let $y_1(x), y_2(x), \ldots, y_n(x)$ be n functions defined in $[x_0-a, x_0+a]$ such that $|y_i(x) - y_i^{(0)}| < b$, and consider the functional transformation:

$$\phi_i(x) = y_i^{(0)} + \int_{x_0}^{x} f_i(t, y(t))\,dt. \tag{1.12}$$

This produces a new system of functions $\phi_n(x)$ corresponding to $y_1(x),\dots,y_n(x)$ such that

$$\phi_i(x_0) = y_i^{(0)}.$$

The problem of finding a solution of the integral system (1.11) is thus equivalent to that of finding a system of continuous functions that is mapped into itself by the functional transformation (1.12). We shall, under the hypotheses in the statement of the theorem, prove, by means of the method of *successive approximations,* the existence of this system of functions "fixed" by the transformation (1.12). If we begin with a system of functions and transform it by means of (1.12) so as to obtain a new system of functions and then transform this new system itself by means of (1.12), we obtain yet another system; this process may be continued indefinitely. The sequence of such systems of functions converges to a system that turns out to be exactly the "fixed" system of functions that we want.

Proof of the Theorem: Let us consider any system of continuous functions with values in R, for example, the following system of constant functions:

$$y_1^{(0)}(x) \equiv y_1^{(0)},\ y_2^{(0)}(x) \equiv y_2^{(0)},\dots,y_n^{(0)}(x) \equiv y_n^{(0)}$$

and put

$$y_i^{(1)}(x) = y_i^{(0)} + \int_{x_0}^{x} f_i(t,y_1^{(0)},y_2^{(0)},\dots,y_n^{(0)})dt \qquad (i = 1,2,\dots,n). \tag{1.13$_1$}$$

One has for $x_0-\delta \le x \le x_0+\delta$

$$|y_i^{(1)}(x)-y_i^{(0)}| \le M|x-x_0| \le M\delta \le b. \tag{1.14$_1$}$$

Thus, the composed functions $f_i(x,y^{(1)}(x))$ are defined for $x_0-\delta \le x \le x_0+\delta$. We now put

$$y_i^{(2)}(x) = y_i^{(0)} + \int_{x_0}^{x} f_i(t,y^{(1)}(t))dt \qquad (i = 1,2,\dots,n) \tag{1.13$_2$}$$

and observe that one also has, for $x_0-\delta \le x \le x_0+\delta$,

$$|y_i^{(2)}(x)-y_i^{(0)}| \le M|x-x_0| \le M\delta \le b. \tag{1.14$_2$}$$

We continue in this way and put

$$y_i^{(m)}(x) = y_i^{(0)} + \int_{x_0}^{x} f_i(t,y^{(m-1)}(t))dt \qquad (i = 1,2,\dots,n). \tag{1.13$_m$}$$

If we suppose that for $x_0-\delta \le x \le x_0+\delta$ we have $|y_i^{(m-1)}(x)-y_i^{(0)}| \le b$, it follows that in the same interval

$$|y_i^{(m)}(x)-y_i^{(0)}| \le M|x-x_0| \le M\delta \le b. \tag{1.14$_m$}$$

The sequences of continuous functions

$$y_i^{(0)}(x), y_i^{(1)}(x), \dots, y_i^{(m)}(x), \dots \qquad (i = 1,2,\dots,n) \tag{1.15}$$

are therefore defined for $x_0-\delta \le x \le x_0+\delta$. Let us prove that such sequences converge uniformly (even totally) in the interval $[x_0-\delta, x_0+\delta]$. To do this, we consider the series of functions

$$y_i^{(0)}(x) + [y_i^{(1)}(x)-y_i^{(0)}(x)] + [y_i^{(2)}(x)-y_i^{(1)}(x)] + \dots\dots + [y_i^{(m)}(x)-y_i^{(m-1)}(x)] + \dots \quad (i = 1,2,\dots,n) \tag{1.16}$$

whose partial sums coincide with the terms of the sequence (1.15), and we prove that these series converge uniformly in the interval $[x_0-\delta, x_0+\delta]$. To this end we observe that from $(1.14)_1$ we have

$$|y_i^{(1)}(x)-y_i^{(0)}| \le M|x-x_0|.$$

From $(1.13)_1$ and $(1.13)_2$, taking note also of (1.7)', we deduce:

$$\begin{aligned}|y_i^{(2)}(x)-y_i^{(1)}(x)| &\le \left|\int_{x_0}^{x} |f_i(t,y^{(1)}(t))-f_i(t,y^{(0)}(t))|dt\right| \\ &\le A\left|\int_{x_0}^{x} \left\{\sum_{s=1}^{n} |y_s^{(1)}(t)-y_s^{(0)}|\right\}dt\right| \le AnM\left|\int_{x_0}^{x} |t-x_0|dt\right| \\ &= AnM\,\frac{|x-x_0|^2}{2}.\end{aligned}$$

If we suppose, by induction, that we have in $[x_0-\delta, x_0+\delta]$

$$|y_i^{(m-1)}(x) - y_i^{(m-2)}(x)| \le A^{m-2}\, n^{m-2}\, M\,\frac{|x-x_0|^{m-1}}{(m-1)!}$$

then, in an analogous manner we get

$$|y_i^{(m)}(x)-y_i^{(m-1)}(x)| \leq \left|\int_{x_0}^{x} |f_i(t,y_1^{(m-1)}(t),\ldots,y_n^{(m-1)}(t))\right.$$

$$\left. - f_i(t,y_1^{(m-2)}(t),\ldots,y_n^{(m-2)}(t))|dt\right|$$

$$\leq A\left|\int_{x_0}^{x} \left\{\sum_{s=1}^{n} |y_s^{(m-1)}(t) - y_s^{(m-2)}(t)|\right\}dt\right|$$

$$\leq An\left|\int_{x_0}^{x} A^{m-2}n^{m-2}M \frac{|t-x_0|^{m-1}}{(m-1)!}\, dt\right| = A^{m-1}n^{m-1}M \frac{|x-x_0|^m}{m!}.$$

Hence, this last condition is valid for arbitrary m, that is, the general term of the series (1.16) is, for $x_0-\delta \leq x \leq x_0+\delta$, dominated by the term

$$\frac{M}{nA} \cdot \frac{(n\cdot A\cdot \delta)^m}{m!} \tag{1.17}$$

of an exponential series. The series (1.16), and therefore the sequences (1.15), are consequently uniformly convergent in the interval $[x_0-\delta,x_0+\delta]$. Let us denote by $y_1(x),\ldots,y_n(x)$ the sums of (1.16) and, therefore, the functions which are the limits of the sequence (1.15). Such functions are continuous in the interval $[x_0-\delta,x_0+\delta]$ and satisfy there the conditions: $|y_i(x)-y_i^{(0)}| \leq b$ $(i = 1,2,\ldots,n)$.

It remains to prove that such a system of functions is precisely the system of continuous functions that satisfies the integral system (1.11), that is, the system of functions "fixed" by the transformation (1.12). This follows from $(1.13)_m$ if we succeed in showing that it is permissible to pass to the limit under the integral sign. It is therefore a matter of proving that the sequences

$$f_i(x,y^{(m)}(x))$$

converge uniformly in $[x_0-\delta,x_0+\delta]$ to the functions $f_i(x,y(x))$. One has, in fact, recalling (1.7)':

$$|f_i(x,y^{(m)}(x)) - f_i(x,y(x))| \leq A \sum_{s=1}^{n} \cdot|y_s^{(m)}(x) - y_s(x)|.$$

But the expression $|y_i^{(m)}(x) - y_i(x)|$ is the (m+1)st remainder of series (1.16); this remainder is, as we have seen, less than the (m+1)st remainder of the series whose general term is given by (1.17) and hence tends to 0 uniformly with respect to x as $m \to \infty$.

The existence theorem is thus completely proved.

Exercise. Prove that if $f\colon [a,b] \times [0,c] \to [0,+\infty]$ satisfies a Lipschitz condition, then the initial value problem

$$y' = f(x,y),\ y(a) = 0$$

has a solution in a certain interval, even though the initial point $y(a)$ does not belong to the interior of the set in which the second variable of f varies.

1.5. Uniqueness Theorem

We now propose to show that under the same hypotheses as in the preceding theorem, the solution to the initial value problem is unique.

Uniqueness Theorem. Under the same hypotheses as in the existence theorem, there is not more than one system of differentiable functions satisfying (1.9) and (1.10).

Let us suppose that there are in fact two solutions of the given initial value problem, namely $y \equiv y(x)$ and $Y \equiv Y(x)$, so that we have

$$y_i(x) = y_i^{(0)} + \int_{x_0}^{x} f_i(t,y(t))dt$$

$$Y_i(x) = y_i^{(0)} + \int_{x_0}^{x} f_i(t,Y(t))dt.$$

Upon subtraction, one obtains, taking note of (1.7)':

$$|y_i(x) - Y_i(x)| \leq A\left|\int_{x_0}^{x} \sum_{s=1}^{n} |y_s(t) - Y_s(t)|dt\right|. \tag{1.18}$$

We now consider in the interval $[x_0,x_0+\delta]$ the function $\phi(x)$ defined by

$$\phi(x) = \sum_{s=1}^{n} \max_{x_0 \leq t \leq x} |y_s(t) - Y_s(t)|$$

which is evidently non-negative and non-decreasing, and if it is identically 0, the two systems of solutions coincide. If it is not identically 0, then a value x_1 exists in the interval $[x_0,x_0+\delta]$ such that one has $\phi(x) = 0$ for $x_0 \leq x \leq x_1$ and $\phi(x) > 0$ for $x > x_1$ and $x-x_1$ small enough. But from (1.18) one may deduce

$$|y_i(x) - Y_i(x)| \leq A \int_{x_1}^{x} \sum_{s=1}^{n} |y_s(t) - Y_s(t)|dt,$$

from which follows

$$\max_{x_0 \le t \le x} |y_i(t) - Y_i(t)| \le A \sum_{s=1}^{n} \max_{x_0 \le t \le x} |y_s(t) - Y_s(t)|(x-x_1)$$

that is,

$$\phi(x) \le nA(x-x_1)\phi(x).$$

But this relation is clearly absurd, whence one may conclude that in the whole interval $[x_0,x_0+\delta]$ one must have $\phi(x) \equiv 0$, that is, the two systems coincide there. In exactly the same way one shows that the functions must coincide even in the interval $[x_0-\delta,x_0]$, and the proof of the theorem is complete.

Alternative Proof of the Uniqueness Theorem. We now give another proof of the uniqueness theorem which also sheds some light upon the magnitude of the error in the process of successive approximations.

The solution, whose existence has been demonstrated in the theorem of Sec. 1.4, is the limit of the sequence of successive approximations defined inductively by $(1.13)_m$. We shall now prove that any other possible solution of the same problem of initial values is a limit of the same sequence and therefore must coincide with that found in the proof of the existence theorem. To be specific, if $Y \equiv Y(x)$ is some solution of the same problem of initial values, that is, a system of functions for which one has in $[x_0-\delta,x_0+\delta]$:

$$Y_i(x) = y_i^{(0)} + \int_{x_0}^{x} f_i(t,Y_1(t),\ldots,Y_n(t))dt, \quad (i = 1,2,\ldots,n), \tag{1.19}$$

one has also

$$|Y_i(x) - y_i^{(0)}| \le M|x - x_0|.$$

Upon subtracting $(1.13)_1$ from (1.19) side from side, one has, taking note of (1.7)':

$$|Y_i(x)-y_i^{(1)}(x)| \le \int_{x_0}^{x} |f_i(t,Y_1(t),\ldots,Y_n(t)) - f_i(t,y_1^{(0)}(t),\ldots,y_n^{(0)}(t))|dt|$$

$$\le A\left|\int_{x_0}^{x} \sum_{s=1}^{n} |Y_s(t) - y_s^{(0)}(t)|dt\right| \le nAM \frac{|x-x_0|^2}{2!}.$$

Proceeding by induction, we find, in a way analogous to that used in the proof of the existence theorem,

$$|Y_i(x) - y_i^{(m)}(x)| \leq \frac{M}{nA}\frac{(nA|x-x_0|)^{m+1}}{(m+1)!} \tag{1.20}$$

Since the second side tends to 0 as m goes to infinity, we are able to make the desired conclusion.

From what we have so far shown we conclude that the successive approximations obtained by $(1.13)_m$ provide approximate values of the solution and that inequality (1.20) also defines the error one makes in substituting the mth approximation given by $(1.13)_m$ for the solution.

1.6. Continuous Dependence on Initial Conditions and Parameters

The lower estimate of the end-points of the interval in which the functions $y_1(x), y_2(x), \ldots, y_n(x)$ satisfying the initial value problem are defined is furnished by the existence theorem under the hypothesis that the initial conditions coincide with the center of the rectangular domain defined by (1.8). Let us now consider a point $(\xi_0, \eta_1^{(0)}, \ldots, \eta_n^{(0)})$ of R that satisfies the additional conditions

$$|\xi_0 - x_0| < \sigma_1; \quad |\eta_i^{(0)} - y_i^{(0)}| \leq \sigma_2 \quad (i = 1,2,\ldots,n) \tag{1.21}$$

and the solution of the initial value problem relative to the system (1.9) with the initial conditions:

$$y_i(\xi_0) = \eta_i^{(0)} \quad (i = 1,2,\ldots,n). \tag{1.22}$$

This solution will, because of the existence theorem, surely be defined in the interval $[\xi_0-h, \xi_0+h]$, where $h = \min[a - \sigma_1, (b-\sigma_2)/M]$, and, therefore, in $[x_0-\delta', x_0+\delta']$ where $\delta' = h - \sigma_1$, as long as, of course, σ_1 and σ_2 are sufficiently small.

Let us consider the sequence of successive approximations relative to the initial value problem (1.22). Each term of this sequence is a continuous function not only of x but also of the other variables $(\xi_0, \eta_1^{(0)}, \eta_2^{(0)}, \ldots, \eta_n^{(0)})$, where x varies in the interval and the others vary in the rectangular domain defined by (1.21). This fact, which is easy to see for the first approximations, may be proved by induction for the others. The functions which are limits of this sequence, that is, the solution of the problem, are consequently continuous functions of $\xi_0, \eta_1^{(0)}, \eta_2^{(0)}, \ldots, \eta_n^{(0)}$, that is to say, of the initial conditions. We are thus able to state:

Theorem. The integrals of a system of differential equations satisfying the hypotheses of the existence theorem are continuous functions of the initial conditions.

Even more generally, let us suppose that the right hand sides of the system (1.9) depend not only on $x, y_1, y_2, \ldots, y_n$ but also on one or more parameters $\lambda_1, \lambda_2, \ldots, \lambda_p$ and are continuous with respect to all these variables in a rectangular domain in $n + p + 1$ dimensional space and satisfy the Lipschitz condition with respect to the y_i. The integrals will evidently be functions not only of x but also of the parameters $\lambda_1, \lambda_2, \ldots, \lambda_p$. We may then prove, as above, the following:

Theorem. The integrals of a system of differential equations satisfying the given conditions are continuous functions of any parameters on which the system depends continuously.

1.7. Interval of Definition and Extension of Solutions

The domain of the functions $y_1(x), y_2(x), \ldots, y_n(x)$ that constitute the solution of the given initial value problem can be extended beyond the bounds guaranteed by the existence theorem. In fact, the functions $y_1(x), y_2(x), \ldots, y_n(x)$ determined by the theorem in the interval $[x_0-\delta, x_0+\delta]$ assume for $x = x_0+\delta$ values $y_1(x_0+\delta), y_2(x_0+\delta), \ldots, y_n(x_0+\delta)$, and these can be considered initial conditions of an initial value problem relative to the same system. The solution of this problem will, by the same theorem, be defined in an interval $[x_0+\delta-\delta_1,\ x_0+\delta+\delta_1]$ where δ_1 will depend on the maximum value that the functions $|f_i(x, y_1, y_2, \ldots, y_n)|$ assume in a rectangular domain of R^{n+1} of the form

$$x_0 + \delta - a_1 \le x \le x_0 + \delta + a_1$$
$$y_i(x_0+\delta) - b_1 \le y_i \le y_i(x_0+\delta) + b_1 \qquad (i = 1, 2, \ldots, n).$$

This domain is contained in the set I where the functions $f_i(x, y)$ are defined. The two solutions, both the primitive one and the one just now obtained, must - by the uniqueness theorem - coincide in the interval $[x_0+\delta-\delta_1,\ x_0+\delta]$, and so the new solution is an *extension* of the primitive one into a bigger interval. By repeating this extension again and again, one arrives at an interval larger than the original one in which the solution is defined; in special cases one may arrive, for example, at the interval $[x_0-a,\ x_0+a]$ which appears in (1.8). That this may actually not happen is clear from the case of

$$y' = y^2.$$

The integral of this for which $y(x_0) = y_0 > 0$ is the function $y(x) = y_0/(1-y_0(x-x_0))$. It is defined in the interval $[x_0, x_0 + \frac{1}{y_0}[$ and may not be extended beyond this interval to the right because at the right-hand end point the integral becomes infinite.

Let us now suppose that the functions $f_i(x,y)$ $(i = 1,2,\ldots,n)$ are defined in the *cylinder* $S = [x_0-a, x_0+a] \times R^n$ and that they satisfy there the conditions indicated in Sec. 1.4, that is, that they are continuous and satisfy a Lipschitz condition in every rectangular domain contained in S. Not even in this case is one able to assert that the integrals are defined in the interval $[x_0-a, x_0+a]$, as is also clear from the preceding example. We therefore propose to indicate certain cases in which the integrals are defined for $x_0-a \le x \le x_0+a$. We begin by supposing that in all S we have

$$|f_i(x,y)| \le M.$$

Then the existence theorem applied with R as in (1.8) and $b > aM$ assures us that the interval in which the functions $y_1(x), y_2(x), \ldots, y_n(x)$ that furnish the solution are defined in $[x_0-a, x_0+a]$.

Another important case in which this happens is that in which (1.7)' holds in the whole cylinder S, that is, when the functions satisfy a Lipschitz condition not only in every rectangular domain contained in S but in all of S. In such a case, if L is a positive number such that

$$|f_i(x,y^0)| < L$$

for $|x-x_0| < a$, $i = 1,2,\ldots,n$ we take R defined by (1.8) with $b > L$. We now observe that there certainly is a point $(\tilde{x}, \tilde{y}_1, \ldots, \tilde{y}_n)$ in R and an index i such that

$$M = |f_i(\tilde{x}, \tilde{y}_1, \ldots, \tilde{y}_n)|,$$

where M is the greatest of the n numbers $M_i = \max|f_i|$ $(i = 1,\ldots,n)$. Thus

$$\frac{M}{b} \le \frac{|f_i(\tilde{x}, \tilde{y}_1, \tilde{y}_2, \ldots, \tilde{y}_n) - f_i(\tilde{x}, y_1^{(0)}, \ldots, y_n^{(0)})|}{b} + \frac{|f_i(\tilde{x}, y_1^{(0)}, \ldots, y_n^{(0)})|}{b}$$

$$\le \frac{A \sum_{i=1}^{n} |\tilde{y}_i - y_i^{(0)}|}{b} + \frac{L}{b} < nA + 1.$$

The existence theorem now assures us that the integral is defined in the interval $[x_0-\delta, x_0+\delta]$, where the number $\delta > 1/nA+1$ is independent of the initial values $(x_0, y_1^{(0)}, \ldots, y_n^{(0)})$ and depends only on the Lipschitz coefficient A that appears in (1.7)'. One may therefore conclude that after a finite number of extensions, the solution is defined in

$$[x_0-a, x_0+a].$$

An important case in which the preceding situation applies is that of the linear system

$$y_i' \equiv a_{i1}(x)y_1 + a_{i2}(x)y_2 + \ldots + a_{in}(x)y_n + b_i(x) \quad (i = 1,2,\ldots,n).$$

As a matter of fact, the functions on the right satisfy (1.7)' in all S, A being the greatest of the maxima of the absolute values of the functions $a_{ij}(x)$, $(i,j = 1,2,\ldots,n)$ for $x_0-a \le x \le x_0+a$. We may thus say that the *integrals of the linear systems are defined in the whole interval where the coefficients and the functions* b_i *are continuous.*

When the domain of the solutions y_i of the given initial value problem does not coincide with the interval $[x_0-a, x_0+a]$, there is a maximal interval of existence J (which may be open or half-open) in the sense that the y_i cannot be extended beyond J and still provide a solution to the given system. One establishes this result by going back to the argument made at the beginning of this section. Specifically, we shall consider the maximal extension to the right of x_0, since that to the left can be considered analogously. Let B be the set of all the $x \in [x_0, x_0+a]$ such that all the y_i are already defined or can be extended as solutions into the interval $[x_0, x]$. Put $\alpha = \sup B$. Let us suppose that $\alpha < x_0 + a$. If all the limits

$$\lim_{x\to\alpha} y_i(x) = \ell_i \qquad (i = 1,\ldots,n)$$

exist and are finite, then we can extend that y_i to α (if they are not already defined) in such a way that the extensions are differentiable and satisfy the given system at α. (It is enough to use the mean value theorem: cf. Exercise 1). We now consider the initial value problem

$$y_i' = f_i(x,y), \quad y_i(\alpha) = \ell_i \qquad (i = 1,\ldots,n)$$

and find a solution to the right of α. With this, one is able to construct a solution to the given initial value problem that is defined in an interval $[x_0, x]$ with $x > \alpha$ by the method of "joining" (cf. Exercise 2); but this contradicts $\alpha = \sup B$. There therefore must exist at least one i such that the limit ℓ_i either does not exist or is infinite. But in that case there is no extension of y_i to the right of

α that is continuous and therefore no solution to the given initial value problem. Finally, let us suppose that $\alpha = x_0+a$. Then there are two cases: $a \in B$ or $a \notin B$. In the first case, the y_i are solutions in $[x_0,x_0+a]$, while in the second case they are solutions only in $[x_0,x_0+a[$. We are thus able to conclude with the following theorem:

Theorem. Every solution of a given system has its graph contained in that of a solution that cannot be extended as a solution of the given system.

Before closing this section, we observe that the integrals' property of depending continuously on the initial conditions and parameters on which the equations depend holds in every closed interval contained in the interval in which the integrals are defined. In fact, these intervals can be covered by a finite number of extensions of the type indicated at the beginning of this section.

Exercise 1. Prove that if the f_i are continuous and the y_i are solutions of the system

$$y_i' = f_i(x,y)$$

in the interval $[a,b[$, if the limits

$$\ell_i = \lim_{x \to b} y_i(x)$$

exist and are finite, and if the point $(b,\ell_1,\ldots,\ell_n)$ belongs to the domain of the f_i, then the y_i can be extended into $[a,b]$ as differentiable functions such that

$$y_i'(b) = f_i(b,y(b)).$$

Therefore they are solutions of the given system in $[a,b]$. An analogous result holds if we consider $]a,b]$ in place of $[a,b[$.

Exercise 2. Let the f_i be continuous and the y_i solutions of the system in the interval $[a,b]$, and let the z_i be solutions of the same system in $[b,c]$. Prove that if $y_i(b) = z_i(b)$, then the function u defined by

$$u_i(t) = \begin{cases} y_i(t), & t \leq b \\ z_i(t), & t \geq b \end{cases}$$

is a solution of the given system in $[a,c]$.

Exercise 3. Use Zorn's Lemma to prove the theorem on the existence of non-extendable solutions.

1.8. Gronwall's Lemma

In the four preceding sections, we have seen that in order to obtain the desired conclusions, we must proceed to the study of appropriate upper bounds of the solutions and of their differences. One may therefore suppose that it may be convenient to make a study in itself of the possible upper bounds of the solutions in order to be able to apply the results later to the theory of ordinary differential equations. We shall now prove a theorem which historically was one of the first results of this kind, while in Chapter III we shall come back to this question with more general methods and results.

Gronwall's Lemma. Let $I \subseteq R$ be an interval, let $a \in I$, and let $u,v\colon I \to R$ be continuous functions with $u \geq 0$. Let c be a constant ≥ 0. From

$$v(t) \leq c + \int_a^t u(s)v(s)ds \qquad (t \geq a)$$

it follows that

$$v(t) \leq ce^{\int_a^t u(s)ds} \qquad (t \geq a),$$

while from

$$v(t) \leq c + \int_t^a u(s)v(s)ds \qquad (t \leq a)$$

it follows that

$$v(t) \leq ce^{\int_t^a u(s)ds} \qquad (t \leq a).$$

Proof of the Lemma. Let us first of all consider the case $t \geq a$. We fix $\varepsilon > 0$ and define a positive function by

$$V_\varepsilon(t) = c + \varepsilon + \int_a^t u(s)v(s)ds.$$

By differentiating the two sides, one obtains

$$V_\varepsilon'(t) = u(t)v(t).$$

From the hypothesis of the lemma it follows that

$$V_\varepsilon'(t) \leq u(t)V_\varepsilon(t)$$

and therefore that

$$\frac{V'_\varepsilon(t)}{V_\varepsilon(t)} \le u(t).$$

By integrating the two sides from a to t, one obtains

$$\ell g\, V_\varepsilon(t) - \ell g\, V_\varepsilon(a) \le \int_a^t u(s)\,ds. \tag{1.23}$$

Since the first side is equal to $\ell g \dfrac{V_\varepsilon(t)}{V_\varepsilon(a)}$, we obtain, by applying the exponential function to the two sides of (1.23),

$$\frac{V_\varepsilon(t)}{V_\varepsilon(a)} \le e^{\int_a^t u(s)ds}.$$

Since $v \le V_\varepsilon$ and $V_\varepsilon(a) = c + \varepsilon$, we deduce

$$v(t) \le (c+\varepsilon)e^{\int_a^t u(s)ds} \qquad (t \ge a;\ \varepsilon > 0).$$

Upon taking the limit as $\varepsilon \downarrow 0$, we obtain the desired formula. The case $t \le a$ reduces to the preceding case after a change of variables. In fact, let us consider the two functions defined for those $s \ge 0$ such that $a - s \in I$ by

$$\overline{v}(s) = v(a - s), \quad \overline{u}(s) = u(a - s).$$

By virtue of the hypotheses of the lemma, it follows that

$$\overline{v}(s) \le c + \int_{a-s}^a u(\xi)y(\xi)d\xi = c - \int_s^0 u(a-\zeta)v(a-\zeta)d\zeta$$

(now we make the change of variables $\phi(\zeta) = a - \zeta$)

$$= c + \int_0^s \overline{u}(\zeta)\overline{v}(\zeta)d\zeta.$$

Since $s \ge 0$, the preceding case (already proved for every interval I and for every $a \in I$) implies that

$$\overline{v}(s) \le ce^{\int_0^s \overline{u}(\zeta)d\zeta} \qquad (s \ge 0). \tag{1.24}$$

Since $\int_0^s \overline{u}(\zeta)d\zeta = \int_{a-s}^a u(\xi)d\xi$ (because of the change of variables $\phi(\zeta) = a - \zeta$) and since it is possible to find for every point $t \in I$, $t \le a$, an $s \ge 0$ such that $t = a - s$, (1.24) implies the second of the desired formulas. The lemma of Gronwall has thus been completely proved.

Exercise 1. Prove the following generalization of Gronwall's Lemma. If $k: I \to [0,\infty[$ is of class C^1, and if

$$v(t) \leq k(t) + \int_a^t u(s)v(s)ds \qquad (t \geq a),$$

then we have

$$v(t) \leq k(a)e^{\int_a^t u(s)ds} + e^{\int_a^t u(s)ds}\int_a^t e^{-\int_a^s u(\xi)d\xi} k'(s)ds.$$

Hint: For any positive integer n, calculate the solution u_n of the linear initial value problem

$$w' = k'(t) + \frac{1}{n} + u(t)w, \qquad w(a) = k(a),$$

and then prove that $v \leq u_n$. Conclude by taking the limit.

Exercise 2. Prove the following additional variant of Gronwall's Lemma. If $k: I \to [0,\infty[$ is continuous, it follows from

$$v(t) \leq k(t) + \int_a^t u(s)v(s)ds \qquad (t \geq a)$$

that

$$v(t) \leq k(t) + \int_0^t k(s)u(s)e^{\int_s^t u(\xi)d\xi} ds.$$

Hint: Consider the function $y(t) = \int_0^t u(s)v(s)ds$, and calculate its derivative; use the hypotheses to find something that dominates it, then differentiate $z(t) = y(t)e^{-\int_0^t u(s)ds}$, and try to reach the conclusion.

1.9. Application of Gronwall's Lemma to the Cauchy Problem

As the first application of the preceding lemma, we are able to obtain an upper bound for the solutions of the system (1.9) satisfying the conditions (1.7)'.

If in fact $y(x)$ is the solution of the system (1.9) satisfying the initial conditions (1.10), one has

$$y_i(x) - y_i^{(0)} = \int_{x_0}^x \{f_i(t,y(t))-f_i(t,y^{(0)})\}dt + \int_{x_0}^x f_i(t,y^{(0)})dt,$$

whence, if we put L_i equal to the maximum value of $|f_i(x,y^{(0)})|$ for $x_0 \leq x \leq x_0+a$, we have:

$$|y_i(x) - y_i^{(0)}| \leq A \int_{x_0}^{x} \sum_{j=1}^{n} |y_j(x) - y_j^{(0)}| dx + L_i a$$

and, even more,

$$\sum_{i=1}^{n} |y_i(x) - y_i^{(0)}| \leq nA \int_{x_0}^{x} \sum_{i=1}^{n} |y_i(x) - y_i^{(0)}| dx + a \sum_{i=1}^{n} L_i .$$

From this point, it follows from Gronwall's Lemma that

$$|y_i(x) - y_i^{(0)}| \leq a \sum_{i=1}^{n} L_i e^{nAa} .$$

We now give another noteworthy application of Gronwall's Lemma. Let us consider the two systems of differential equations

$$y_i' = f_i(x,y) \qquad (i = 1,2,\dots,n)$$

$$Y_i' = g_i(x,Y) \qquad (i = 1,2,\dots,n),$$

where the functions $f_i(x,y)$ and $g_i(x,y)$ are supposed to be continuous in both variables and satisfy a Lipschitz condition with respect to y in the rectangle

$$R \equiv (x_0 - a \leq x \leq x_0 + a;\ y_i^{(0)} - b \leq y_i \leq y_i^{(0)} + b; \quad (i = 1,\dots,n)).$$

Now let $y(x)$, $Y(x)$ be the solutions of the two systems satisfying respectively the initial conditions

$$y_i(x_0) = y_i^{(0)} \quad ; \quad Y_i(\alpha) = \beta_i .$$

We propose to prove that the differences

$$|y_i(x) - Y_i(x)| \qquad (i = 1,2,\dots,n)$$

may be made as small as we please uniformly in x by making sufficiently small the quantities

$$|f_i(x,y) - g_i(x,y)|;\ |x_0 - \alpha|,\ |y_i^{(0)} - \beta_i|;$$

in other words, we propose to show that the solutions depend continuously on all the quantities given in the problem, thereby generalizing a preceding result which examined the dependence of the solutions on the initial conditions.

We may write

$$y_i(x) = y_i^{(0)} + \int_{x_0}^{x} f_i(t,y(t))dt$$

$$Y_i(x) = \beta_i + \int_{\alpha}^{x} g_i(t,Y(t))dt,$$

since the existence theorem and the discussion in Sec. 1.7 insure that the functions $y_i(x)$, $Y_i(x)$ are defined in the interval $[x_0-\delta, x_0+\delta]$, where

$$\delta = \min\left\{a - \sigma_1, \frac{b}{M}, \frac{b - \sigma_2}{M}\right\}$$

with M greater than the maxima of $|f_i|$ and $|g_i|$ $(i = 1,2,\ldots,n)$ in R, σ_1 and σ_2 sufficiently small, and

$$|x_0 - \alpha| < \sigma_1, \quad |y_i^{(0)} - \beta_i| \le \sigma_2.$$

Upon subtracting side from side in the equalities above, we obtain

$$y_i(x)-Y_i(x) = y_i^{(0)} - \beta_i - \int_{\alpha}^{x_0} g_i(t,Y(t))dt$$
$$+ \int_{x_0}^{x} \{f_i(t,y(t)) - g_i(t,Y(t))\}dt$$

and also

$$y_i(x) - Y_i(x) = y_i^{(0)} - \beta_i - \int_{\alpha}^{x_0} g_i(t,Y(t))dt$$
$$+ \int_{x_0}^{x} \{f_i(t,y(t)) - g_i(t,y(t))\}dt$$
$$+ \int_{x_0}^{x} \{g_i(t,y(t)) - g_i(t,Y(t))\}dt.$$

From this one easily deduces

$$|y_i(x) - Y_i(x)| \le |y_i^{(0)} - \beta_i| + M|x_0 - \alpha|$$
$$+ \int_{x_0}^{x} |f_i(t,y(t)) - g_i(t,y(t))|dt + A\left|\int_{x_0}^{x} \sum_{j=1}^{n} |y_j(t) - Y_j(t)|dt\right|.$$

If we assume that the conditions

$$|x_0 - \alpha| < \sigma, \quad |y_i^{(0)} - \beta_i| < \sigma$$

are satisfied with $\sigma < \min\{\sigma_1,\sigma_2\}$ and that in R we have

$$|f_i(x,y) - g_i(x,y)| < \varepsilon,$$

it follows that, upon summing with respect to i,

$$\sum_{i=1}^{n} |y_i(x) - Y_i(x)| < n\sigma + nM\sigma + n\varepsilon a + nA\left|\int_{x_0}^{x} \sum_{j=1}^{n} |y_j(t) - Y_j(t)|dt\right|.$$

Therefore, if $x > x_0$, one has, by the lemma of Gronwall:

$$\sum_{i=1}^{n} |y_i(x) - Y_i(x)| \leq n[\sigma(1+M) + \varepsilon a]e^{nAa}.$$

If $x < x_0$, one has, in the same manner,

$$-\sum_{i=1}^{n} |y_i(x) - Y_i(x)| > -n[\sigma(1+M) + \varepsilon a] + nA\int_{x_0}^{x} \sum_{j=1}^{n} |y_j(t) - Y_j(t)|dt,$$

and, therefore, again by Gronwall's Lemma,

$$\sum_{i=1}^{n} |y_i(x) - Y_i(x)| \leq n[\sigma(1+M) + \varepsilon a]e^{nAa}.$$

The proposition has thus been completely proved, and we obtain the following:

Kamke's Theorem. Under the Lipschitz condition, the solutions depend uniformly continuously on all the data of the problem: the initial values x_0, y_0 and the functions f.

Following the two applications given here, the reader will easily be able to use Gronwall's Lemma to find another way to solve the problems in the previous sections. We attach some exercises to serve as guides.

Exercise 1. Use Gronwall's Lemma to prove the uniqueness theorem under the Lipschitz condition.

Exercise 2. Use Gronwall's Lemma to prove that if f_i: $[a,b] \times R^n \to R$ satisfy a Lipschitz condition with respect to the variable y, then the initial value problem

$$y_i' = f_i(x,y), \quad y_i(a) = y_i^0$$

has a solution in the whole interval $[a,b]$.

Exercise 3. Use Gronwall's Lemma to extend Kamke's theorem in the following way. Let $f_i\colon [a,b] \times R^n \to R$ satisfy a Lipschitz condition in the variable y, and let $g_{m,i}\colon [a,b] \times R^n \to R$ be continuous (not necessarily satisfying a Lipschitz condition) for every positive integer m. Prove that if $\lim_m g_{m,i} = f_i$ uniformly and if $\lim_m y_m^0 = y^0$, then the solutions y_m of the initial value problems

$$y'_{m,i} = g_{m,i}(x,y_m), \quad y_{m,i}(a) = y^0_{m,i}$$

exist in $[a,b]$, if m is sufficiently large, and converge uniformly to the unique solution of

$$y'_i = f_i(x,y), \quad y_i(a) = y^0_i.$$

Hint: Begin by proving that there are two constants A and B such that

$$|f_i(x,y)| \leq A||y|| + B.$$

Then prove that if m is sufficiently large, the $g_{m,i}$ satisfy a similar inequality, though, in general, A and B will be different. Conclude that the relative initial value problems have solutions in $[a,b]$. Then proceed to the second part.

2. QUALITATIVE PROPERTIES OF SOLUTIONS

In the following sections, we shall be concerned with finding the qualitative properties of the solution of the system (1.9) in correspondence with the assumptions about the functions on the right side.

2.1. Differentiability of Solutions

We begin by supposing that the functions $f_i(x,y)$ are defined in the rectangular domain R given by (1.8) and satisfy there the conditions of the uniqueness theorem. This same theorem guarantees that the functions $y_i(x)$, which constitute the solution, and their first derivatives are continuous in the interval $[x_0-\delta, x_0+\delta]$ fixed by the theorem. One may immediately deduce that for these, the inequality

$$\frac{|y_i(x'') - y_i(x')|}{|x'' - x'|} \leq M_i$$

is valid, where M_i is the maximum of the absolute value of the function f_i in R. If, in addition to the given hypotheses, the functions f_i satisfy a Lipschitz condition with respect to the variable x, then the first derivatives of the $y_i(x)$ are also of bounded incremental ratio; one has, in fact, taking note of the preceding inequality and setting M =

$$\left|\frac{y_i'(x'')-y_i'(x')}{x''-x'}\right| = \frac{|f_i(x'',y_1(x''),\ldots,y_n(x''))-f_i(x',y_1(x'),\ldots,y_n(x'))|}{|x''-x'|}$$

$$\leq A\frac{|x''-x'| + \sum_{i=1}^{n}|y_i(x'')-y_i(x')|}{|x''-x'|} = A + A\frac{\sum_{i=1}^{n}|y_i(x'')-y_i(x')|}{|x''-x'|}$$

$$\leq A(1+nM).$$

If we furthermore suppose that the functions f and all their first derivatives are continuous, then the functions $y_i(x)$ constituting the solution of the initial value problem have continuous second derivatives, and one has, upon application of the "chain rule" to the relation $y_i'(x) = f_i(x,y_1(x),\ldots,y_n(x))$:

$$y_i''(x) = \frac{\partial f_i}{\partial x} + \sum_{s=1}^{n}\frac{\partial f_i}{\partial y_s}y_s'(x).$$

One may differentiate once again if the functions f_i and their second derivatives are continuous with respect to all the variables on which they depend, and one finds that

$$y_i'''(x) = \frac{\partial^2 f_i}{\partial x^2} + \sum_{s=1}^{n}\frac{\partial^2 f_i}{\partial y_s \partial x}y_s'(x) + \sum_{s,j}^{1\ldots n}\frac{\partial^2 f_i}{\partial y_s \partial y_j}y_s'(x)y_j'(x) + \sum_{s=1}^{n}\frac{\partial f_i}{\partial y_s}y_s''(x).$$

In general, we may affirm the following:

Theorem. If the functions f_i and their derivatives of the first p orders are continuous, then the functions $y_i(x)$ that constitute a solution to the system (1.9) and their derivatives up to the (p+1)st order are continuous.

2.2. Analyticity of the Solutions

We now suppose that the functions $f_i(x,y_1,\ldots,y_n)$ are *analytic*, that is, expandible about the initial point $(x_0,y_1^{(0)},\ldots,y_n^{(0)})$ in a multiple power series that is absolutely convergent for $|x-x_0| < r$, $|y_i-y_i^{(0)}| < t$:

$$f_i(x,y_1,\ldots,y_n) = \sum_{p_0p_1\ldots p_n}^{0\ldots\infty} a_{p_0p_1\ldots p_n}^{(i)}(x-x_0)^{p_0}(y_1-y_1^{(0)})^{p_1}\ldots(y_n-y_n^{(0)})^{p_n}.$$

Without loss of generality, we suppose that the initial conditions for the solution of the system (1.0) are

$$y_i(0) = 0$$

so that $x_0 = y_1^{(0)} = \dots = y_n^{(0)} = 0$. We shall then have

$$f_i(x,y_1,\dots,y_n) = \sum_{p_0\dots p_n}^{0\dots\infty} a^{(i)}_{p_0\dots p_n} x^{p_0} y_1^{p_1} \dots y_n^{p_n} \tag{2.1}$$

whence

$$a^{(i)}_{p_0\dots p_n} = \frac{1}{p_0!p_1!\dots p_n!}\left(\frac{\partial^{(p_0+\dots+p_n)} f_i}{\partial x^{p_0} \partial y_1^{p_1} \dots \partial y_n^{p_n}}\right)(0,\dots,0),$$

and we shall suppose the multiple series to be convergent for

$$|x| < r; \quad |y_i| < t, \qquad (i = 1,2,\dots,n).$$

The functions $y_i(x)$ that constitute the solution of the system (1.9) with the given initial conditions therefore admit (as we have shown) derivatives of all orders in some interval about $x = 0$; we propose to prove that they are *analytic*, that is, that they can be expanded in power series around $x = 0$, in other words, that one may represent them by

$$y_i(x) = \sum_{s=1}^{\infty} c_s^{(i)} x^s \qquad \left(c_s^{(i)} = \frac{1}{s!} y_i^{(s)}(0)\right) \tag{2.2}$$

in an interval $]-\rho,\rho[$.

We begin now by observing that by means of the procedure indicated for the calculation of $y_i^{(s)}$ in 2.1, one has

$$c_0^{(i)} = y_i(0) = 0$$

$$c_2^{(i)} = y_i'(0) = f_i(0,0\dots0)$$

$$c_2^{(i)} = \frac{y_i''(0)}{2!} = \frac{1}{2!} f_{ix}'(0\dots0) + \sum_{s=1}^{n} f_{iy_s}'(0\dots0)\cdot f_s(0\dots0) \tag{2.3}$$

$$c_3^{(i)} = \frac{y_i'''(0)}{3!} = \frac{1}{2!} f_{ixx}''(0\dots0) + \sum_{s=1}^{n} f_{ixy_s}''(0\dots0) f_s(0\dots0)$$

$$+ \sum_{s,j}^{1\dots n} f_{iy_s y_j}''(0\dots0) f_s(0\dots0) f_j(0\dots0) + \sum_{s=1}^{n} f_{iy_s}(0\dots0) c_2^{(i)} 2!,$$

that is, the coefficients of the expansion (2.2) are determined by the derivatives of the f_i. This does not insure, however, that the series on the right hand side of (2.2) converges and have sums $y_i(x)$. If, though these series converge in an interval of the type $]-\rho,\rho[$, the sums must necessarily coincide with the functions $y_i(x)$ that provide the solution. In fact, if $\phi_i(x)$ is the sum of the series on the right hand side of (2.2), then the function

$$\phi_i'(x) - f_i(x,\phi_1(x),\ldots,\phi_n(x))$$

has value 0, as do all its derivatives, when $x = 0$, as one may easily verify by differentiating, keeping (2.3) in mind. Therefore, it is identically 0 in $]-\rho,\rho[$. This means that $\{\phi_i(x)\}$ is the solution of the system (1.9) satisfying the same initial conditions as $\{y_i(x)\}$ and hence

$$\phi_i(x) \equiv y_i(x).$$

Consequently, in order to prove (2.2), it suffices to show that the series (2.2) has non-zero radius of convergence. To do this, assume for the moment that there is an analytic function $F(x,y_1,\ldots,y_n)$ which is positive together with its partial derivatives at the point $(0\ldots0)$, and satisfies the condition

$$\left|\left(\frac{\partial^{(p_0+p_1+\ldots+p_n)}f_i}{\partial x^{p_0}\partial y_1^{p_1}\ldots\partial y_n^{p_n}}\right)(0,\ldots,0)\right| \leq \left(\frac{\partial^{(p_0+\ldots+p_n)}F}{\partial x^{p_0}\partial y_1^{p_1}\ldots\partial y_n^{p_n}}\right)(0,\ldots,0) \qquad (2.4)$$

for all $p_0,p_1,\ldots,p_n$ and for every i. Such a function is called a majorant function, and one writes briefly

$$f_i(x,y_1,\ldots,y_n) \ll F(x,y_1,\ldots,y_n).$$

Let us consider the system

$$Y_i' = F(x,Y_1,\ldots,Y_n)$$

and its solution that satisfies the initial conditions $Y_i(0) = 0$ $(i = 1,2,\ldots,n)$. We suppose that its solution is analytic and therefore has the expansion

$$Y_i(x) = \sum_{s=1}^{\infty} Y_s^{(i)}x^s \qquad \left(Y_s^{(i)} = \frac{1}{s!}Y_i^{(s)}(0)\right) \qquad (2.5)$$

in an interval of the type $]-\rho,\rho[$. It follows,that, because of (2.3) and (2.4),

$$|y_i'(0)| \leq Y_i'(0),$$

$$|y_i''(0)| \leq Y_i''(0),$$

and, in general

$$|y_i^{(s)}(0)| \leq Y_i^{(s)}(0).$$

Thus series (2.2) is dominated by a convergent series, hence it is convergent in the same open interval $]-\rho,\rho[$.

As a follow-up we proceed to show that it is possible to find a majorant function $F(x,y_1,\ldots,y_n)$ such that the solution of the system $Y_i' = F(x,Y_1,\ldots,Y_m)$ satisfying the conditions $Y_i(0) = 0$ $(i = 1,2,\ldots,n)$ can be expanded in a power series whose radius of convergence is non-zero.

To do this we observe that because of the absolute convergence of series (2.1) for $|x| \leq r$, $|y_i| \leq t$, it follows that there is a constant M such that for every collection of indices $p_0,p_1,\ldots,p_n$ one has

$$|a^{(i)}_{p_0p_1\ldots p_n}|r^{p_0}t^{p_1+\ldots+p_n} \leq M,$$

whence

$$\left|\frac{\partial^{(p_0+p_1+\ldots+p_n)} f_i}{\partial x^{p_0}\partial y_1^{p_1}\ldots\partial y_n^{p_n}}(0,\ldots,0)\right| \leq \frac{p_0!p_1!\ldots p_n!}{r^{p_0}t^{p_1+\ldots+p_n}}M. \tag{2.6}$$

Let us now consider the function

$$F(x,y_1,\ldots,y_n) = \frac{M}{(1-\frac{x}{r})(1-\frac{y_1}{t})\ldots(1-\frac{y_n}{t})}$$

$$= M\sum_{p_0\ldots p_n}^{0\ldots\infty}\left(\frac{x}{r}\right)^{p_0}\left(\frac{y_1}{t}\right)^{p_1}\ldots\left(\frac{y_n}{t}\right)^{p_n}.$$

The series that appears here is convergent for $|x| < r$, $|y_1| < t,\ldots,$ $|y_n| < t$. The function $F(x,y_1,\ldots,y_n)$ is therefore analytic for such values of the variables; one has, furthermore,

$$\frac{\partial^{(p_0+\ldots+p_n)}F(0)}{\partial x^{p_0}\partial y_1^{p_1}\ldots\partial y_n^{p_n}} = \frac{p_0!p_1!\ldots p_n!}{r^{p_0}t^{p_1+\ldots+p_n}}M.$$

It then follows from (2.6) that the function $F(x,y_1,\ldots,y_n)$ is a majorant of the functions $f_i(x,y_1,\ldots,y_n)$. Let us now consider the solution of the system of differential equations

$$Y_i' = \frac{M}{(1 - \frac{x}{r})\left(1 - \frac{Y_1}{t}\right) \cdots \left(1 - \frac{Y_n}{t}\right)}$$

satisfying the initial conditions $Y_i(0) = 0$, $(i = 1,2,\ldots,n)$. Since the right hand sides of the system under consideration do not depend on the index i, it follows, upon taking note of the initial conditions, that $Y_1(x) \equiv Y_2(x) \equiv \ldots \equiv Y_n(x) \equiv Y(x)$.

We are thus led back to solving the single differential equation of the first order

$$Y'(x) = \frac{M}{(1 - \frac{x}{r})(1 - \frac{Y}{t})^n}.$$

By separation of variables, the desired integral is easily found; it is given by the formula

$$Y_1(x) \equiv \ldots \equiv Y_n(x) \equiv Y(x) \equiv t\left[1 - \sqrt[n+1]{1 + \frac{n+1}{t} Mr \log(1 - \frac{x}{r})}\ \right],$$

where the radical is to be taken in the absolute value, and is defined for

$$x < r\left(1 - e^{-\frac{t}{(n+1)Mr}}\right).$$

The radical may be expanded in a binomial series if

$$\frac{n+1}{t} Mr \left|\log(1 - \frac{x}{r})\right| < 1,$$

that is, if

$$r\left(1 - e^{\frac{t}{(n+1)Mr}}\right) < x < r\left(1 - e^{\frac{-t}{(n+1)Mr}}\right)$$

and therefore surely if

$$|x| < r\left(1 - e^{-\frac{t}{(n+1)Mr}}\right) \tag{2.7}$$

since

$$r\left(1 - e^{\frac{t}{(n+1)Mr}}\right) < -r\left(1 - e^{\frac{-t}{(n+1)Mr}}\right).$$

In fact, from

$$e^{\frac{t}{(n+1)Mr}} + e^{-\frac{t}{(n+1)Mr}} - 2 = \left[e^{\frac{t}{2(n+1)Mr}} - e^{-\frac{t}{2(n+1)Mr}}\right]^2 > 0$$

one has

$$1 - e^{\frac{t}{(n+1)Mr}} < -\left(1 - e^{\frac{-t}{(n+1)Mr}}\right).$$

Therefore, in the interval (2.7), one has

$$\sqrt[n+1]{1 + \frac{n+1}{t} Mr \log(1 - \frac{x}{r})} = \sum_{s=0}^{\infty} \binom{\frac{1}{n+1}}{s} \left(\frac{n+1}{t} Mr\right)^s \left[\log(1 - \frac{x}{r})\right]^s,$$

and since $\log(1 - \frac{x}{r})$ can be developed in a MacLaurin series in the same interval, the solution $\{Y_i(x)\}$ can be developed analogously.

We may now state the following theorem. (This theorem is valid even in the complex domain, and the proof there is formally the same. It is due to Cauchy, and the method of proof is called the method of majorant functions.)

Theorem. If the functions $f_i(x,y_1,\ldots,y_n)$ are analytic in an interval about the point $(\alpha,\beta_1,\ldots,\beta_n)$, then the solution $\{y_i(x)\}$ of the system (1.9) satisfying the initial conditions

$$y_i(\alpha) = \beta_i$$

is also analytic in an interval about α.

Observe also that under the stated hypotheses, formulas (2.3) furnish power series expansions of the functions $y_i(x)$ that constitute the solution.

3. SOLUTIONS AS FUNCTIONS OF THE INITIAL DATA

We consider the system of differential equations that depends on one parameter,

$$y_i' = f_i(x,y;\lambda),$$

where the functions $f_i(x,y_1,\ldots,y_n;\lambda)$ are continuous, along with their first partial derivatives with respect to all the variables on which they depend, in the rectangular domain R of the space R^{n+2} defined by the inequalities

$$R \equiv \{a \le x \le b;\ c_i \le y_i \le d_i\ (i = 1,2,\ldots,n);\ \lambda_1 \le \lambda \le \lambda_2\}.$$

If $(x_0,y_1^{(0)},\ldots,y_n^{(0)};\bar{\lambda})$ is a point of R, there is one and only one solution $\bar{y}_i = \bar{y}_i(x)$ of the system under consideration, where we put $\bar{\lambda}$ in place of λ, that satisfies the initial conditions

$$\overline{y}_i(x_0) = y_i^{(0)}.$$

Such a solution depends on the fixed values $(x_0,y_1^{(0)},\ldots,y_n^{(0)};\overline{\lambda})$ and may therefore be written in the form

$$\overline{y}_i = \overline{y}_i(x;x_0;y_1^{(0)},\ldots,y_n^{(0)};\overline{\lambda}).$$

We thereby obtain a transformation T of $[a,b] \times R$ into R^n associating with each point $(x,x_0,y_1^{(0)},\ldots,y_n^{(0)},\overline{\lambda})$ the value at x of the solution that has initial values $y_1^{(0)},\ldots,y_n^{(0)}$ at the point x_0.

We propose to study the dependence of the solution on the arguments $x_0,y_1^{(0)},\ldots,y_n^{(0)},\overline{\lambda}$; the dependence on x has already been studied in the preceding paragraph. In other words, we want to study the qualitative properties of the transformation T.

3.1. Differentiability with Respect to the Parameter

We begin by studying the dependence on the parameter λ. We have already seen in Sec. 1 that this dependence is continuous; we now propose to show that it is differentiable. Let us consider the solution $\{y_i\}$ that satisfies the same initial conditions and corresponds to an arbitrary value λ of the parameter; we are then able to write

$$\overline{y}_i \equiv y_i^{(0)} + \int_{x_0}^{x} f_i(t,\overline{y}_1,\overline{y}_2,\ldots,\overline{y}_n;\overline{\lambda})dt$$

$$y_i \equiv y_i^{(0)} + \int_{x_0}^{x} f_i(t,y_1,\ldots,y_n;\lambda)dt.$$

Subtracting side from side and making use of the Mean Value Theorem, one obtains

$$\frac{y_i - \overline{y}_i}{\lambda - \overline{\lambda}} = \int_{x_0}^{x}\left\{f^*_{i\lambda} + \sum_{s=1}^{n} f^*_{iy_s}\frac{y_s - \overline{y}_s}{\lambda - \overline{\lambda}}\right\}dt, \tag{3.1}$$

where the derivatives $f^*_{i\lambda} = \partial f_i/\partial\lambda$, $f^*_{iy_s} = \partial f_i/\partial y_s$ are intended to be calculated at an interior point of the segment that joins the points $(t,\overline{y}_1,\ldots,\overline{y}_n,\overline{\lambda})$, $(t,y_1,\ldots,y_n,\lambda)$.

We now consider the integral system

$$U_i(x) = \int_{x_0}^{x}\left\{\overline{f}_{i\lambda} + \sum_{s=1}^{n} \overline{f}_{iy_s}U_s\right\}dt, \tag{3.2}$$

where $\overline{f}_{i\lambda}$, $\overline{f}_{iy_s}$ indicate the same derivatives calculated at the point

$(t,\bar{y}_1,\ldots,\bar{y}_n,\bar{\lambda})$.

This integral system admits a unique solution $\{U_i(x)\}$ defined in the whole interval where $\{\bar{y}_i\}$ is defined, for $\{U_i(x)\}$ is the solution of the linear system

$$U_i'(x) = \bar{f}_{i\lambda} + \sum_{s=1}^{n} \bar{f}_{iy_s} U_s \tag{3.2)'$$

that satisfies the initial conditions $U_i(x_0) = 0$ (cf. Sec. 1.7).

Upon subtracting (3.2) from (3.1) side from side, we get

$$\frac{y_i - \bar{y}_i}{\lambda - \bar{\lambda}} - U_i(x) = \int_{x_0}^{x} \left\{ f^*_{i\lambda} - \bar{f}_{i\lambda} + \sum_{s=1}^{n} (f^*_{iy_s} - \bar{f}_{iy_s}) U_s(t) + \sum_{s=1}^{n} f^*_{iy_s} \left[\frac{y_s - \bar{y}_s}{\lambda - \bar{\lambda}} - U_s(t) \right] \right\} dt.$$

Because of the above-mentioned continuous dependence of the solutions on the parameter λ, we may, for every $\varepsilon > 0$, determine a number $\delta_\varepsilon > 0$ in such a way that for $|\lambda - \bar{\lambda}| < \delta_\varepsilon$ we have

$$|f^*_{i\lambda} - \bar{f}_{i\lambda}| = |f_{i\lambda}(x,\bar{y}_1 + \theta(y_1-\bar{y}_1),\ldots,\bar{y}_n + \theta(y_n-\bar{y}_n),\bar{\lambda} + \theta(\lambda-\bar{\lambda})) - f_{i\lambda}(x,\bar{y}_1,\ldots,\bar{y}_n,\bar{\lambda})| < \varepsilon$$

with $0 < \theta < 1$ and

$$|f^*_{iy_s} - \bar{f}_{iy_s}| < \varepsilon \qquad (s = 1,2,\ldots,n).$$

If M is a number greater than the maximum absolute values in R of the functions f_i and of their first derivatives, and if K is a number greater than the maximum absolute values of the functions $U_i(x)$, we obtain, for $|\lambda - \bar{\lambda}| < \delta_\varepsilon$,

$$\left| \frac{y_i - \bar{y}_i}{\lambda - \bar{\lambda}} - U_i \right| \le \varepsilon(1+K)(b-a) + M \left| \int_{x_0}^{x} \sum_{s=1}^{n} \left| \frac{y_s - \bar{y}_s}{\lambda - \bar{\lambda}} - U_s \right| dt \right|$$

and hence

$$\sum_{i=1}^{n} \left| \frac{y_i - \bar{y}_i}{\lambda - \bar{\lambda}} - U_i \right| \le n\varepsilon(1+K)(b-a) + nM \left| \int_{x_0}^{x} \sum_{s=1}^{n} \left| \frac{y_s - \bar{y}_s}{\lambda - \bar{\lambda}} - U_s \right| dt \right|.$$

We then have by Gronwall's Lemma

$$\left| \frac{y_i - \bar{y}_i}{\lambda - \bar{\lambda}} - U_i(x) \right| \le n\varepsilon(1+K)(b-a)e^{nM(b-a)},$$

and this means that

$$\lim_{\lambda \to \overline{\lambda}} \frac{y_i - \overline{y}_i}{\lambda - \overline{\lambda}} = U_i(x),$$

that is, *the derivative of the solution* $y_i(x)$ *with respect to the parameter* λ *exists and is precisely the solution of the linear system* (3.2)'.

The linear system (3.2)' is called the *linear variational equation* of the given system. It is found merely by differentiating the two sides of the given system with respect to λ and calculating the coefficients of the derivatives along the solution $\overline{y}_i(x)$.

3.2. Differentiability with Respect to the Initial Point

We shall now study the behavior of the solutions of the given system as a function of the initial conditions, that is, we shall examine the qualitative properties with respect to the variables $(x_0, y_1^{(0)}, \ldots, y_n^{(0)})$ of the function T that associates with every point $(x_0, y_1^{(0)}, \ldots, y_n^{(0)}, \lambda)$ of $[a,b] \times R$ the value at x of the unique solution of the given system that passes through $(x_0, y_1^{(0)}, \ldots, y_n^{(0)}, \lambda)$:

$$T_i(x, x_0, y_1^{(0)}, \ldots, y_n^{(0)}, \lambda) = \overline{y}_i(x, x_0, y_1^{(0)}, \ldots, y_n^{(0)}, \lambda).$$

We know from Kamke's theorem that T is continuous. We shall therefore now inquire about its differentiability, while in the next section we shall examine its analyticity.

This situation is different from the one studied in the preceding section because an increment in only one of the variables $y_k^{(0)}$ implies an increment in the f_i simultaneously with respect to all the variables $\overline{y}_1, \ldots, \overline{y}_n$. For the purpose of discovering what we should expect, let us consider the simplest case of a system of two equations that does not depend on any parameter,

$$y_i' = f_i(x, y_1, y_2), \qquad (i = 1,2),$$

and let us look for the partial derivatives $\dfrac{\partial}{\partial y_1^{(0)}} \overline{y}_i(x, x_0, y_1^{(0)}, y_2^{(0)})$.

The given system can then be rewritten in the following way:

$$\frac{\partial}{\partial x} \overline{y}_i(x, x_0, y_1^{(0)}, y_2^{(0)}) = f_i(x, \overline{y}_1(x, x_0, y_1^{(0)}, y_2^{(0)}), \overline{y}_2(x, x_0, y_1^{(0)}, y_2^{(0)})). \tag{3.3}$$

If we assume that the derivatives $\dfrac{\partial}{\partial y_1^{(0)}} y_i$ are continuous, then, by differentiating the two sides of (3.3), we obtain

$$\frac{\partial}{\partial y_1^{(0)}}\frac{\partial}{\partial x}\bar{y}_i = \sum_{s=1}^{2} f_{iy_s}(x,\bar{y}_1,\bar{y}_2)\frac{\partial}{\partial y_1^{(0)}}\bar{y}_s,$$

or, by the theorem of Schwarz

$$\frac{\partial}{\partial x}\frac{\partial}{\partial y_1^{(0)}}\bar{y}_i = \sum_{s=1}^{2} f_{iy_s}(x,\bar{y}_1,\bar{y}_2)\frac{\partial}{\partial y_1^{(0)}}\bar{y}_s.$$

If we apply the same argument to $\frac{\partial}{\partial y_2^{(0)}}\bar{y}_i$, we see that the functions $U_{i1} = \frac{\partial}{\partial y_1^{(0)}}\bar{y}_i$ satisfy the linear system

$$U'_{i1} = \sum_{s=1}^{2} f_{iy_s}(x,\bar{y}_1,\bar{y}_2)U_{s1} \qquad (i = 1,2). \tag{3.4}$$

To determine the $\frac{\partial}{\partial y^{(0)}}\bar{y}_i$ completely, it remains to discover for what initial values we must solve the system (3.4); since the solutions of (3.4) vary with the varying of the initial data $U_{i1}(0)$, we shall have a unique pair of initial values for which the solution of (3.4) coincides with the $\frac{\partial}{\partial y_1^{(0)}}\bar{y}_i$. To this end, we observe that it follows from the definition of the derivative that for each x,

$$\lim_{h\to 0}\left|\frac{\bar{y}_1(x,x_0,y_1^{(0)}+h,y_2^{(0)}) - \bar{y}_1(x,x_0,y_1^{(0)},y_2^{(0)})}{h} - U_{11}(x)\right| = 0$$

$$\lim_{h\to 0}\left|\frac{\bar{y}_2(x,x_0,y_1^{(0)}+h,y_2^{(0)}) - \bar{y}_2(x,x_0,y_1^{(0)},y_2^{(0)})}{h} - U_{21}(x)\right| = 0.$$

In particular, if $x = x_0$, it follows that, since $\bar{y}_i(x_0,x_0,u_1,u_2) = u_i$,

$$\lim_{h\to 0}|1 - U_{11}(x_0)| = 0$$

$$\lim_{h\to 0}|0 - U_{21}(x_0)| = 0,$$

and, therefore, the initial values must be

$$U_{11}(x_0) = 1, \quad U_{21}(x_0) = 0.$$

Similarly, we see that the candidates to be the derivatives $\frac{\partial}{\partial y_2}y_i$ and $\frac{\partial}{\partial x_0}y_i$ are the solutions of certain initial value problems for linear systems. These "probable" derivatives are really the derivatives, as is proved in the following

Theorem. If the partial derivatives $f_{ik} = \frac{\partial}{\partial y_k} f_i$ are continuous, then the partial derivatives $\frac{\partial}{\partial y_k^{(0)}} \bar{y}_i$ and $\frac{\partial}{\partial x_0} \bar{y}_i$ of the solutions of the given system considered as functions of the initial data exist, are continuous, and are the solutions of two Cauchy problems for linear systems. Precisely, if one sets

$$U_{ik}(x) = \frac{\partial}{\partial y_k^{(0)}} \bar{y}_i(x,x_0,y_1^{(0)},\ldots,y_n^{(0)},\lambda),$$

$$V_i(x) = \frac{\partial}{\partial x_0} \bar{y}_i(x,x_0,y_1^{(0)},\ldots,y_n^{(0)},\lambda),$$

then one has

(a)

$$U'_{ik} = \sum_{s=1}^{n} f_{iy_s}(x,x_0,\bar{y}(x,x_0,y^{(0)},\lambda),\lambda)\cdot U_{sk}, \quad U_{ik}(x_0) = \begin{cases} 1 & \text{if } i = k \\ 0 & \text{if } i \neq k \end{cases};$$

(b)

$$V'_i = \sum_{s=1}^{n} f_{iy_s}(x,x_0,\bar{y}(x,x_0,y^{(0)},\lambda),\lambda)V_s, \quad V_i(x_0) = -f_i(x_0,y_0,\lambda).$$

Among the U_{ik} and the V_i there is the following relation:

$$V_i = -\sum_{k=1}^{n} U_{ik}\, f_k(x_0,y_0,\lambda).$$

The matrices $U = (U_{ik})$, $V = (V_i)$ are the Jacobian matrices of $(\bar{y}_1,\ldots,\bar{y}_n)$ considered as functions of $y_1^{(0)},\ldots,y_n^{(0)}$ alone or of x_0 alone. If we denote by f_{y_s} the Jacobian matrix of f with respect to the $y_1,\ldots,y_n$, we may interpret (a) and (b) symbolically as first order equations between matrices

$$U' = f_{y_s}\cdot U, \qquad V' = f_{y_s}\cdot V$$

with the initial data $U(0) = I$ the identity matrix, and $V(0) = -f(x_0,y_0,\lambda)$. Equations between matrices are meaningful and will be studied in Chapter II. Such a reformulation has the advantage of showing that the systems in (a) and (b) are formally similar: they are therefore called the linear variational equation associated with the given system.

Proof of the Theorem: We begin with (a). Let U_{ik} be the unique solution of the Cauchy problem (a). First of all, let us observe that the U_{ik} are defined in the whole interval because of what was established in Sec. 1.7 for linear systems. Let us fix i and k and prove that $\partial \bar{y}_i/\partial y_k^{(0)} = U_{ik}$. We must show that the function

$$\frac{y_i(x,x_0,y_1^{(0)},\ldots,y_{k-1}^{(0)},y_k^{(0)}+h,y_{k+1}^{(0)},\ldots,y_n^{(0)},\lambda)-y_i(x,x_0,y_1^{(0)},\ldots,y_n^{(0)},\lambda)}{h} - U_{ik}(x) \tag{3.5}$$

tends to 0 as $h \to 0$. To do this, we fix for the moment the increment h of the variable $y_k^{(0)}$ and consider the integral representations of the solutions of the given system and of that in (a). We put, for the sake of brevity,

$$\bar{\bar{y}}_i(x) = \bar{y}_i(x,x_0,y_1^{(0)},\ldots,y_{k-1}^{(0)},y_k^{(0)}+h,y_{k+1}^{(0)},\ldots,y_n^{(0)},\lambda)$$

and

$$\tilde{y}_i(x) = \bar{y}_i(x,x_0,y_1^{(0)},\ldots,y_n^{(0)},\lambda).$$

Then

$$\bar{\bar{y}}_i(x) = \left\{\begin{matrix} y_i^{(0)}+h & \text{if } k = i \\ y_i^{(0)} & \text{if } k \neq i \end{matrix}\right\} + \int_{x_0}^{x} f_i(t,\bar{\bar{y}}(t),\lambda)dt,$$

$$\tilde{y}_i(x) = y_i^{(0)} + \int_{x_0}^{x} f_i(t,\tilde{y}(t),\lambda)dt,$$

$$U_{ik}(x) = \left\{\begin{matrix} 1 & \text{if } k = i \\ 0 & \text{if } k \neq i \end{matrix}\right\} + \int_{x_0}^{x} \sum_{s=1}^{n} f_{iy_s}(t,\tilde{y}(t),\lambda)\cdot U_{sk}(t)dt.$$

By substitution in (3.5) one obtains

$$\left|\frac{\bar{\bar{y}}_i(x) - \tilde{y}_i(x)}{h} - U_{ik}(x)\right| = \left|\int_{x_0}^{x}\left[\frac{f_i(t,\bar{\bar{y}}(t),\lambda) - f_i(t,\tilde{y}(t),\lambda)}{h} - \sum_{s=1}^{n} f_{iy_s}(t,\tilde{y}(t),\lambda)U_{sk}(t)\right]dt\right|.$$

At this point, by adding and subtracting $\Sigma_s f_{iy_s}(t,\bar{y}(t),\lambda)\frac{\bar{\bar{y}}_s(t)-\tilde{y}_s(t)}{h}$ under the integral sign and making the obvious bounds, one obtains:

$$\left|\frac{\bar{\bar{y}}_i(x) - y_i(x)}{h} - U_{ik}(x)\right| \le \int_{x_0}^{x} \sum_s |f^*_{iy_s} - \bar{f}_{iy_s}| \frac{|\bar{\bar{y}}_s(t) - \tilde{y}_s(t)|}{|h|}\,dt + \int_{x_0}^{x} \sum_s |\bar{f}_{iy_s}| \left|\frac{\bar{\bar{y}}_s(t) - \tilde{y}_s(t)}{h} - U_{sk}(t)\right| dt, \tag{3.6}$$

where the derivatives $f^*_{iy_s}$ are intended to be calculated at an interior point of the segment that joins the points $(t,\bar{\bar{y}}(t),\lambda)$ and $(t,\bar{y}(t),\lambda)$, and the derivatives $\bar{f}_{iy_s}$ are to be calculated at $(t,\bar{y}(t),\lambda)$.

The quantity $\sum_s \frac{|\bar{\bar{y}}_s(x) - \tilde{y}_s(x)|}{|h|}$ is bounded uniformly in x; in fact, one has

$$|\bar{\bar{y}}_s(x) - \tilde{y}_s(x)| \le \begin{Bmatrix} |h| \text{ if } i = k \\ 0 \text{ if } i \ne k \end{Bmatrix} + \int_{x_0}^{x} |f_i(t,\bar{\bar{y}}(t),\lambda) - f_i(t,\tilde{y}(t),\lambda)|\,dt$$

$$\le |h| + \int_{x_0}^{x} \sum_s |f^*_{iy_s}|\, |\bar{\bar{y}}_s(t) - \tilde{y}(t)|\,dt$$

(by the Mean Value Theorem). Summing on the left with respect to s and observing that there exists a constant L such that

$$|f_{iy_s}|| \le L \tag{3.7}$$

in R, one obtains

$$\sum_s |\bar{\bar{y}}_s(x) - \tilde{y}_s(x)| \le n|h| + \int_{x_0}^{x} nL \sum_s |\bar{\bar{y}}_s(t) - \tilde{y}_s(t)|\,dt.$$

We are consequently able to apply Gronwall's Lemma to the function $v(x) = \sum_s |\bar{\bar{y}}_s(x) - \tilde{y}_s(x)|$ and obtain

$$\frac{\sum_s |\bar{\bar{y}}_s(x) - \tilde{y}_s(x)|}{|h|} \le ne^{nL(b-a)},$$

which proves that $\sum_s \frac{|\bar{\bar{y}}_s(x) - \tilde{y}_s(x)|}{|h|}$ is bounded. From this and from the fact that $\lim_{h\to 0} (f^*_{iy_s} - \bar{f}_{iy_s}) = 0$ because of the continuous dependence on the data, the first term in the second part of (3.6) tends to 0 as $h \to 0$. Thus, for every $\varepsilon > 0$, there exists $\delta > 0$ such that

$$\int_{x_0}^{x} \sum_s |f^*_{iy_s} - \bar{f}_{iy_s}| \frac{|\bar{\bar{y}}_s(t) - \tilde{y}_s(t)|}{|h|}\,dt \le \varepsilon \qquad (|h| \le \delta).$$

For $|h| < \delta$, one obtains from (3.6) and (3.7)

$$\left|\frac{\bar{\bar{y}}_i(x) - \tilde{y}_i(x)}{h} - U_{ik}(x)\right| \leq \varepsilon + \int_{x_0}^{x} L \sum_s \left|\frac{\bar{\bar{y}}_s(t) - \tilde{y}_s(t)}{h} - U_{sk}(t)\right| dt.$$

Upon summing the first side with respect to i, one obtains

$$\sum_i \left|\frac{\bar{\bar{y}}_i(x) - \tilde{y}_i(x)}{h} - U_{ik}(x)\right| \leq n\varepsilon + \int_{x_0}^{x} nL \sum_s \left|\frac{\bar{\bar{y}}_s(t) - \tilde{y}_s(t)}{h} - U_{sk}(t)\right| dt.$$

We here apply Gronwall's Lemma to

$$v(x) = \sum_i \left|\frac{\bar{\bar{y}}_i(x) - \tilde{y}_i(x)}{h} - U_{ik}(x)\right|,$$

and obtain

$$\sum_i \left|\frac{\bar{\bar{y}}_i(x) - \tilde{y}_i(x)}{h} - U_{ik}(x)\right| \leq n\varepsilon e^{nL(b-a)}, \qquad (|h| \leq \delta).$$

Since the quantity on the right tends to zero as $\varepsilon \to 0$, we must have

$$\sum_i \left|\frac{\bar{\bar{y}}_i(x) - \tilde{y}_i(x)}{h} - U_{ik}(x)\right| = 0$$

and (a) is thereby proved.

The proof of (b) is similar and is therefore left as an exercise for the interested reader. As for the last formula, it can be easily verified by observing that the function defined by

$$W_i(x) = -\sum_k U_{ik}(x)\cdot f_k(x_0,y_0,\lambda)$$

is a solution of the initial value problem (a). This completes the proof of the theorem.

Exercise 1. Prove (b) in all detail. Also prove the last formula in the theorem according to the directions given in the proof above.

Exercise 2. In this as in the preceding section, we have used the following fact in the proofs: If $f: [a,b] \to R$ is of class C^1 and $x,y: [a,b] \to [a,b]$ are continuous, and if, for every t, we choose a ξ_t for which the relation $f(x(t)) - f(y(t)) = f'(\xi_t)(x(t) - y(t))$ of the Mean Value Theorem holds, then the function g defined by $g(t) = f'(\xi_t)$ is summable. Why is this true? Hint: Consider the set of t for which $x(t) \neq y(t)$.

Exercise 3. The reader who knows the concept of derivative in Banach spaces as linear operator should use this idea to prove the theorems of Secs. 3.1 and 3.2. (Some formal simplification will be obtained.)

Exercise 4. Using the notation introduced in this section, prove that when the given system has at most one solution for every choice of initial data (as in the case in the systems that were considered), then the following formula holds:

$$\bar{y}_i(x,x_0+h,y^{(0)},\lambda) = \bar{y}_i(x,x_0,\bar{y}(x_0,x_0+h,y^{(0)},\lambda),\lambda).$$

Exercise 5. Prove Alekseev's formula for a non-linear change of variables: Let $f,F\colon [a,b] \times [c,d] \to \mathrm{R}$ be continuous functions with continuous partial derivatives with respect to the second variable. Denote by $x(t,t_0,x_0)$ the unique solution of

$$x' = f(t,x), \quad x(t_0) = x_0 \tag{3.8}$$

and by $y(t,t_0,x_0)$ the unique solution of

$$y' = f(t,y) + F(t,y), \quad y(t_0) = x_0. \tag{3.9}$$

Then the following relation holds:

$$y(t,t_0,x_0) = x(t,t_0,x_0) + \int_{t_0}^{t} \frac{\partial}{\partial x_0} x(t,s,y(s,t_0,x_0))F(s,y(s,t_0,x_0))ds.$$

Hint: Write $y(t) = y(t,t_0,x_0)$, and differentiate the function $x(t,s,y(s))$ with respect to s. Then integrate the result from t_0 to t. (The formula holds in R^n also. State and prove it in that case.)

Exercise 6. Let $f,F\colon [a,\infty[\ \times \mathrm{R} \to \mathrm{R}$ be continuous functions with continuous partial derivatives with respect to the second variable. Prove that

$$y' = f(t,y) + F(t,y)$$

has a bounded solution if $f(t,0) \equiv 0$, $f_x \le m < 0$ and $|F(t,y)| \le h(t)$ with $\int_a^{+\infty} h(s)ds < \infty$.

Exercise 7. With the notation of Exercise 5, prove that

$$x(t,t_0,y_0) - x(t,t_0,x_0) = \int_0^1 \frac{\partial}{\partial x_0} x(t,t_0,x_0+\lambda(y_0-x_0))\cdot(y_0-x_0)d\lambda.$$

3.3 Higher Order Differentiability and Analyticity

We now suppose that in the rectangular domain R, the functions $f_i(x,y_1,\ldots,y_n,\lambda)$ $(i = 1,2,\ldots,n)$ as well as their partial derivatives up to the pth order with respect to all the variables are continuous. Under these conditions, the functions

$$y_i = y_i(x;x_0,y_1^{(0)},\ldots,y_n^{(0)},\lambda) \tag{3.10}$$

and their partial derivatives up to order p with respect to the variables $y_1^{(0)},\ldots,y_n^{(0)},\lambda$ are also continuous; differentiability with respect to x was already studied in Sec. 1.

This follows immediately from what was shown in the previous sections. In fact, the derivatives of y_i with respect to each variable satisfies a system of linear equations, the variational system, whose coefficients depend on the variables $y_1^{(0)},\ldots,y_n^{(0)},\lambda$ via the composite functions

$$f'_{iy_s}(x,y_1,\ldots,y_n,\lambda),\ f'_i\,(x,y_1,\ldots,y_n,\lambda)$$

and are therefore differentiable with respect to the parameter λ and the variables $y_1,\ldots,y_n$. From this, one may deduce the existence of the second derivatives of the y_i. Proceeding in this way, we are able to verify our assertion.

One may prove - by a procedure similar to that used in Sec. 2 - that the functions (3.10) are analytic if the functions $f_i(x,y_1,\ldots,y_n,\lambda)$ are, but we omit the proof.

Instead, we observe that, in this case, there is an expression

$$y_i \equiv y_i(x,x_0;\beta_1,\ldots,\beta_n)$$

which - at least in a neighborhood of the point $(x_0,y_1^{(0)},\ldots,y_n^{(0)})$ - furnishes all the solutions of the system. This expression depends in an essential way on n arbitrary constants $\beta_1,\beta_2,\ldots,\beta_n$ (the coordinates of the initial point) and constitutes - at least locally - the *general solution* or *general integral* of the system (1.9).

This observation brings us back to what was said - in general - in Sec. 1.2; but we make clear that the problem of finding the analytic solution of the general integral is quite difficult - and up to now more or less abandoned - because this integral may be complicated even when the functions f_i have very simple analytic expressions. In the second chapter, we shall study this problem for the case when the given system is linear.

3.4. Remark about a More General Point of View

Up to now, we have studied the qualitative properties of the function T that associates with each point $(x,x_0,y_1^{(0)},\ldots,y_n^{(0)},\lambda)$ of $[a,b] \times R$ the value at x of the unique solution that passes through $(x_0,y_1^{(0)},\ldots,y_n^{(0)})$. Instead of T, one could study another function closely connected with it: the function which associates with each point $(x_0,x_0,y_1^{(0)},\ldots,y_n^{(0)},\lambda)$ the solution of the given system that passes through $(x_0,y_1^{(0)},\ldots,y_n^{(0)})$. We would then have a function S from $[a,b] \times R$ into the set $C([a,b],R^n)$ of continuous functions from $[a,b]$ to R^n. We may extend all the previous results to S using the Banach space structure of $C([a,b],R^n)$ which will be introduced in Sec. 4.2 below. The proofs are the same as those which are given to establish the properties of T.

4. SYSTEMS OF EQUATIONS AS PARTICULAR TRANSFORMATIONS BETWEEN FUNCTION SPACES

Let us consider an ordinary differential equation

$$y' = f(x,y)$$

with $f\colon I \times R \to R$ and look for the solution that satisfies the initial condition $y(x_0) = y_0$. We know that the solution u that we want can be written in the following form

$$u(t) = y_0 + \int_{x_0}^{x} f(s,u(s))ds. \tag{4.1}$$

If we look at the second side of this equality, we note that the formula

$$y_0 + \int_{x_0}^{x} f(s,u(s))ds$$

associates with each continuous function u defined on the interval I another continuous function defined on I. (4.1) therefore defines a transformation T from the set $C(I)$ of continuous real-valued functions defined on I into itself:

$$T\colon C(I) \to C(I).$$

In this section, we will examine the properties of C(I) and of T that are of importance for the study of the Cauchy problem. To determine the properties of T, we observe that u is a solution of the given initial value problem if and only if u satisfies identity (4.1). This means that u is invariant under T, that is, u must satisfy the relation

$$u = Tu.$$

Points enjoying this property are called *fixed points* of T. In Sec. 4.1 we shall prove a sufficient condition for the existence of fixed points and we shall use it to study Cauchy's problem.

4.1. Review of Metric Spaces

In this section, we shall go over those notions of general topology which will prove to be useful in the sequel.

There are on the real line two different but equivalent ways of defining the concept of limit: by relying on the notion of interval or on that of absolute value. In trying to generalize these two definitions to an arbitrary set, one arrives respectively at topological spaces and metric spaces.

If we associate with every real number x the family $\mathcal{U}_x$ of all subsets of $\mathbb{R}$ containing an open interval to which x belongs, one obtains a collection of subsets of $\mathbb{R}$ that has special properties with respect to the operations of union, intersection, inclusion, and "is an element of": $\cup$, $\cap$, $\subseteq$, $\in$. If these are taken as axioms of a structure, one obtains the concept of *topological space*. In other words, a topological space is a set X endowed with the structure defined by associating with each point $x \in X$ a family $\mathcal{V}_x$ of subsets of X in such a way that $(\mathcal{V}_x)_{x\in X}$ has the same properties as $(\mathcal{U}_x)_{x\in \mathbb{R}}$ with respect to the operations of union, intersection, inclusion, and "is an element of." The family $(\mathcal{V}_x)_{x\in X}$ is called the *system of neighborhoods of the topological space* X, $\mathcal{V}_x$ is called the *system of neighborhoods of* x, and every $V \in \mathcal{V}_x$ is called a *neighborhood of* x. We do not stop here to discuss what these properties are formally because we are interested only in *using* topology, and in the cases when we meet it, we shall always be able to make use of the more convenient and manageable notion of metric space which we shall introduce further below.

We begin by examining certain subsets and certain points of a topological space which are privileged with respect to the topological structure because of their special properties. Let X be a topological space with the topology defined by the system of neighborhoods $(\mathcal{V}_x)_{x\in X}$, let A be a subset of X, and let x be a point in X. We say that A is *open* if it is a neighborhood of each of its points. We say that A is *closed* if its complement is open. We say that x is an *adherent point* of A if each of its neighborhoods meets A, that is, if

$$A \cap V \neq \emptyset, \qquad V \in \mathcal{V}_x.$$

The set of adherent points of A is called the *closure of* A and denoted by $\bar{A}$. We say that x is an *interior point* of A if A is a neighborhood of x. The set of interior points of A is called the *interior of* A and denoted A^0. We say that x is a *boundary point* of A if x is in the interior neither of A nor of the complement of A, that is, if $x \in \bar{A} \cap \overline{(X-A)}$. The set of boundary points of A is called the *boundary of* A and denoted by ∂A. We say that A is *dense* in B if $A \subseteq B$ and $\bar{A} = B$.

A subset C of a topological space X is called *compact* if every open covering has a finite subcovering, that is, if from every family $(U_\alpha)_{\alpha\in A}$ of open sets of X whose union contains C, we can extract a finite family whose union contains C; this means that there exist $\alpha_1,\ldots,\alpha_n$ such that $C \subseteq U_{\alpha_1} \cup\cdots\cup U_{\alpha_n}$. A subset C of a topological space X is called *connected* if there are no two open sets U_1 and U_2 of X such that $C \subseteq U_1 \cup U_2$, $C \cap U_i \neq \emptyset$ for $i = 1,2$ and $(C \cap U_1) \cap (C \cap U_2) = \emptyset$. This definition can be reformulated more simply in the following manner.

If A is a subset of the topological space X with system of neighborhoods $(\mathcal{V}_x)_{x\in X}$, one may define a topology in A in a very natural way by saying that the neighborhoods of the points in A are the "traces" in A of the points' neighborhoods in X, that is, one takes for the system of neighborhoods of a point x in A the family of sets

$$A \cap U, \qquad (U \in \mathcal{V}_x).$$

The topological structure thus obtained is called the *topology induced* by X on A. In this terminology, a subset C of a topological space X is connected if and only if it cannot be broken up into the union of two disjoint non-empty sets which are open in the topology induced by X on A.

We come at last to the concept of limit in topological spaces. Let X be a topological space with system of neighborhoods $(\mathcal{V}_x)_{x\in X}$ and let $(x_n)_{n=1}^\infty$ be a sequence in X. We say that $(x_n)_{n=1}^\infty$ converges to the point x_0 of X, written $\lim_n x_n = x_0$, if, for every neighborhood U of x_0, there is an n_U such that $x_n \in U$ for every $n \geq n_U$. The condition

$$x_n \in U, \qquad (n \geq n_U)$$

is expressed by saying that $(x_n)_{n=1}^{\infty}$ is eventually in U. Therefore, $\lim_n x_n = x_0$ if and only if $(x_n)_{n=1}^{\infty}$ is eventually in every neighborhood of x_0.

We now consider two topological spaces X and Y. Among all the functions from X to Y, there are some which respect the topological structure of the two sets. These are the continuous functions. More precisely, let us consider a function $f: X \to Y$. We say that f *is continuous at the point* $x \in X$ if, for every neighborhood V of f(x) in Y, there is a neighborhood U of x in X such that $f(U) \subseteq V$. We say that f is continuous if it is continuous at every point $x \in X$. The fact that a *continuous* function "respects" the topological structures of the two systems depends on the following property, which is easy to prove: f is *continuous* if and only if the inverse image $f^{-1}(V)$ of every open (closed) set V of Y is an open (closed) set of X. It is just as easy to prove that if C is compact (connected), and f is continuous, then f(C) is also compact (connected).

We now come to the notion of metric space, in which the intuitive notion of the distance between two points in the plane or in three dimensional space is extended into abstract spaces. Precisely, a *metric space* is a set X for which there exists a function $d: X \times X \to [0,\infty[$ such that

(i) $d(x,y) = 0 \Longleftrightarrow x = y$

(ii) $d(x,y) = d(y,x)$

(iii) $d(x,y) \leq d(x,z) + d(z,y)$.

Such a function d is called a *distance* or a *metric*. (ii) expresses the symmetry property; (iii) the triangle inequality. One denotes by (X,d) the metric space consisting of the set X together with the metric d. One may define a metric in the set of real numbers quite easily by putting

$$d(x,y) = |x - y|,$$

whereas in $\mathbb{R}^n$ one may define various metrics in a natural way:

$$d(x,y) = \sqrt{\sum_{i=1}^{n} |x_i - y_i|^2}, \quad d(x,y) = \sup_{i=1,\dots,n} |x_i - y_i|,$$

$$d(x,y) = \sum_{i=1}^{n} |x_i - y_i|.$$

The first of these is called the Euclidean metric because it is the one that corresponds to the notion of distance in Euclidean geometry.

Let (X,d) be a metric space. If $x \in X$ and $\varepsilon > 0$, the set

$$\{y \in X \mid d(x,y) < \varepsilon\}$$

is called the *ball with center* x *and radius* ε and is denoted by $B(x,\varepsilon)$. One proves in general topology that if for every $x \in X$ we define $\mathcal{V}_x$ to be the set of all $V \subseteq X$ for which there is an $\varepsilon > 0$ such that $B(x,\varepsilon) \subseteq V$, then $(\mathcal{V}_x)_{x \in X}$ is the system of neighborhoods for a topology on X called the *topology associated with the metric space* (X,d). We now consider the notion of limit. If we recall the definition of the system of neighborhoods of a point $x_0 \in (X,d)$, we easily see that $\lim_n x_n = x_0$ in the topology of (X,d) if and only if $\lim_n d(x_n,x_0) = 0$ in $\mathbb{R}$. Thus the notion of limit in metric spaces is reduced to that of limit on the real line.

The notion of the limit of a sequence can be used to characterize all the topological concepts in a metric space. We give two common cases as examples.

If $A \subseteq (X,d)$, then $x \in \overline{A}$ if and only if there exists a sequence $(x_n)_{n=1}^{\infty}$ of points of A that converge to x.

A subset A of a metric space is compact if and only if every sequence of points of A has a subsequence that converges to a point of A.

As is known from the theorem of Bolzano and Weierstrass, the compact sets of $\mathbb{R}^n$ are the closed and bounded ones.

Let us consider a convergent sequence in a metric space (X,d): $\lim_n x_n = x_0$. Upon applying the triangle inequality, we get

$$d(x_n,x_m) \le d(x_n,x_0) + d(x_0,x_m)$$

and therefore

$$\lim_{n,m} d(x_n,x_m) = 0. \tag{4.2}$$

We have thus found a necessary condition for the convergence of a sequence in a metric space. It is not, however, sufficient. For example, let us consider the metric space consisting of the set $\mathbb{Q}$ of rational numbers with the same metric

$$d(x,y) = |x - y|$$

already defined for the real numbers. The sequence of rational numbers with general term

$$a_n = \left(1 + \frac{1}{n}\right)^n$$

converges in the set of real numbers to the irrational e. (a_n) thus satisfies condition (4.2) relative to the distance in $\mathbb{R}$ and so also relative to that in $\mathbb{Q}$, since the two are equal. But (a_n) does not converge in $\mathbb{Q}$. (a_n) is therefore an example of a sequence in $(\mathbb{Q},d)$ which satisfies (4.2) but does not converge to any point of $(\mathbb{Q},d)$.

Sequences $(x_n)_{n=1}^{\infty}$ satisfying condition (4.2) are called *Cauchy sequences*. Metric spaces in which all Cauchy sequences converge are called *complete spaces*. It is known that $\mathbb{R}$ and $\mathbb{R}^n$ are complete metric spaces. The following property is a direct result of the definition: A closed subset of a complete metric space is complete in the induced metric. As appears natural, *the metric induced by* (X,d) on its subset A is the restriction $d|_{A\times A}$, which is clearly a metric.

In metric spaces, we can define some special types of functions: uniformly continuous functions, functions satisfying a Lipschitz condition, and functions satisfying a Lipschitz condition locally. Suppose (X,d) and (Y,d') are two metric spaces. We say $f\colon X \to Y$ is *uniformly continuous* if, for every $\varepsilon > 0$ there exists a $\delta > 0$ such that

$$d'(f(x),f(y)) \le \varepsilon$$

for $d(x,y) < \delta$. We say that $f\colon X \to Y$ *satisfies a Lipschitz condition with Lipschitz constant* L if $L > 0$ and

$$d'(f(x),f(y)) \le Ld(x,y), \qquad (x,y \in X).$$

When the Lipschitz constant is less than 1, the function f is called a *contraction*. We say that $f\colon X \to Y$ *satisfies a Lipschitz condition locally* if, for every $x \in X$, there exists a neighborhood V_x and $L_x \ge 0$ such that $f|_{V_x}$ satisfies a Lipschitz condition with constant L_x. The general properties of functions satisfying a Lipschitz condition and their relationship to uniformly continuous functions are gathered in the exercises of this section and in Exercises 3 and 4 of the following section.

We now prove the following theorem, which deals with the existence of fixed points, and is of interest to us because of its impact on ordinary differential equations.

Theorem of Banach and Caccioppoli on the Fixed Point of Contractions. If (X,d) is a complete metric space and $f\colon X \to Y$ is a contraction, then f has a unique fixed point, that is, there exists a unique point

$x_0 \in X$ such that $x_0 = f(x_0)$. Moreover, if y_0 is any point whatsoever in X, the sequence defined inductively by

$$y_{n+1} = f(y_n) = \underbrace{f\circ \ldots \circ}_{n \text{ times}} f(y_0)$$

converges to x_0.

The sequence $(y_n)_{n=1}^{\infty}$ is called the *sequence of successive approximations* for f with initial point y_0.

Proof of the theorem: We follow the argument used to prove the existence theorem of Sec. 1. Let k be a Lipschitz constant for f. Since f is a contraction, we may take $k < 1$. We fix y_0 and prove that the sequence $(y_n)_{n=1}^{\infty}$ converges. From the Lipschitz property of f, we have, for every n,

$$d(y_n,y_{n-1}) = d(f(y_{n-1}),f(y_{n-2})) \leq kd(y_{n-1},y_{n-2}).$$

Applying this formula again and again yields

$$d(y_n,y_{n-1}) \leq k^{n-1}d(y_1,y_0).$$

From this and the triangle inequality it follows that

$$d(y_{n+p},y_n) \leq d(y_{n+p},y_{n+p-1}) + d(y_{n+p-1},y_{n+p-2}) + \ldots + d(y_{n+1},y_n)$$
$$\leq (k^{n+p-1} + \ldots + k^n)d(y_1,y_0).$$

The expression $k^n + \ldots + k^{n+p-1}$ is a partial remainder of the geometric series with ratio k, $0 < k < 1$, and therefore tends to zero uniformly with respect to p as $n \to \infty$. This implies

$$\lim_{n,m} d(y_n,y_m) = 0.$$

$(y_n)_{n=1}^{\infty}$ is therefore a Cauchy sequence. Because of the completeness of X there exists $y_\infty \in X$ such that $\lim_n y_n = y_\infty$. Since f is continuous (cf. Exercise 1), we may pass to the limit on both sides of

$$y_{n+1} = f(y_n)$$

and get $y_\infty = f(y_\infty)$, so that f has at least one fixed point. To show that the fixed point is unique, we suppose that there is another fixed point x_1 and observe that

$$d(x_1,y_\infty) = d(f(x_1),f(y_\infty)) \leq kd(x_1,y_\infty) < d(x_1,y_\infty),$$

since $d(x_1,y_\infty) > 0$ for $x_1 \neq y_\infty$. The theorem is thus proved.

We give a series of exercises that illustrate the properties of functions that satisfy a Lipschitz condition.

Exercise 1. Prove that a function satisfying a Lipschitz condition is uniformly continuous, whereas one satisfying a Lipschitz condition locally is continuous.

Exercise 2. Prove that if $A \subseteq \mathbb{R}^n$ is open and $f: A \to \mathbb{R}^m$ is of class C^1, then f satisfies a Lipschitz condition locally.

Exercise 3. If A is a rectangle in $\mathbb{R}^n$ and $f: A \to \mathbb{R}^m$ is of class C^1, prove that f satisfies a Lipschitz condition if and only if all its partial derivatives are bounded.

Exercise 4. Let (X,d) be a compact metric space, (Y,d') an arbitrary metric space, and $f: X \to Y$. Prove that if f satisfies a Lipschitz condition locally, then f satisfies a Lipschitz condition. Hint: For every $x \in X$, there is a ball $B(x,\varepsilon_x)$ in which f satisfies a Lipschitz condition with Lipschitz constant L_x. There exist finitely many points $x_1,\ldots,x_n \in X$ such that X is the union of the balls $B(x_i,\varepsilon x_i/3)$. Put $\delta = \min \varepsilon_{x_i}/3$, $L_0 = \max L_{x_i}$, $M \geq d'(f(x),f(y))$ (why is M finite?) and $L = \max\{\frac{M}{\delta},L_0\}$. In order to prove that

$$d'(f(x),f(y)) \leq Ld(x,y)$$

for every $x,y \in X$, let us fix x and y. Let i be such that $x \in B(x_i,\varepsilon x_i/3)$. If $y \notin B(x_i,\varepsilon x_i/3)$, then $d(x,y) > \delta$ and therefore $d'(f(x),f(y)) \leq M$.

Exercise 5. Extend Exercise 4 to the case $f: I \times X \to Y$ with $I \subseteq \mathbb{R}$ compact, X and Y as in Exercise 4, and f locally satisfying a Lipschitz condition with respect to the variable $x \in X$.

Exercise 6. f is called a Hölder function if

$$d'(f(x),f(y)) \leq L(d(x,y))^{\alpha}$$

with $L, \alpha > 0$. State and prove a result analogous to that of Exercise 4 for Hölder functions.

4.2. Review of Banach Spaces

In this section, we shall summarize those notions of linear functional analysis that will be used in the sequel. The reader who wishes to go deeper into these questions may consult the books of Banach [4], Dunford and Schwartz [10], and Yoshida [36].

Banach spaces are the most commonly found examples of metric spaces in mathematical analysis. They have the particular attribute of combining with the topological structure derived from the metric the algebraic structure of a vector space, the two structures being linked together. We recall that a *vector space* is a set X with the following two operations defined on it:

(α) *sum*, which is a function from $X \times X \to X$ whose values at (x,y) is denoted by $x + y$ and enjoys the following properties:

(α_1) associativity: $(x+y) + z = x + (y+z)$

(α_2) commutativity: $x + y = y + x$

(α_3) existence of zero: there is an element 0 such that $x + 0 = x$ for every $x \in X$;

(α_4) existence of the inverse: for every $x \in X$, there is an element $-x$ such that $x + (-x) = 0$.

(β) *scalar product,* which is a function from $\mathbb{R} \times X$ to X whose value at (λ,x) is denoted by λx and enjoys the properties of

(β_1) distributivity with respect to vectors: $\lambda(x+y) = \lambda x + \lambda y$;

(β_2) distributivity with respect to scalars: $(\mu+\nu)x = \mu x + \nu x$;

(β_3) associativity: $\mu(\nu x) = (\mu\nu)x$;

(β_4) $1x = x$.

Properties (α_i) mean that $(X,+)$ is a commutative group. One can easily show that the element 0 that appears in (α_3) is unique; it is called the *origin* or *zero* of the vector space. Similarly, one can show that the element $-x$ of (α_4) is unique for every x; it is called the *inverse* of x. The elements of X are called *vectors* whereas the real numbers, in the context of vector spaces, are called *scalars*.

The following are the examples of vector spaces that we meet most frequently in the theory of ordinary differential equations:

(1) $\mathbb{R}$ with the usual operations of sum and product.

(2) $\mathbb{R}^n$ with the operations defined coordinate-wise:

$$x + y = (x_1 + y_1, \dots, x_n + y_n)$$

$$\lambda x = (\lambda x_1, \dots, \lambda x_n).$$

(3) The set of $m \times n$ matrices with the operations defined element-wise:

$$(\alpha_{ij})_{ij} + (\beta_{ij})_{ij} = (\alpha_{ij} + \beta_{ij})_{ij}$$

$$\lambda(\alpha_{ij})_{ij} = (\lambda\alpha_{ij})_{ij}.$$

(4) The set of continuous functions (resp. C^p functions) from $I \subseteq \mathbf{R}$ into $\mathbf{R}^n$, with operations defined as follows: if f and g are functions and $\lambda \in \mathbf{R}$, then $f + g$ and λf are the functions whose values at $x \in I$ are given by the formulas:

$$(f+g)(x) = f(x) + g(x)$$
$$(\lambda f)(x) = \lambda f(x).$$

This space will be denoted by $C(I,\mathbf{R}^n)$ (resp. $C^p(I,\mathbf{R}^n)$), or simply $C(I)$ (resp. $C^p(I)$) when $n = 1$. Observe that the usual properties of continuous functions and of class C^p functions guarantee that $f + g$ and λf are themselves continuous or of class C^p respectively.

(5) The set $C(X,Y)$ of continuous functions between two topological spaces X and Y with the operations defined as in (4), Y being in addition a vector space.

In a vector space X the operation

$$\sum_{i=1}^{n} \lambda_i x_i$$

is well defined for every n-tuple $x_1,\ldots,x_n$ of vectors and n-tuple $\lambda_1,\ldots,\lambda_n$ of scalars:

$$\sum_i \lambda_i x_i = (\ldots((\lambda_1 x_1 + \lambda_2 x_2) + \lambda_3 x_3) + \ldots) + \lambda_n x_n).$$

This operation is called *linear combination*. A set $A \subseteq X$ is called a *convex set* if whenever $x_1,\ldots,x_n \in A$ and $\lambda_1,\ldots,\lambda_n \geq 0$ with $\Sigma_{i=1}^n \lambda_i = 1$ then $\Sigma_{i=1}^n \lambda_i x_i \in A$. One may easily prove that this is equivalent to saying that if $x,y \in A$ and $\lambda \in [0,1]$, then $\lambda x + (1-\lambda)y \in A$.

A *vector subspace* of X is a set $A \subseteq X$ such that if $x_1,\ldots,x_n \in A$ and $\lambda_1,\ldots,\lambda_n \in \mathbf{R}$, then $\Sigma_i \lambda_i x_i \in A$. One may easily prove that this is equivalent to saying that if $x,y \in A$ and $\alpha,\beta \in \mathbf{R}$, then $\alpha x + \beta y \in A$. One may prove just as easily that A is a vector subspace of X if and only if A is a vector space under the induced operations.

There are functions between two vector spaces that respect the algebraic structure of each; these are called linear functions or, even better, linear operators. Precisely, $f\colon X \to Y$ is a linear operator if

$$f(\alpha x + \beta y) = \alpha f(x) + \beta f(y), \qquad (\alpha,\beta \in \mathbf{R};\ x,y \in X).$$

The structure of a metric space may be introduced into a vector space through the notion of a norm. A *norm* on a vector space X is a function, generally denoted by $||\cdot||$, from X into $[0,\infty[$ that has the following properties:

(i) $||x|| = 0 \Leftrightarrow x = 0 \qquad (x \in X);$

(ii) $||\lambda x|| = |\lambda| \; ||x|| \qquad (\lambda \in R;\; x \in X);$

(iii) $||x + y|| \le ||x|| + ||y|| \quad (x,y \in X).$

(ii) is called the homogeneity property, and (iii) the triangle inequality. From the three characteristic properties of the norm, it follows that

$$\big| ||x|| - ||y|| \big| \le ||x - y|| \qquad (x,y \in X).$$

A normed space is a vector space with a norm.

In the examples (1),...,(4) of vector spaces, various norms may be defined.

(a) In R, the absolute value $|x|$ is a norm.

(b) In R^n, the following functions are norms:

$$||x|| = \sum_i |x_i|$$

$$||x|| = \sqrt{\Sigma_i |x_i|^2} \qquad \text{(Euclidean norm)}$$

$$||x|| = \max_i |x_i| \qquad \text{(sup norm)}.$$

(c) In the space of $m \times n$ matrices norms analogous to those in (b) may be defined:

$$||(\alpha_{ij})|| = \sum_{ij} |\alpha_{ij}|$$

$$||(\alpha_{ij})|| = \sqrt{\Sigma_{ij} |\alpha_{ij}|^2}$$

$$||(\alpha_{ij})|| = \max_{ij} |\alpha_{ij}|.$$

(d) In $C^p(I,R^n)$, where I is a compact interval in R, one may define a norm in the following manner:

$$||f|| = \sup_{x \in I}\{||f(x)|| + ||f'(x)|| + \ldots + ||f^{(p)}(x)||\},$$

or as follows

$$||f|| = \sup_{x \in I}||f(x)|| + \max_{k=1,\ldots,p}\; \sup_{x \in I}||f^{(k)}(x)||.$$

(e) In $C(X,Y)$, where X is a compact space and Y a normed space, a norm may be defined by

$$||f|| = \sup_{x \in X}||f(x)||.$$

This is called the *sup norm*. The topology associated with its metric is that of uniform convergence. We are interested in the case when X is a compact interval in R and $Y = R^n$.

(f) Let X be a normed space, I a closed interval (or half-open on the right) and $a = \inf I$, $b = \sup I$. Let $h: I \to [0,\infty)$ be a continuous function such that

$$\int_a^b h(t)dt < +\infty$$

where $\int_a^b$ represents the improper integral when I is not closed. Let $\varepsilon > 0$ and $X_{\varepsilon,h}$ be the set of all continuous functions $f: I \to X$ such that the number $||f||_{\varepsilon,h}$ defined by

$$||f||_{\varepsilon,h} = \sup_{t\in I} e^{-\varepsilon\int_a^t h(s)ds} ||f(t)||$$

is finite. Then $X_{\varepsilon,h}$ is a vector space and $||\cdot||_{\varepsilon,h}$ is its norm.

A metric may be associated with a normed space in a way that is similar to what is done in the real line by means of the absolute value. If X is a normed space with norm $||\cdot||$, then one defines a metric on X by putting

$$d(x,y) = ||x - y||.$$

Such a metric is called the metric or distance associated with the norm $||\cdot||$ or with X. Requiring that such a metric space be complete leads to the concept of a Banach space. Precisely, a *Banach space* is a normed space which, as a vector space, is complete in the metric associated with its norm. All the examples (a),...,(f) of normed spaces are also examples of Banach spaces.

Remark I. We saw in (b) that we can define various norms in R^n. These in turn have various associated metrics and therefore, one would think, various topologies. One can prove, however, that all the norms on R^n define the same topology, that is, are equivalent according to the following definition. We say that two norms $||\cdot||_1$, $||\cdot||_2$ on a vector space are *equivalent* when they have the same topology in the associated metrics. One may prove that this is equivalent to the existence of two constants A and B such that

$$A||x||_1 \leq ||x||_2 \leq B||x||_1, \qquad (x \in X).$$

Remark II. If X is a vector space, $A \subseteq X$ and $f: A \to X$, then x_0 is a fixed point of f if and only if $x_0 - f(x_0) = 0$. In Banach spaces, therefore, the search for the fixed points of a function f can be replaced by the search for the zeroes of the function $I - f$, where I is the identity transformation on X. In particular, f has a fixed point if $I - f$ is surjective. This point of view will be used in Chapters III and IV.

Exercise 1. Prove the following property of the norm:

$$\big|\,||x|| - ||y||\,\big| \leq ||x-y||.$$

Exercise 2. Prove that in examples (a),...,(f) the normed spaces are Banach spaces, that is, that we are dealing in each case with vector spaces with a norm, and that the associated metric is complete.

Exercise 3. Let X be a metric space and Y a normed space. Prove that if $f: X \to Y$ is a function satisfying a Lipschitz condition then it does so for every equivalent norm on X and Y. When, therefore, we deal with ordinary differential equations whose second member satisfies a Lipschitz condition, it is of no particular importance what norm we select on R^n, except for special circumstances.

Exercise 4. Let X be a metric space. Prove that $f: X \to \mathsf{R}^n$ satisfies a Lipschitz condition if and only if every coordinate function $f_i: X \to \mathsf{R}$ does.

Exercise 5. Let X be a compact metric space and Y a normed space. Prove that if $f_n \in C(X,Y)$, then $\lim_n f_n = f_0$ in the sup norm if and only if $\lim_n f_n = f_0$ uniformly in X.

Exercise 6. Prove that when the interval I is compact, the set X defined in example (f) of normed spaces coincides with $C(I,X)$, and that the sup norm and the norm $||\cdot||_{\varepsilon,h}$ are equivalent, i.e., $\lim_n x_n = x$ holds in $X_{\varepsilon,h}$ if and only if $\lim_n x_n = x$ uniformly.

Exercise 7. Prove that the set of real-valued functions that are bounded and continuous on $[a,\infty[$ constitutes a Banach space with the sup norm.

4.3. The Cauchy Problem and Fixed Points of Certain Transformations in Banach Spaces

In this section, we intend to study Cauchy's problem for a system

$$y_i' = f_i(x,y), \quad y_i(a) = y_i^{(0)} \tag{4.3}$$

using the method sketched in the introduction of the paragraph; we want to associate with it a Banach space of functions and a transformation T defined on it in such a way that T has fixed points and these are the solutions of (4.3). Let us therefore suppose that the functions f_i are functions from $[a,b] \times \mathbb{R}^n$ with values in $\mathbb{R}$ and satisfy a Lipschitz condition with respect to the variable y, that is, there exists a constant L such that (cf. Exercise 4 of the preceding section)

$$|f_i(x,y) - f_i(x,\bar{y})| \leq L||y - \bar{y}||,$$

where $||\cdot||$ is any norm on $\mathbb{R}^n$. In agreement with Remark I and Exercise 3 of the preceding section, we may assume that

$$||x|| = \max_i |x_i|.$$

Let us consider the Banach space $C([a,b],\mathbb{R}^n)$ with the sup norm. We associate with every function $y \in C(I,\mathbb{R}^n)$ a new function Ty whose i-th coordinate assumes at $x \in [a,b]$ the value

$$y_i^{(0)} + \int_a^x f_i(t,y(t))dt,$$

that is,

$$Ty(x) = \left(y_1^{(0)} + \int_a^x f_1(t,y(t))dt,\ldots,y_n^{(0)} + \int_a^x f_n(t,y(t))dt\right).$$

We deduce from the integral representation of the solutions that the solution of (4.3) is a fixed point of T and vice versa. The question that interests us is, therefore, whether or not T has fixed points. To find out, we shall see whether T is a contraction. To this end, let us consider $u,v \in C([a,b],\mathbb{R}^n)$ and the inequalities

$$\left|y_i^{(0)} + \int_a^x f_i(t,u(t))dt - y_i^{(0)} - \int_a^x f_i(t,y(t))dt\right|$$
$$\leq \int_a^x |f_i(t,u(t)) - f_i(t,y(t))|dt \leq \int_a^x L||u(t) - v(t)||dt$$
$$\leq (x-a)L||u - v||,$$

where $||u - v||$ is the sup norm of $u - v$. By taking first the max and then the sup on both sides we get

$$||Tu - Tv|| \leq (a-b)L||u - v||,$$

where $||\cdot||$ represents the sup norm. T is thus a contraction if

$(b-a)L < 1.$

In case this condition cannot be verified, one may restrict the interval to a smaller one $[a,b_0] \subseteq [a,b]$ with length $\delta = b_0 - a$ satisfying the condition $0 < \delta < 1/L$. Upon repeating for $C([a,b_0],R^n)$ the steps already made, we find that T is a contraction of $C([a,b_0],R^n)$ into itself and therefore has a fixed point $\bar{y}$ by the theorem on contractions. Such a function $\bar{y}$ is a solution of (4.3) in $[a,b_0]$. From the fixed point theorem for contractions, one also obtains that $\bar{y}$ is the unique solution and is the uniform limit of the sequence of successive approximations

$$y_{n+1i} = y_i^{(0)} + \int_a^x f_i(t,y_n(t))dt, \tag{4.4}$$

where y_0 is any continuous function from $[a,b_0]$ into R^n.

In this manner, one may also prove the similar result for the case in which the domain of the f_i is a rectangle in R^{n+1}; see Exercise 4.

Although the proof we have just given is remarkably simple, it does not lead to new results because the proof of the fixed point theorem for contractions is in substance that of the existence theorem given in Section 1. We shall, however, show that by changing the sup norm, we succeed in obtaining directly global existence, uniqueness, and convergence of the successive approximations. (Recall that in Section 1 local and global existence were treated separately using different methods.)

Let $\varepsilon > 0$ be a number which we shall fix later. By means of the formula

$$||x||_\varepsilon = \sup_{a \le t \le b} e^{-\varepsilon L(t-a)}||x(t)||$$

we can define a Banach space norm on $C([a,b],R^n)$ whose topology coincides with that of uniform convergence. (Cf. Example (f) and Exercise 6 of the preceeding section and take h to be the function with constant value L.) We shall show that the operator T is a contraction if we equip $C([a,b], R^n)$ with the norm $||\cdot||_\varepsilon$. Then the fixed point theorem for contractions will provide global existence, uniqueness, and convergence of the successive approximations.

Let us fix $u,v \in C([a,b],R^n)$. From the inequalities

$$\left|y_i^{(0)} + \int_a^x f_i(t,u(t))dt - y_i^{(0)} - \int_a^x f_i(t,v(t))dt\right|$$

$$\le \int_a^x |f_i(t,u(t)) - f_i(t,v(t))|dt \le \int_a^x L||u(t) - v(t)||dt$$

$$= L\int_a^x e^{\varepsilon L(t-a)}e^{-\varepsilon L(t-a)}||u(t) - v(t)||dt$$

$$\le L\int_a^x e^{\varepsilon L(t-a)}||u - v||_\varepsilon dt$$

$$= L||u - v||_\varepsilon \int_a^x e^{\varepsilon L(t-a)}dt$$

$$= L||u - v||_\varepsilon \frac{e^{\varepsilon L(x-a)} - 1}{\varepsilon L}$$

$$\le \frac{||u-v||_\varepsilon}{\varepsilon} e^{\varepsilon L(x-a)}$$

we obtain, by taking the max,

$$||Tu(x) - Tv(x)|| \le \frac{e^{\varepsilon L(x-a)}}{\varepsilon} ||u - v||_\varepsilon.$$

If we divide the two sides by $e^{\varepsilon L(x-a)}$ and take $\sup_x$, we get

$$||Tu - Tv||_\varepsilon \le \frac{1}{\varepsilon} ||x - y||_\varepsilon.$$

This is true for every $\varepsilon > 0$. In particular, it is true for $\varepsilon > 1$. But then T is a contraction for the metric associated with $||\cdot||_\varepsilon$, as we wanted to show.

Exercise 1. Using the ideas of the last proof, treat the case in which

$$|f_i(x,y) - f_i(x,\bar{y})| \le h(x)||y - \bar{y}||$$

with $\int_a^{+\infty} h(t)dt < +\infty$. Is this condition sufficient to assure existence in $[a,\infty[$?

Exercise 2. Let $f\colon [a,b] \times \mathbb{R}^n \to \mathbb{R}^n$ be a continuous function. Prove that there is $0 < \varepsilon \le b-a$ such that the successive approximations (4.4) converge to the solution of (4.3) uniformly in $[a,a+\varepsilon]$ if and only if $\lim_n (y_{n+1} - y_n) = 0$ uniformly in $[a,a+\varepsilon]$. Hint: Make use of the theorem of Arzelà-Ascoli.

Exercise 3. Let $U \subseteq \mathbb{R}^n$ be open, let $f\colon [a,b] \times U \to \mathbb{R}^n$ be continuous, and let $x_0 \in U$. Prove that there exists $0 < \varepsilon < b-a$ such that the successive approximations (4.4) converge to the solution of (4.3) uniformly in $[a,a+\varepsilon]$ if there exists a continuous function $\omega: [a,b] \times [0,c] \to [0,\infty[$ such that

$$||f_i(x,y) - f_i(x,\bar{y})|| \le \omega(x,||y - \bar{y}||),$$

where $\omega(t,\cdot)$ is increasing, and the unique solution of the initial value problem

$$u' = \omega(x,u), \quad u(a) = 0$$

is $u \equiv 0$. Hint: For a suitable $M > 0$, consider the successive approximations

$$u_0(x) = M(x-a), \quad u_{n+1}(x) = \int_a^x \omega(t,u_n(t))dt$$

and prove that $\lim_n u_n = 0$ and

$$\sup_{n,m\ge k} ||y_n(x) - y_m(x)|| \le u_k(x), \qquad (k \ge 1;\ a \le x \le a+\varepsilon).$$

Exercise 4. Prove the theorem of local existence (that of Section 1.4) using contractions. Hint: Use the fact that if $A \subseteq \mathbb{R}^n$ is closed, then $C(I,A)$ is a complete metric subspace of the Banach space $C(I,\mathbb{R}^n)$ with the sup norm.

5. EXERCISES

We propose to the reader a series of exercises whose solution will require elementary methods that are usually introduced in calculus courses.

Determine the solutions of the following equations.

5.1. Variables Separable Equations

1. $y' = xy$ general solution: $y(x) = C\exp(x^2/2)$.
2. $y' = 1 - y^2$ general solution: $y(x) = \dfrac{Ce^{2x} - 1}{Ce^{2x} + 1}$.
3. $y' = 1 + y^2$ general solution: $y(x) = \tan(x + C)$.
4. $y' = \dfrac{1}{y}$ general solution: $\begin{cases} y(x) = \sqrt{2x + c} \\ y(x) = -\sqrt{2x + c} \end{cases}$.
5. $y' = \dfrac{x}{y}$ general solution: $\begin{cases} y(x) = \sqrt{x^2 + c} \\ y(x) = -\sqrt{x^2 + c} \end{cases}$.

6. $y' = e^{-(x+y)}$ general solution: $y(x) = \lg(c - e^{-x})$.

7. $y' = \sqrt{1 + y^2}$ general solution: $y(x) = \cosh(x + c)$.

In the next exercise, use the trigonometric identity for the sine of the sum and difference of two angles.

8. $y' = \sin(x+y) + \sin(x-y)$

general solution: $y(x) = 2[\arctan(ce^{-2\cos x}) - \frac{\pi}{4}]$.

5.2. Equations Reducible to Separable Equations

Equations whose second parts are quotients of homogeneous polynomials of the same order are reducible to separable equations by means of the substitution $z = y/x$. Example:

$$y' = \frac{x^2 + y^2}{x^2 - y^2}, \quad y = zx; \quad y' = z'x + z$$

$$z'x + z = \frac{1 + z^2}{1 - z^2}, \quad z' = \left(\frac{1 + z^2}{1 - z^2} - z\right)\frac{1}{x}. \tag{5.1}$$

In general, such a transformation does not allow us to give a solution in terms of elementary functions; nevertheless, it reduces the problem to one of integration and inversion of a function, problems which are, among other things, much more simple from the numerical point of view. We shall now give some exercises in which the solutions are obtained in an explicit form.

1. $y' = \frac{1}{4}\frac{3x - 4y}{x - y}$ general solution: $\begin{cases} y = x + \frac{1}{2}\sqrt{x^2 + c} \\ y = x - \frac{1}{2}\sqrt{x^2 + c} \end{cases}$

2. $y' = \frac{y}{x + y}$. If we put $z = y/x$, we obtain $-z^{-1} + \ell g|z| = -\ell g|x| + c$, whence $-\frac{x}{y} + \ell g|y| - \ell g|x| = -\ell g|x| + c$. Therefore, the solution in implicit form is given by

$$x = y\,\ell g|y| + c'y.$$

Observe that the same result is obtained by studying the linear equation $\frac{dx}{dy} = \frac{x + y}{y} = 1 + \frac{1}{y}\cdot x$.

3. $y' = \frac{xy}{x^2 - y^2}$. By putting $z = y/x$, we are able to get the solution

$$-\frac{1}{2}\frac{1}{z^2} - \ell g|z| = \ell g|x| + c, \text{ whence } x^2 = c'y^2 - 2y^2\ell g|y|.$$

We now observe that (5.1), after an integration, allows us to express x explicitly as a function of z. In such a case we obtain an explicit representation of the solution in terms of polar coordinates with $\rho = \rho(\theta)$. Specifically, $z = \tan\theta$; $\rho^2 = x^2 + z^2x^2 = (1 + \tan^2\theta)\cdot[x(\tan\theta)]^2$. For examples we take the two following exercises:

4. $y' = \frac{x + y}{x - y}$. Putting $z = y/x$ yields $\arctan z - \ell g\sqrt{1 + z^2} = \ell g\, x + c$, whence we get $x = \frac{1}{K}\frac{1}{\sqrt{1 + z^2}}\cdot\exp(\arctan z)$, so

$$\rho^2 = x^2 + y^2 = \frac{1}{K^2}\exp(2\theta); \quad \rho = \frac{1}{|K|}\exp(\theta).$$

In conjunction with Exercises 1, 2 and 4 one should also consult Sec. 1.3 of Chapter V, in which the interpretation of the solutions of homogeneous problems of order 1 is clarified.

5. $y' = 2\cdot\frac{x^2 + xy + y^2}{x^2 - y^2}$. The solution is $cx^5 = \frac{\exp(4\arctan z)}{(1 + z^2)|z + 2|^3}$,

from which, setting $\rho = \sqrt{x^2 + y^2} = \sqrt{1 + z^2}\,x$, we obtain

$$\rho = C'\left(\frac{\sqrt{1 + z^2}}{|z + 2|}\right)^{3/5}\exp\left(\frac{4}{5}\arctan z\right) = C'\left(\frac{\sqrt{1 + \tan^2}}{|2 + \tan\theta|}\right)^{3/5}\exp\left(\frac{4}{5}\theta\right).$$

The solutions are therefore of the type indicated in the figure.

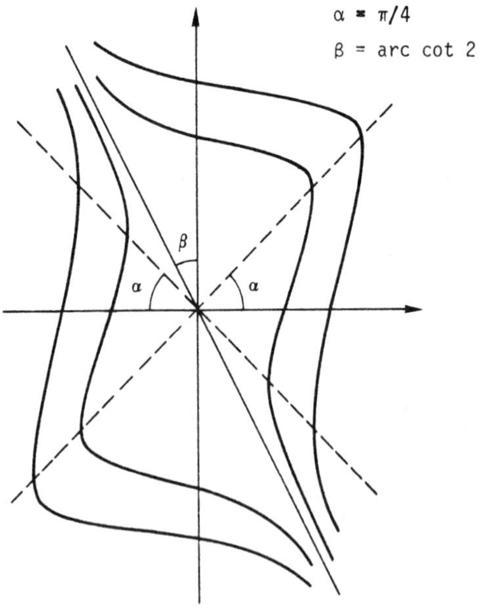

Figure 1.1

5.3. Linear Equations of the First Order

1. $y' = -xy$ general solution: $y = e\ \exp\left(-\frac{x^2}{2}\right)$.

2. $y' = 3y + \sin x$ general solution: $y = e\ \exp(3x) + \frac{3}{10}\sin x - \frac{1}{10}\cos x$.

3. $y' = 2y + e^{2x}$ general solution: $y = e\ \exp(2x) + x\ \exp(2x)$.

4. $y' = -y + x^2 - 1$ general solution: $y = e\ \exp(-x) + x^2 - 2x + 1$.

5. $y' = -xy + x^3$ general solution: $y = e\ \exp\left(-\frac{x^2}{2}\right) + x^2 - 2$.

6. $y' = \sin x \cdot y + \sin 2x$

   general solution: $y = e\ \exp(-\cos x) - 2\cos x + 2$.

5.4. Linear Equations of Order Higher than the First with Constant Coefficients

1. $y'' = y$ general solution: $y = Ae^x + Be^{-x}$.

2. $y'' + 4y = 0$ general solution: $y = A\cos(2x) + B\sin(2x)$.

3. $y'' - 3y' + 2y = 0$ general solution: $y = Ae^{2x} + Be^x$.

4. $y'' + y' + \frac{5}{4}y = 0$ general solution: $y = e^{-x/2}(A\cos x + B\sin x)$.

5. $y'' - 2y' + y = 0$ general solution: $y = (A + Bx)e^x$.

6. $y^{IV} + 4y = 0$ general solution: $y = e^x(A\cos x + B\sin x)$
   $+ e^{-x}(C\cos x + D\sin x)$.

7. $y''' - y'' + y' - y = 0$ general solution: $y = Ae^x + B\cos x + C\sin x$.

8. $y^{IV} + 9y'' = 0$ general solution: $y = A + Bx + C\cos(3x)$
   $+ D\sin(3x)$.

9. $y'' + 3y' + 2y = x + 1$ general solution: $y = Ae^{-2x} + Be^{-x} + \frac{1}{2}x - \frac{1}{4}$.

10. $y'' - y = xe^x$ general solution: $y = Ae^x + Be^{-x} + \frac{1}{4}x^2e^x + \frac{1}{4}xe^x$.

(Observe that since e^x is a solution of the associated homogeneous equation, a particular solution of the complete equation is sought in the form $(Ax^2 + Bx)e^x$ and not in the form $(Ax + B)e^x$. Compare with the next exercises. Keep this observation in mind even in the subsequent exercises.)

11. $y'' - y = xe^{2x}$ general solution: $y = Ae^x + Be^{-x} + (\frac{1}{3}x - \frac{4}{9})e^{2x}$.

12. $y'' + y = \sin(2x)$ general solution: $y = A\cos x + B\sin x - \frac{1}{3}\sin(2x)$.

13. We now examine a more general case of Exercise 12 relative to the phenomena that occur when the frequency of the forcing term is near that of the solution to the homogeneous problem. First of all, determine the solutions of the equation $y'' + y = \sin[(1+\varepsilon)x]$, $\varepsilon > 0$, with the general solution

$$y = A\cos x + B\sin x - \frac{1}{\varepsilon(2+\varepsilon)}\sin[(1+\varepsilon)x].$$

Such a solution may be written in the form

$$\begin{aligned} y &= A\cos x + \left[B + \frac{1}{\varepsilon(2+\varepsilon)}\right]\sin x - \frac{1}{\varepsilon(2+\varepsilon)}[\sin x + \sin(1+\varepsilon)x]] \\ &= A\cos x + \left[B + \frac{1}{\varepsilon(2+\varepsilon)}\right]\sin x - \frac{2}{\varepsilon(2+\varepsilon)}\sin\left[(\frac{2+\varepsilon}{2})x\right]\cos\frac{\varepsilon x}{2}. \end{aligned}$$

The interpretation is that a solution results from the overlapping of the pure sinusoidal wave

$$A\cos x + \left[B + \frac{1}{\varepsilon(2+\varepsilon)}\right]\sin x$$

and of the term

$$-\frac{2}{\varepsilon(2+\varepsilon)}\sin\left[\left(\frac{2+\varepsilon}{2}\right)x\right]\cos\frac{\varepsilon x}{2},$$

on account of which the maximum amplitudes of $y(x)$ are increased or decreased in comparison with the mean amplitude by a factor of amplitude $\frac{2}{\varepsilon(2+\varepsilon)}$ and period $\frac{2}{\varepsilon}2\pi$.

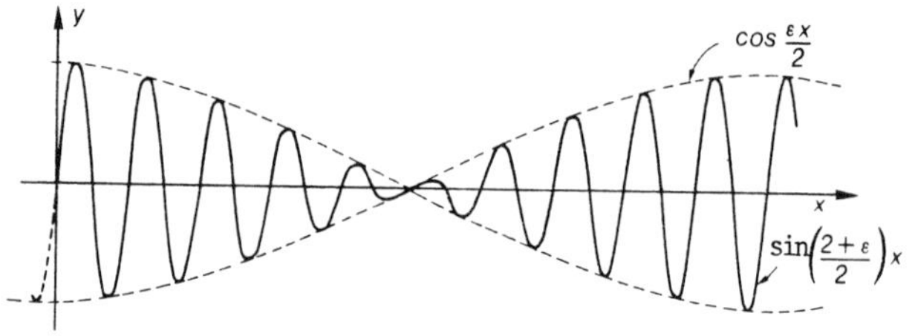

Figure 1.2

14. In the equation of exercise 13, determine the solution to the initial value problem for $y(0) = 0$, $y'(0) = 0$, and calculate, for fixed x, $\lim_{\varepsilon\to 0} y(x)$.

Solution: $y(x) = \frac{1}{\varepsilon(2+\varepsilon)}[\sin x - \sin(1+\varepsilon)x]] + \frac{1}{2+\varepsilon}\sin x$

Limit: $\tilde{y}(x) = -\frac{1}{2}x\cos x + \frac{1}{2}\sin x.$

15. $y'' + y = \sin x$ general solution: $y = A\cos x + B\sin x - \frac{1}{2}x\cos x.$

Observe that the solution relative to the initial data $y(0) = 0$, $y'(0) = 0$ coincides with the limit of solutions of the previous exercise.

16. $y'' - 2y' + 4y = e^x$ general solution: $y = e^x[A \cos(\sqrt{3}\, x) + B \sin(\sqrt{3}\, x) + \frac{1}{3}]$.

17. $y^{IV} + y'' = x^2$ general solution: $y = A \cos x + B \sin x + C + Dx + x^2 + \frac{1}{12} x^4$.

18. $y'' + y = \frac{1}{\tan x}$ general solution: $y = A \cos x + B \sin x + \sin x \cdot \ell g|\tan \frac{x}{2}|$.

5.5. Euler Equations

1. $x^2y'' + xy' - y = 0$ general solution: $y = Ax + \frac{B}{x}$.
2. $x^2y'' = 2y = 0$ general solution: $y = Ax^2 + \frac{B}{x}$.
3. $x^2y'' + xy' + y = 0$ general solution: $y = A \cos \ell g\, x + B \sin \ell g\, x$
4. $x^2y'' - xy' + y = 0$ general solution: $y = Ax + Bx\, \ell g\, x$
5. $x^2y'' - xy' + 5y = 0$ general solution: $y = x(A \cos \ell g(x^2) + B \sin \ell g(x^2))$.
6. $(x+1)y'' + y' - \frac{4y}{(x+1)} = 0$

   general solution: $y = A(x+1)^2 + B(x+1)^{-2}$.
7. $x^2y'' + 4xy' + 2y = x$ general solution: $y = Ax^{-2} + Bx^{-1} + \frac{1}{6}x$.
8. $x^2y'' + 4xy' + 2y = \frac{1}{x}$ general solution: $y = Ax^{-2} + Bx^{-1} + \frac{\ell g\, x}{x}$.
9. $x^2y'' - 2xy' + 2y = \ g\, x$

   general solution: $y = Ax + Bx^2 + \frac{1}{2}\ell g\, x + \frac{3}{2}$.

5.6. Envelopes and Differential Equations (cf. Sec. 1.2)

1. Consider the following family of curves depending on the parameter C:

$$y - Cx + \frac{C^2}{2} = 0.$$

Determine the differential equation whose solutions this family represents and calculate the envelope solution of the same differential equation.

Solution: The equation is $y'^2 - 2xy' + 2y = 0$.

To determine the equation of the envelope, differentiate the equation and obtain

$$2y'2y'' - 2xy'' = 0.$$

We see that the equation of the envelope is $y' = x$, $y = \frac{x^2}{2} + K$. Upon substituting into the equation it follows that $K = 0$.

2. Obtain the differential equation that the curves $y - x\exp(Cx) = 0$ satisfy. What can one say about the solutions of the equation so obtained?

Solution: Since $Cx = \ln\frac{y}{x}$, it follows that the equation is

$$y' = \frac{y}{x} + \ln\frac{y}{x}.$$

Such an equation is defined for $x \neq 0$ and $y \neq 0$, so the solutions are given by $y = xe^{Cx}$, $x > 0$ and $y = xe^{Cx}$, $x < 0$ respectively. It therefore follows that the initial values problem admits a solution only if $y_0 \cdot x_0 > 0$, and the solutions are defined either on the positive x semiaxis or on the negative.

3. Given the family of two-parameter curves

$$y - Ae^x - Bx = 0,$$

determine the differential equation having these curves as solutions.

Solution: $(x - 1)y'' = xy' - y$.

4. Given the family of curves

$$\begin{cases} y - Ae^t - Be^{2t} = 0 \\ x - 2Ae^t + Be^{2t} = 0 \end{cases}$$

determine the system of differential equations having these curves as solutions.

Answer: $x' = \frac{4}{3}x - \frac{2}{3}y$, $y' = -\frac{1}{3}x + \frac{5}{3}y$.

5. Given the family of curves

$$y + x^2 - 4Cx + 2C^2 = 0,$$

determine its envelope; determine the differential equation (in non-normal form) having these curves as solutions.

Answer: envelope: $y = x^2$, equation: $(y' - 2x)^2 = 8(x^2 - y)$. Observe that there exist solutions only if $x^2 - y \geq 0$. If one substitutes $z = y - x^2$, the equation is reduced to one with variables separable.

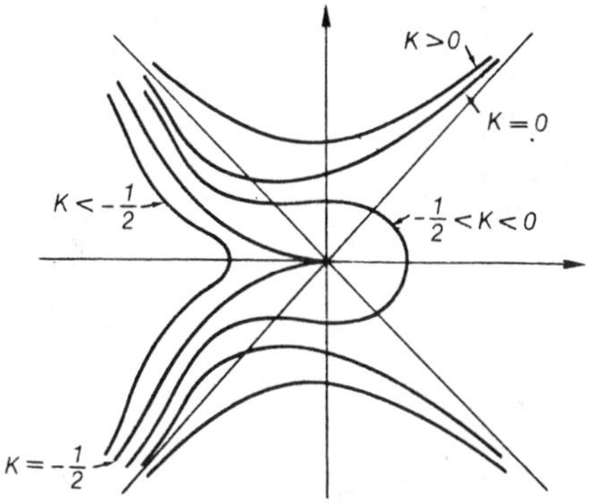

Figure 1.3

5.7. Various Exercises

1. Solve the equation

$$z' = z - \frac{x^2}{z},$$

and study its integrals.

Answer: (Multiply by z and use the unknown z^2.)

$$z = \pm\sqrt{Ke^{2x} + x^2 + x + \frac{1}{2}}.$$

2. Let the equation

$$z' = -(x+1)z^2 + x$$

be given. Prove that the solution satisfying $z(-1) = 1$ is defined from $-\infty$ to $+\infty$ and has an absolute maximum and minimum. Also prove that the solutions that pass through the first quadrant are defined up to $+\infty$ and that $\lim_{x\to\infty} z(x) = 1$ holds.

Solution: The curve of the zeros is $z^2 = \frac{x}{x+1}$; where necessary, compare the solutions with those of $z' = -1$.

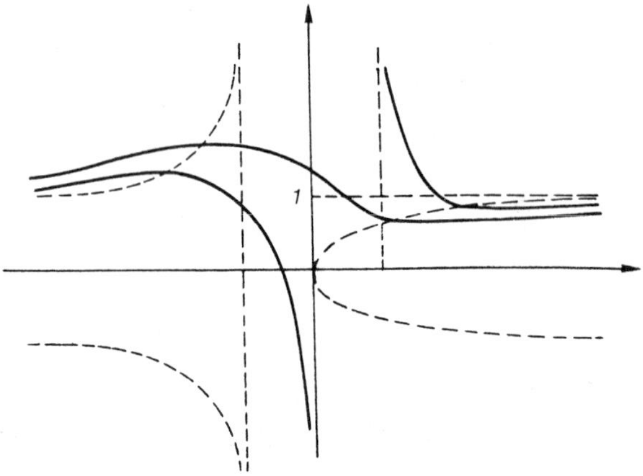

Figure 1.4

5.8. Selected Exercises

Exercise 1. Given the initial value problem

$$\begin{cases} y' = \dfrac{1}{x^2 + y^2} \\ y(0) = y_0 \neq 0, \end{cases}$$

prove

1. that the solutions of this problem are all defined from $-\infty$ to ∞,
2. that $\lim_{x\to-\infty} y(x)$, $\lim_{x\to+\infty} y(x)$ exist and are finite, and
3. give bounds depending on y_0 for $\lim_{x\to+\infty} y(x) - \lim_{x\to-\infty} y(x)$.

Outline of Proof:

1. The theorem on the existence of the solutions assures us that there exists a $\delta > 0$ such that the solution exists in $-\delta \leq x \leq \delta$. For $|x| > \delta$, the right side is uniformly bounded in the whole plane by $1/\delta^2$.

2. Since the right side is positive, y is increasing; hence the limits exist.

We show as an example that $\lim_{x\to\infty} y(x) < +\infty$. Let us suppose that from a certain x_0 on, $y \geq 1$. We then have $y' \leq 1/1+x^2$ so

$$y(x) \leq y(x_0) + \int_{x_0}^{x} \frac{1}{1+\xi^2}\, d\xi = 1 + (\arctan x - \arctan x_0)$$

and so for all x

$$y(x) \leq 1 + \pi.$$

The proof for $\lim_{x\to-\infty} y(x)$ is done similarly.

3. Let us call the two limits L^+ and L^- and give estimates for $L^+ - y_0$ and $y_0 - L^-$ separately. In each case we have, for $y_0 > 0$,

$$L^+ - y_0 \leq \int_0^{+\infty} \frac{1}{y_0^2 + \xi^2}\, d\xi = \frac{1}{y_0}\frac{\pi}{2}. \tag{5.2}$$

This is a good estimate for a "large" y_0. If y_0 is "small", let us consider some value $a > y_0$; we shall have

$$L^+ < a + \int_0^{+\infty} \frac{d\xi}{a^2 + \xi^2} \leq a + \frac{1}{a}\frac{\pi}{2}.$$

That is, if $0 < y_0 \leq \sqrt{\pi/2}$,

$$L^+ - y_0 < L^+ < 2\sqrt{\pi/2}. \tag{5.3}$$

We now consider the case $y < 0$. If $y_0 > -\sqrt{2\pi}$ we may modify (5.3) and obtain

$$L^+ - y_0 < 2\sqrt{2\pi}. \tag{5.4}$$

If $y_0 \leq -\sqrt{2\pi}$, then, as long as x is such that $y < y_0/2$, we have

$$y - y_0 \leq \int_0^x \frac{1}{x^2 + (y_0/2)}\, dx \leq -\frac{\pi}{y_0}.$$

On the other hand, if $y_0 \leq -\sqrt{2\pi}$, we have $-\frac{\pi}{y_0} + y_0 \leq \frac{y_0}{2}$. In this case, we have the estimate

$$L - y_0 \leq \frac{\pi}{|y_0|}. \tag{5.5}$$

Analogous estimates hold for $y_0 - L'$. We thus have the desired bounds.

Exercise 2. Given the equation

$$y'' = a^2 - y^2, \quad a > 0, \tag{5.6}$$

study the behavior of the solutions, paying particular attention to periodic and non-periodic solutions defined from $-\infty$ to ∞.

Outline of the Proof: From (5.6) one may conclude that the solutions are concave up if $|y| < a$ and concave down if $|y| > a$. There are, furthermore, the two constant solutions $y = a$ and $y = -a$. One

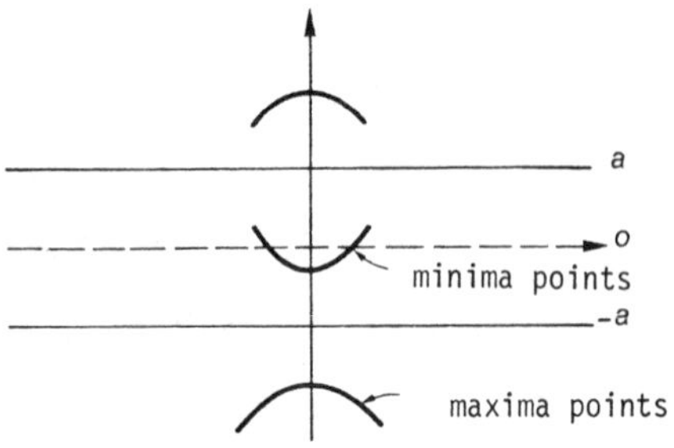

Figure 1.5

may at once conclude that whatever periodic solutions there are must have a maximum value $> a$ and a minimum value between $-a$ and a. If we multiply (1) by $2y'$ and integrate, we produce

$$y'^2 = y_0'^2 - \frac{2}{3}(y^3 - y_0^3) + 2a^2(y - y_0). \tag{5.7}$$

Let us suppose that a solution has minimum value y_0 between $-a$ and a. The maximum $\bar{y}$ will in this case be given, because of (5.7), the the root of

$$0 = -\frac{2}{3}(\bar{y}^2 - y_0^3) + 2a^2(\bar{y} - y_0) \tag{5.8}$$

and is therefore greater than a. The three roots are

$$y_0; \; -\frac{y_0}{2} - \frac{\sqrt{12a^2 - 3y_0^2}}{2} < -a; \; -\frac{y_0}{2} + \frac{\sqrt{12a^2 - 3y_0^2}}{2} > a.$$

The maximum for a periodic solution corresponding to a minimum of y_0 is therefore

$$-\frac{y_0}{2} + \frac{\sqrt{12a^2 - 3y_0^2}}{2}.$$

The amplitude is less than $3a$, so the periodic solutions do not have maxima bigger than $2a$. The solution having maximum $2a$ satisfies the equation

$$y'^2 = -\frac{2}{3}(y^3 - 8a^3) + 2a^2(y - 2a) = -\frac{2}{3}(y + a)^2(y - 2a); \tag{5.9}$$

the only other value of y for which (5.9) vanishes is $-a$. This cannot be a minimum since $y \equiv -a$ is a solution and we have uniqueness. We therefore see that the solution with maximum $2a$ tends asymptotically to $-a$ as $x \to \infty$. It is given by

$$y = a\left[2 - 3\left(\frac{e^{\sqrt{2a}|t-t_0|} - 1}{e^{\sqrt{2a}|t-t_0|} + 1}\right)^2\right].$$

We now examine the shape of the non-periodic solutions. They have maximum value $> 2a$ or $< -a$. If y_0 is such a maximum, Eq. (5.7) becomes

$$y'^2 = \left[-\frac{2}{3}(y^2 + yy_0 + y_0^2) + 2a^2\right](y-y_0); \tag{5.10}$$

if $|y_0| > 2a$, there are no other solutions except y_0. If $-2a \le -y_0 < -a$, the solutions y^* of (5.10) exist but $y^* > -a$ which is unacceptable. (They are maxima and minima of the periodic solutions discussed previously.) As a result, if a solution has maximum value greater than $2a$ or less than $-a$, the values it assumes are not bounded below. From this it easily follows that if x_0 gives a maximum, the solutions are defined only in an open interval $]-b + x_0, b + x_0[$. In fact, if $y < -2|y_0|$, one has from (5.10) $y'^2 \ge -\frac{1}{12}y^3$, whence the assertion follows.

It remains finally to consider the solutions without maxima. Since (5.7) always has at least one root, a solution without a maximum has a horizontal asymptote. From (5.6) it follows that the only possible asymptote is $y = -a$. In (5.7), the condition of having root $-a$ also translates into Eq. (5.9), and in this case the solution is

$$y = a\left[2 - 3\left(\frac{e^{\sqrt{2a}|t-t_0|} + 1}{e^{\sqrt{2a}|t-t_0|} - 1}\right)^2\right].$$

Exercise 3. Prove that the problem

$$\begin{cases} y'' = -y^2 \\ y(-c) = y(c) = 0. \end{cases} \tag{5.11}$$

admits exactly two solutions for each real number c.

Outline of Proof: Consider the results of the previous exercise with $a = 0$. One solution of the problem (5.11) is given by $y \equiv 0$. Consider now the initial value problem relative to the initial data

$$y(-c) = 0, \qquad y'(-c) = \alpha. \tag{5.12}$$

Since problem (5.11) has an even solution, it is sufficient to prove that among all the solutions of the initial value problem with data (5.12) as α varies, there is exactly one (non-trivial) solution with maximum at 0.

If $\ell(\alpha)$ is the point where the maximum occurs as α varies ($\alpha > 0$ since otherwise the solution could not exist), we shall prove that $\ell(\alpha)$ is a continuous, decreasing function and that $\lim_{\alpha\to 0^+} \ell(\alpha) = +\infty$, $\lim_{\alpha\to\infty} \ell(\alpha) = -c$. The assertion follows from this.

The continuity of $\ell(\alpha)$ is a consequence of the general theorems concerning the continuous dependence on the data. For monotonicity, note first of all that from the equation obtained by integrating (5.11)

$$y'^2 = y_0'^2 - \frac{2}{3}(y - y_0^3) \tag{5.13}$$

it follows that if $y_1'(-c) > y_2'(c)$, then $y_1(x) > y_2(x)$ for $x > c$ until y_1 or y_2 attains the maximum. If we set $g_1(x) = y_1(x)/y_1'(x)$, $g_2(x) = y_2(x)/y_2'(x)$, we obtain

$$g_i' = 1 + y_i(x)g_i^2, \qquad g_i(-c) = 0, \tag{5.14}$$

and, therefore, since $y_1 > y_2$ from (5.14), it follows that

$$g_1(x) > g_2(x). \tag{5.15}$$

Thus $g_1(x) \to \infty$ for $x \to x_1$, and $g_2(x) \to \infty$ for $x \to x_2$ (which corresponds to the maximum point) and $x_1 \leq x_2$. Suppose, on the contrary, that $x_1 = x_2$. If M_1 and M_2 are the respective maxima, it then follows that $M_1 > M_2$, and so from certain $\bar{x} < x_1 = x_2$ up to $x_1 = x_2$ we would have

$$g_1' \geq 1 + \frac{M_2 + M_1}{2} g_1^2, \qquad g_2' \leq 1 + M_1 g_1^2 \tag{5.16}$$

with $g_1(\bar{x}) > g_2(\bar{x})$. Upon integrating the inequalities (5.16), we arrive at the contradiction. This proves the monotonicity of $\ell(\alpha)$. We now prove the last point. One first of all gets from (5.13) the maximum value $m(\alpha)$;

$$0 = \alpha^2 - \frac{2}{3} m(\alpha)^3; \qquad m(\alpha) = 3\sqrt{3/2}\,\alpha^2.$$

We consider Eq. (5.13) from a maximum point on. We get

$$y'^2 = +\frac{2}{3}[m(\alpha) - y][m^2(\alpha) + m(\alpha)y + y^2].$$

Since $0 \leq y \leq m(\alpha)$, we then have the two inequalities

$$+2[m(\alpha) - y]m^2(\alpha) \leq y'^2 \leq + \frac{2}{3}[m(\alpha) - y]m^2(\alpha).$$

which, upon being integrated, yield

$$m(\alpha) - \frac{1}{2}(m(\alpha))^2|x - x_M|^2 \leq y \leq m(\alpha) - \frac{1}{6}|m(\alpha)|^2|x - x_M|^2,$$

where x_M is the point that gives the maximum. Since $y(c) = 0$, we get at once

$$(\frac{16}{3})^{1/6}, \quad \alpha^{-1/3} < \ell(\alpha) < 16^{1/3}\alpha^{-1/3}$$

and the proof is complete.

Exercise 4. Given the initial value problem

$$\begin{cases} y' = \sin(xy) \\ y(0) = a, \quad a > 0 \end{cases}$$

prove that there is a solution in $-\infty < x < \infty$, that $y(x) > 0$ for all x, and that $\lim_{x\to\pm\infty} y(x) = 0$.

Solution:

a. $\sin xy$ is a Lipschitz function in the whole plane, and this insures local existence and the uniqueness of the solutions. Since $\sin xy$ is defined in all of R^2 and is bounded there, the solutions are defined in $-\infty < x < \infty$.

b. $y = 0$ is a solution of the equation. Since the solution to the initial value problem is unique, if a solution is positive at a point, it is always positive.

c. Note first of all that the locus where the solutions have their maxima is given by the hyperbolas

$$xy = (2k + 1)\pi, \tag{5.17}$$

and the location of the minima by the hyperbolas

$$xy = 2k\pi. \tag{5.18}$$

In the sets between the hyperbolas

$$xy = 2k\pi; \quad xy = (2k + 1)\pi$$

the solutions are increasing, whereas between the hyperbolas

$$xy = (2k - 1); \quad xy = 2k\pi$$

they are decreasing.

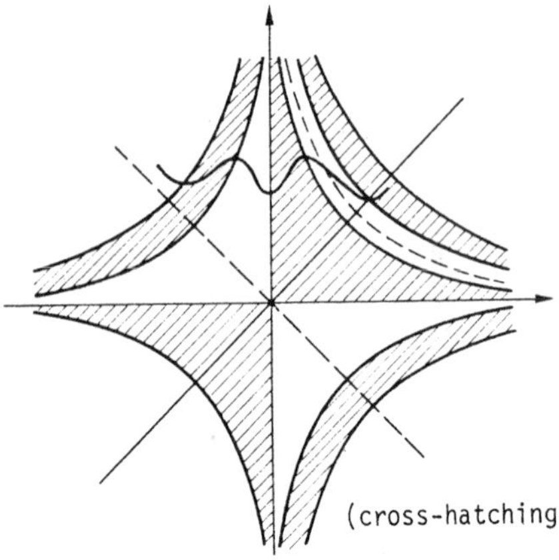

Figure 1.6

Since, furthermore, $\sin(-xy) = -\sin xy$, the solution must also be such that $y(-x) = y(x)$. It is therefore sufficient to study the solution for x positive. Consider now the hyperbolas $xy = (2k - \frac{1}{2})\pi$. On these, $y' = -1$. If the point (x,y) is such that $|y| < |x|$, the tangent to the hyperbola forms an angle less than $\frac{1}{4}\pi$ with the x-axis. We now suppose that a solution y of the given problem crosses the line $y = x$ at x_0. Since $|y'| < 1$, the solution will, for $x > x_0$, satisfy the relation

$$0 < y(x) < x,$$

and therefore it also follows that

$$0 \leq y(x) < \frac{(2k - \frac{1}{2})\pi}{x},$$

where k is the smallest natural number for which $x_0^2 < (2k - \frac{1}{2})\pi$. It therefore remains to prove that every solution crosses the line $y = x$.

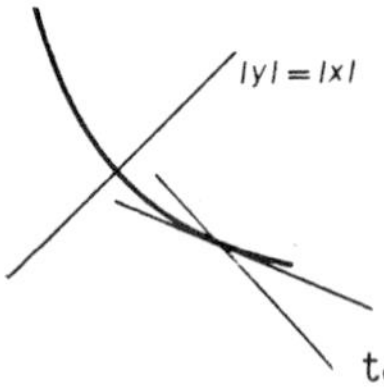

Figure 1.7

Consider the function defined by

$$\tilde{f}(x,y) = 0 \quad \text{if} \quad \sin(xy) \le 0$$
$$\tilde{f}(x,y) = 1 \quad \text{if} \quad \sin(xy) > 0$$

The function y which solves the equation $y' = \tilde{f}(x,y)$ with the initial condition $y(0) = a > 0$ dominates the solution y of our initial problem (since $\tilde{f} \ge f$).

The relation

$$\tilde{y} \le \left(a + \frac{\sqrt{\pi}}{2}\right) + \frac{1}{2} x \tag{5.19}$$

holds, and so every solution crosses the line $y = x$.

To prove (5.19), observe that the graph of $\tilde{y}$ is a piecewise linear arc with vertices on the hyperbolas (5.17) and (5.18). If (x_k, y_k) are the vertices, then

$$\frac{y_{k+2} - y_k}{x_{k+2} - x_k} < \frac{1}{2}$$

whence the conclusion. (See the figure; observe that $x_{k-2} - x_{k-1} = x_{k-1} - \bar{x} > x_{k+1} - x_k$.)

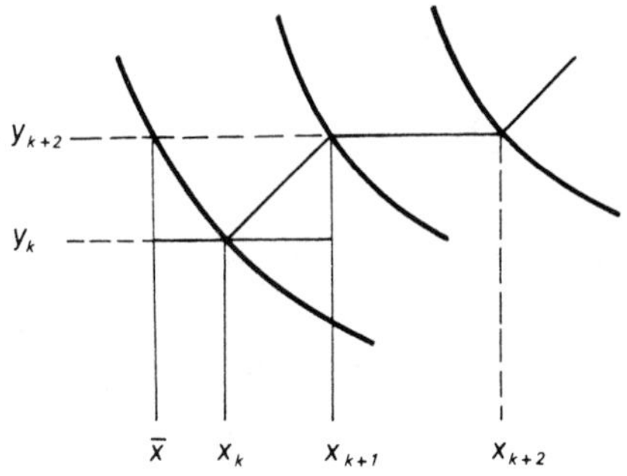

Figure 1.8

Exercise 5. Consider the problem

$$\begin{cases} y'' + y^2 = 0 \\ y(-a) = y(a) = b. \end{cases} \tag{5.20}$$

Prove that there is a continuous, monotonically decreasing function ϕ such that $\lim_{x \to 0^+} \phi(x) = +\infty$ and $\lim_{x \to +\infty} \phi(x) = 0$ for which problem (5.20) has one solution if $b = \phi(a)$, two solutions if $b < \phi(a)$, and no solution if $b > \phi(a)$.

Sketch of Solution: Keep in mind the results and notation of Exercises 2 and 3. The problem has already been solved for $b = 0$; observe that by extending the solutions of the problem for $y(a) = y(-a) = 0$ beyond the interval $|x| \le a$ we get a family of curves that covers the subspace $y < 0$, $|x| > 0$. What is more, for each point with coordinates $(-a,b)$, $a > 0$, $b < 0$, there passes one and only one such solution. If in fact there were another, it follows from what we have shown that one, let us say y_2, would always be greater than the other, y_1, right up to where the maximum occurs. There would therefore exist $x_2 < x_1$ such that $y_2(x_2) = y_1(x_1) = 0$; we would then also have $y_2'(x_2) > y_1'(x_1)$. But in this case the solution y_2 with maximum $m_2 > m_1$ would achieve its maximum before the point $x = 0$, which contradicts the fact that the solutions are even concave functions and must therefore take on the only maximum for $x = 0$.

We now consider the function $\ell(\alpha)$ relative to the Cauchy problem with initial data $y(-a) = b$, $y'(-a) = \alpha$, with $b < 0$. Observe that in correspondence with the solution

$$y = -\frac{6}{(x+1)^2} \quad \text{with} \quad \gamma = +a + \sqrt{-(6/b)} \left[\alpha = \alpha_0 = 12(-\tfrac{6}{b})^{-3/2}\right],$$

$\lim_{\alpha \to \alpha_0} \ell(\alpha) = +\infty$. It thus follows that the graph of $\ell(\alpha)$ is as follows:

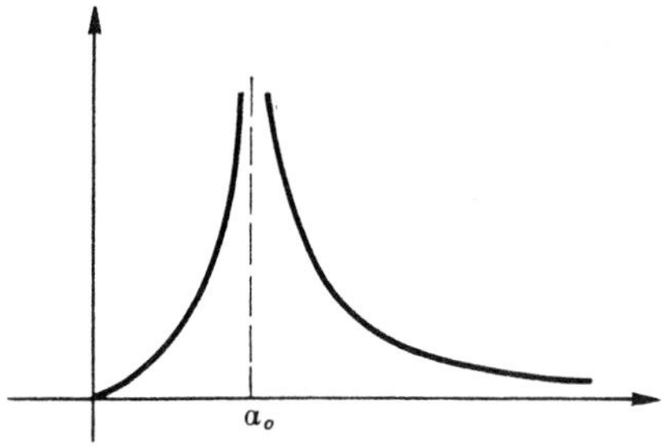

Figure 1.9

Monotonicity between 0 and α_0 is guaranteed by the monotonicity of $-y^2$, which holds for solutions $y < 0$. Between α_0 and ∞ it follows from the preceding discussion. Thus, for every value a, the equation $\ell(\alpha) = a$ admits exactly two solutions.

We now pass to the case $b > 0$. We recall that all the possible solutions of problem (5.20) have a maximum at $x = 0$ and therefore satisfy, as y_0 varies, the equation obtained by integrating (5.20), which corresponds to (5.7) with $a = 0$, $y_0' = 0$:

$$y'^2 = \frac{2}{3}(y_0^3 - y^3).$$

If we consider only $x \geq 0$, we see that they also satisfy

$$y' = -\sqrt{2/3}\sqrt{y_0^3 - y^3}. \tag{5.21}$$

We now construct the inverse $\psi(y)$ of the function $\phi(x)$ that we want. If $\bar{y}$ is fixed, we propose to find

$$I_{\bar{y}} = \{x > 0\colon \text{ problem (5.20) has a solution for } a = x,\ b = \bar{y}\}. \tag{5.22}$$

If we integrate (5.21), we find that the set (5.22) consists of those x such that

$$x(y_0) = \sqrt{3/2}\int_{\bar{y}}^{y_0} \frac{1}{\sqrt{y_0^3 - y^3}}\,dy \tag{5.23}$$

for $y > \bar{y}$. Let us prove that the function (5.23) is continuous, that it has a unique maximum, and that $x(\bar{y}) = 0$, $\lim_{y_0 \to +\infty} x(y_0) = 0$. The continuity is obvious. $x(\bar{y}) = 0$ and $\lim_{y_0 \to +\infty} x(y_0) = 0$ follow from what we saw in Exercise 3. If we calculate the derivative of (5.23), we get (ignoring for the moment the factor $\sqrt{3/2}$)

$$\frac{1}{h}\left\{\int_{\bar{y}}^{y_0+h} \frac{1}{\sqrt{(y_0+h)^3 - y^3}}\,dy - \int_{\bar{y}}^{y_0} \frac{1}{\sqrt{y_0^3 - y^3}}\,dy\right\}$$

$$= \frac{1}{h}\left\{\int_{\bar{y}}^{y_0+h} \left(\frac{y_0}{y_0+h}\right)^{3/2} \frac{1}{\sqrt{y_0^3 - y^3(y_0/y_0+h)^3}}\,dy - \int_{\bar{y}}^{y_0} \frac{1}{\sqrt{y_0^3 - y^3}}\,dy\right\}$$

$$= \frac{1}{h}\left\{\int_{\bar{y}\cdot\frac{y_0}{y_0+h}}^{y_0} \sqrt{y_0/y_0+h}\,\frac{1}{\sqrt{y_0^3 - y^3}}\,dy - \int_{\bar{y}}^{y_0} \frac{1}{\sqrt{y_0^3 - y^3}}\,dy\right\}$$

$$= \frac{1}{h}\left\{\int_{\bar{y}}^{y_0} \frac{1}{\sqrt{y_0^3 - y^3}}\left(\sqrt{y_0/y_0+h} - 1\right)dy \right.$$

$$\left. + \int_{\bar{y}\cdot\frac{y_0}{y_0+h}}^{\bar{y}} \sqrt{y_0/y_0+h}\,\frac{1}{\sqrt{y_0^3 - y^3}}\,dy\right\};$$

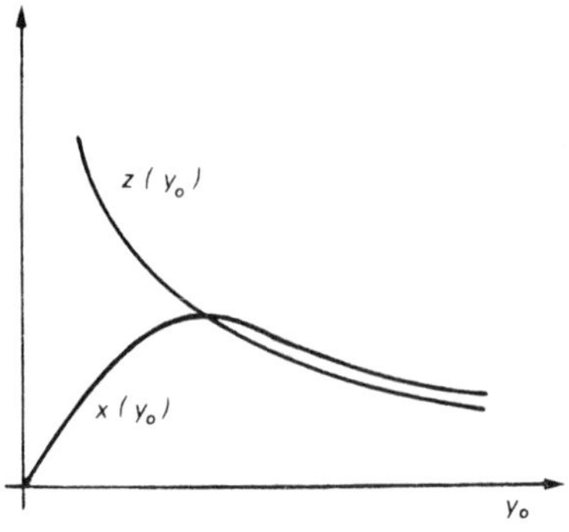

Figure 1.10

whence, passing to the limit as $h \to 0$, we find that

$$\frac{dx}{dy_0} = -\frac{1}{2y_0}\cdot x(y_0) + \sqrt{3/2}\,\frac{\bar{y}}{y_0}\,\frac{1}{\sqrt{y_0^3 - \bar{y}^3}}$$

$$= \frac{1}{2y_0}\left[-x(y_0) = \sqrt{6}\,\bar{y}\,\frac{1}{\sqrt{y_0^3 - \bar{y}^3}}\right]. \qquad (5.24)$$

To find out the sign of the derivative, it is sufficient, since $y_0 \geq \bar{y} > 0$, to examine the inequality

$$\sqrt{6}\,\bar{y}\,\frac{1}{\sqrt{y_0^3 - \bar{y}^3}} = z(y_0) > x(y_0). \qquad (5.25)$$

The function $z(y_0)$ has the graph given in the above figure while, in a neighborhood of $\bar{y}$, $x(y_0)$ is less than $z(y_0)$, so that by (5.24) it must be increasing. Since $\lim_{y_0 \to \infty} z(y_0) = 0$, it follows that there exists a first point y^* for which equality holds in (5.25). There cannot be a second point y^{**} for which equality holds because (5.24) would require that $x'(y_0) = 0$ at that point, which is impossible, since $z'(x_0) < 0$.

We observe that if we put $\psi(\bar{y}) = x(y^*) = \max I_{\bar{y}}$, it follows from the study of $x(y_0)$ that for $a < \psi(b)$ there are two solutions of problem (5.20), for $a = \psi(b)$ one solution, and for $a > \psi(b)$ no solution.

We leave to the reader to prove the continuity and strict monotonicity of $\psi(\bar{y})$, which implies invertibility. (Note that $\phi(x)$ is the envelope curve of the solutions of (5.21); monotonicity follows.)

6. BIBLIOGRAPHICAL NOTES

Questions regarding uniqueness and existence in the large will be taken up again in greater generality in Chapter III. We therefore put off until then the portion of the bibliography that concerns them.

For systems that are not in normal form, certain results can be found in Pontryagin [28].

The convergence of the successive approximations has claimed the attention of many mathematicians, from Picard, who was the first, to Lipschitz, Dieudonné, LaSalle, Coddington and Levinson, Brauer, Krasnoselski, Wazewski, and others. The appropriate references are to be found in Lakshmikantham-Leela [19]. The result of Exercise 2 in Sec. 4.3 is due to Dieudonné, while that of Exercise 3 of the same section is due to Coddington-Levinson. It was for many years an open problem to determine whether the theorem of Coddington and Levinson was true without assuming that $\omega(t,\cdot)$ was increasing, the other hypotheses being retained. This question was solved recently by Evans and Feroe [12] who showed that the answer is yes if $n > 2$ but no if $n = 1$. See Vidossich [34] and [35] for the global convergence of the successive approximations under hypotheses different from those of Sec. 4.3 and for the proof that the systems of ordinary differential equations for which the successive approximations converge are "more" than those for which they do not.

Various authors have generalized Gronwall's Lemma; see, for example, Eisenfeld-Lakshmikantham [11], Pachpatte [26], Rasmussen [30] and the works cited therein.

The analyticity of the solutions is very important. For a more complete treatment, see Friedrichs [13], while for a panorama of the most recent results see Diliberto [8], Hsieh [15], Kaplan [18], Mawhin [23], and the works cited there. For an application of the theory of functions of a complex variable to the theory of ordinary equations, cf. Morris-Feldstein-Bowen [24]; for an application of the existence theorems of analytic functions for ordinary equations to other areas of mathematics, cf. Nussbaum [25].

For a different proof of the theorem on differentiability of Sec. 3.2 based on a fixed point theorem, cf. Sotomayor [33]. The theorem on differentiability can be used in boundary value problems; cf. Friedrichs [13], Proctor [29], and Schmitt [31]. The formula of exercise 5 in Sec. 3.2 is due to Alekseev [1] and has been much used in the study of the asymptotic properties of the solutions; cf. the references in Chapter V.

For generalizations of the fixed point theorem for contractions, see the recent book of Martin [21], which also contains the essential references to the other works on the subject.

The space $X_{\varepsilon,h}$ of the example (f) in Sec. 4.2 was introduced by Bielecki [6].

In the last century, Lie tried to construct a general theory of integration by quadratures of ordinary differential equations; his method was to study the invariance of the ordinary differential equations for certain groups of transformations called infinitesimal groups. He showed that if there exists an infinitesimal group of transformations which leaves invariant a given equation of the first order, then the equation is integrable by quadratures. Unfortunately, there does not exist a general criterion to find the groups that leave a given equation invariant or to prove that such groups do not exist. The interested reader should consult Bluman-Cole [7] and Matsuda [22]. For a collection of the various methods of quadratures, see Kamke [17].

If we consider an autonomous system with uniqueness,

$$y_i' = f_i(y),$$

and if $\delta_1(t,x),\ldots,\delta_n(t,x)$ is the value at t of the solution of the system with initial data $y_i(0) = x_i$, then the function $\delta\colon \mathsf{R}^n \times \mathsf{R}^n \to \mathsf{R}^n$ has the following properties:

(i) $\delta(0,x) = x$

(ii) $\delta(t+s,x) = \delta(t,\delta(s,x))$

(iii) δ is continuous.

These three properties may be taken as the axioms defining the concept of a dynamical system. This notion was introduced by Poincare and has been the object of very profound study, the results of which were then applied to the study both of ordinary and partial differential equations. A general treatment may be found in Bhatia-Szëgo [5], and a panorama of the more recent results of Peixoto [27]. For applications to ordinary equations, consult Andronov-Leontovich-Gordon-Maier [2] and [3], Lasota [20], and Sell [32].

[1] V. M. Alekseev, An upper bound for the perturbations of the solutions of ordinary differential equations (in Russian), *Vestnik Moscov. Univ. I Mat. Mech.*, *2*(1961), 28-36.

[2] A. A. Andronov, E. A. Leontovich, II. Gordon and A. G. Maier, *Theory of bifurcations of dynamic systems on a plane,* Israel Program Scient. Translation, Jerusalem, 1971.

[3] A. A. Andronov, E. A. Leontovich, I. I. Gordon and A. G. Maier, *Qualitative theory of second-order dynamic systems,* Israel Program Scient. Translation, Jerusalem, 1973.

[4] S. Banach, *Théorie des Opérations Linéaires,* Chelsea, New York, 1955.

[5] N. P. Bhatia and G. P. Szëgo, *Stability Theory of Dynamical Systems,* Springer-Verlag, Berlin, 1970.

[6] A. Bielecki, Une remarque sur la méthode de Banach-Caccioppoli-Tikhonov dans la théorie des équations differentielles ordinaires, *Bull. Acad. Polon. Sci. Cl., III, 4*(1956), 261-264.

[7] G. W. Bluman and J. D. Cole, *Similarity Methods for Differential Equations,* Springer-Verlag, Berlin, 1974.

[8] S. P. Diliberto, A new technique for proving the existence of analytic functions in WEISS, *Ordinary Differential Equations,* Academic Press, New York, 1972.

[9] J. Dugundji, *Topology,* Allyn and Bacon, Boston, 1966.

[10] N. Dunford and J. T. Schwartz, *Linear Operators, Part I,* Interscience, New York, 1964.

[11] J. Eisenfeld and V. Lakshmikantham. Comparison principle and nonlinear contractions in abstract Spaces, *J. Math. Anal. Appl., 49* (1975), 504-511.

[12] J. W. Evans and J. A. Feroe, Successive approximations and the general uniqueness theorem, *Amer. J. Math., 96*(1974), 505-510.

[13] K. O. Friedrichs, *Lectures on Advanced Ordinary Differential Equations,* Gordon and Breach, London, 1967.

[14] G. E. O. Giacaglia, *Perturbations Methods in Nonlinear Systems,* Springer-Verlag, Berlin, 1972.

[15] P. F. Hsieh, Recent advances in the analytic theory of nonlinear ordinary differential equations with an irregular type singularity, in ANTOSIEWICZ, *International Conference on Differential Equations,* Academic Press, New York, 1975.

[16] J. R. Isbell, *Uniform Spaces,* Amer. Math. Soc., Providence, R. I., 1964.

[17] E. Kamke, *Differentialgleichungen: Lösungsmethoden und Lösungen, Vol. I,* Leipzig Akademische V., Leipzig, 1942.

[18] W. Kaplan, Analytic ordinary differential equations in the large, in HARRIS and SIBUYA, *Proceedings United States-Japan Seminar on Differential Equations,* Benjamin, New York, 1967.

[19] V. Lakshmikantham and S. Leela, *Differential and Integral Inequalities, Vol. I,* Academic Press, New York, 1969.

[20] A. Lasota, Relaxation of oscillations and turbulence, in WEISS, *Ordinary Differential Equations,* Academic Press, New York, 1972.

[21] R. H. Martin, Jr., *Nonlinear Operators and Differential Equations in Banach Spaces,* John Wiley and Sons, New York, 1976.

[22] M. Matsuda, Integration of ordinary differential equations of the first order by quadratures, *Osaka J. Math., 11*(1974), 23-36.

[23] J. Mawhin, Fredholm mappings and solutions of linear differential equations at singular points, *Ann. Mat. Pure Appl. 108*(1976), 329-335

[24] G. R. Morris, A. Feldstein and E. W. Bowen, The Phragmén-Lindelöf principle and a class of functional differential equations, in WEISS, *Ordinary Differential Equations,* Academic Press, New York, 1972.

[25] R. D. Nussbaum, Periodic solutions of analytic functional differential equations are analytic, *Michigan Math. J., 20*(1973), 249-255.

[26] B. G. Pachpatte, On some integral inequalities similar to Bellman-Bihari inequalities, *J. Math. Anal. Appl., 49*(1975), 794-802.

[27] M. M. Peixoto, *Dynamical Systems,* Academic Press, New York, 1973.

[28] L. S. Pontryagin, *Equations Différentielles Ordinaires,* MIR, Moscow, 1969.

[29] T. G. Proctor, Periodic solutions for perturbed differential equations, *J. Math. Anal. Appl., 47*(1974), 310-323.

[30] D. L. Rasmussen, Gronwall's inequality for functions of two independent variables, *J. Math. Anal. Appl., 55*(1976), 407-417.

[31] K. Schmitt, Applications of variational equations to ordinary, and partial differential equations - Multiple solutions of boundary value problems, *J. Diff. Eq., 17*(1975), 154-186.

[32] G. R. Sell, Topological dynamical techniques for differential and integral equations, in WEISS, *Ordinary Differential Equations,* Academic Press, New York, 1972.

[33] J. Sotomayor, Smooth dependence of solutions of differential equations on initial data: a simple proof, *Boletim Soc. Brasil. Mat., 4*(1973), 55-59.

[34] G. Vidossich, Global convergence of successive approximations, *J. Math. Anal. Appl., 45*(1974), 285-292.

[35] G. Vidossich, Most of the successive approximations do converge, *Math. Anal. Appl., 45*(1974), 127-131.

[36] K. Yosida, *Functional Analysis,* Springer-Verlag, Berlin, 1971.

Chapter II
Linear Systems

In the preceding chapter, we saw that linear systems of differential equations were used to calculate the partial derivatives of the solutions of non-linear systems of differential equations taken as functions of the initial data and of the parameters. This is not the only example in which linear systems are used to study non-linear systems; one of the most frequently used techniques for studying the non-linear system

$$y_i' = f_i(x,y)$$

is that of decomposing f_i into the sum of a linear and a non-linear part

$$f_i(x,y) = \sum_{j=1}^{n} h_{ij}(x)y_j + g_i(x,y),$$

using, for example, Taylor's formula, and comparing the solutions of the linear system

$$y_i' = \sum_j h_{ij}(x)y_j$$

with those of the original system. Other examples of this procedure will be treated in Chapters IV and V. It therefore seems useful to study the properties of the solutions of linear systems in greater depth.

1. ELEMENTS OF LINEAR ALGEBRA

1.1. Matrices and Eigenvalues

To study differential systems and finite difference systems of linear type, it is worthwhile to discuss first some notions concerning matrices and operators between Banach spaces. In this section, we shall briefly review certain elementary properties of matrices.

a) The vector space of $m \times n$ matrices.

We indicate by $A = (a_{ij})$ a matrix

$$A = \begin{pmatrix} a_{11} & a_{12} & \cdots & a_{1n} \\ a_{21} & a_{22} & \cdots & a_{2n} \\ \vdots & & & \\ a_{m1} & a_{m2} & \cdots & a_{mn} \end{pmatrix}.$$

The matrix is of type $m \times n$, where m is the number of rows and n the number of columns. The real or complex numbers a_{ij} are called the entries, elements, or components of the matrix. Two matrices are equal if and only if their elements are equal. The matrix whose elements are all zero is called the null matrix and is indicated by 0. The vector space structure is given by the operation of sum

$$A + B = (a_{ij} + b_{ij})$$

and that of scalar product

$$\alpha A = (\alpha a_{ij}) = A\alpha.$$

As we pointed out in Sec. 4.2 of Chapter I, the $m \times n$ matrices form a Banach space under the norm

$$||A|| = \left(\sum_{i=1}^{m} \sum_{j=1}^{n} |a_{ij}|^2\right)^{1/2}. \tag{1.1}$$

This satisfies all the properties required of a norm. One sees this by identifying the space of $m \times n$ matrices with the Euclidean space of dimension $d = mn$.(*)

(*) In this space, the "triangle" inequality $||x+y|| \le ||x|| + ||y||$ may be proved as follows:

$$||x+y||^2 = \sum_{i=1}^{d} |x_i+y_i|^2 \le \sum_{i=1}^{d}\left(|x_i|+|y_i|\right)^2 \le ||x||^2 + ||y||^2 + 2\sum_{i=1}^{d} |x_i|\cdot|y_i|.$$

We recall that

$$\left(\sum_{i=1}^{d} |x_i|\cdot|y_i|\right)^2 \le \sum_{i=1}^{d} |x_i|^2\cdot\sum_{i=1}^{d} |y_i|^2 = ||x||^2\cdot||y||^2.$$

since, for every λ, $(\Sigma_{i=1}^{d}|x_i| + \lambda|y_i|)^2$ is nonnegative. It therefore follows that $||x+y||^2 \le (||x|| + ||y||)^2$.

Another operation that may be defined on matrices is that of product. Let $A = (a_{is})$ and $B = (b_{sj})$ be an $m \times n$ and an $n \times r$ matrix respectively. The $m \times r$ matrix $C = (c_{ij})$ whose ij-th element c_{ij} is the product of the ith row of A with the jth column of B,

$$c_{ij} = \sum_{s=1}^{n} a_{is} b_{sj}, \tag{1.2}$$

is the product of A and B. Even when both products AB and BC are defined, they may not be the same. In fact, AB may be null when neither A nor B is. For examples, consider the two cases

$$\begin{pmatrix} 0 & 0 \\ 1 & 1 \end{pmatrix}\begin{pmatrix} 1 & 0 \\ 1 & 0 \end{pmatrix} = \begin{pmatrix} 0 & 0 \\ 2 & 0 \end{pmatrix}, \quad BA \neq AB$$

$$\begin{pmatrix} 1 & 0 \\ 1 & 0 \end{pmatrix}\begin{pmatrix} 0 & 0 \\ 1 & 1 \end{pmatrix} = \begin{pmatrix} 0 & 0 \\ 0 & 0 \end{pmatrix}, \quad AB = 0.$$

In general, we have

$$||AB|| \leq ||A|| \; ||B||, \tag{1.3}$$

since

$$\begin{aligned} ||AB||^2 &= \sum_{i=1}^{m} \sum_{j=1}^{r} \left(\sum_{s=1}^{n} a_{is} b_{sj} \right)^2 \leq \sum_{i=1}^{m} \sum_{j=1}^{r} \left(\sum_{s=1}^{n} |a_{is}| \cdot |b_{sj}| \right)^2 \\ &\leq \sum_{i=1}^{m} \sum_{j=1}^{r} \left(\sum_{s=1}^{n} |a_{is}|^2 \cdot \sum_{s=1}^{n} |b_{sj}|^2 \right) = \sum_{i=1}^{m} \sum_{s=1}^{n} |a_{is}|^2 \cdot \sum_{j=1}^{r} \sum_{s=1}^{n} |b_{sj}|^2 \\ &= ||A||^2 ||B||^2. \end{aligned}$$

If A is a $m \times n$ matrix and $v \in \mathbb{R}^n$ we denote by Av the product of A and the column matrix $(\beta_{i1})_i$ such that $\beta_{i1} = v_i$. Obviously, one can canonically identify Av with a vector in R^m.

b) Square matrices.

These are the matrices of the type $m \times m$. They have a determinant, which we denote by det A. A square matrix is called *regular* or *non-singular* if $\det A \neq 0$ and *singular* or *degenerate* if $\det A = 0$. Note that $\det A = 0$ does not imply that A is null.

A matrix whose nondiagonal entries are all zero is called a *diagonal* matrix. If, in particular, all these diagonal entries are 1, the matrix is called the *unit matrix* and is denoted by $I = (\delta_{ij})$ where δ_{ij} is Kronecker's delta.

Observe that if one of two $m \times m$ matrices whose product is 0 is regular, then the other must be null, for if $\det A \neq 0$, the linear homogeneous system in the unknowns $b_{1j}, b_{2j}, \ldots, b_{nj}$,

$$a_{i1}b_{1j} + a_{i2}b_{2j} + \ldots + a_{im}b_{mj} = 0, \qquad i = 1,2,\ldots,m$$

has only the null solution.

Similarly, if A is regular, there is a unique matrix B such that $AB = I$, since for every i, the linear system

$$a_{i1}b_{1j} + a_{i2}b_{2j} + \ldots + a_{im}b_{mj} = \delta_{ij}, \qquad i = 1,2,\ldots,m$$

in the unknowns $b_{1j}, b_{2j}, \ldots, b_{nj}$ admits one and only one solution. The matrix B is called the *inverse* of A and is denoted by A^{-1}.

From the theorem of Binet, saying that

$$\det(AB) = \det A \cdot \det B, \tag{1.4}$$

we may conclude that *a matrix is invertible if and only if it is nonsingular.*

We finally observe that the square matrices are closed under matrix multiplication, which is an associative operation, $(AB)C = A(BC)$. We may therefore, in summary, state the following theorem.

Theorem 1.1. The nonsingular square matrices of order $m \times m$ form a group under the operation of multiplication. This group is noncommutative.

Corollary 1.2. The right inverse equals the left inverse, that is, $BA = I = AB'$ implies that $B = B'$.

We prove this by observing that if we multiply $BA = I$ on the right by B', we get, from associativity, $B(AB') = B'$, whence $B = B'$.

c) Eigenvalues and Eigenvectors

We may ask whether, for an $m \times m$ square matrix, there are scalars λ and nonzero vectors v in R^m for which

$$Av = \lambda v.$$

In this case, the λ are called *eigenvalues* and the v *eigenvectors* associated with λ.

In general, even if the entries of A are real, we cannot say that real λ and vectors v exist satisfying (1.5), for what (1.5) implies in the real field is that vectors exist in R^m which are not rotated by the transformation A from R^m to R^m but merely multiplied by the number λ. It is clear that a matrix like

$$\begin{pmatrix} 0 & 1 \\ -1 & 0 \end{pmatrix}$$

rotates every vector by $\frac{1}{2}\pi$. For this reason, λ and v will, in this section, stand for scalars and vectors which may be complex.

We first of all observe that in order for (1.5) to have nonzero solutions v, it is necessary (and sufficient, if we admit that v may be complex) that λ be a real (or complex) solution of the algebraic equation

$$P(\lambda) = \det(A - \lambda I) = 0; \tag{1.6}$$

$P(\lambda)$ is a polynomial of degree m in the variable λ, and Eq. (1.6) is called the *characteristic* or *secular equation* of the matrix A. Its coefficients, for a reason which we shall soon see, are called the invariants of A; we have

$$P(\lambda) = \lambda^m - I_A\lambda^{m-1} + II_A\lambda^{m-2} + \dots + (-1)^m \mathrm{Inv}_A^m \cdot \lambda^0. \tag{1.7}$$

We recall that the invariants are obtained as follows. To produce the jth invariant, pick a combination of j elements from the principal diagonal and consider the $j \times j$ matrix whose rows and columns are determined by these j elements. Take its determinant, and then form the sum of all such determinants, with their signs, for all possible combinations of j elements from the principal diagonal. This sum is the jth invariant. In particular, $I_A = \Sigma_{i=1}^m a_{ii}$ is called the *trace* of A ($\mathrm{tr}\ A$), while $\mathrm{Inv}_A^m = \det A$. We may now state the following important proposition.

Proposition 1.3. Let T be a nonsingular matrix of the same order as A. Then the characteristic equation of $T^{-1}AT$

$$\det(T^{-1}AT - \lambda I) = 0, \tag{1.8}$$

has the same roots as the characteristic equation of A.

To prove this, observe that

$$T^{-1}(A - \lambda I)T = (T^{-1}AT - \lambda I). \tag{1.9}$$

Since $\det T^{-1} \cdot \det T = 1$, (1.4) implies that we always have

$$\det(T^{-1}AT - \lambda I) = \det(A - \lambda I). \tag{1.10}$$

Note that (1.10) assures that the coefficients of the polynomials $P(\lambda)$ are the same in the two cases, so that they are truly invariants under transformations of the type $T^{-1}AT$.

The matrix $A^t \equiv (a_{ji})$ is called the *transpose* of the matrix $A \equiv (a_{ij})$; $A^* = (\overline{a}_{ji})$ is its *adjoint*, where $\overline{a}_{ji}$ is the complex conjugate of a_{ji}. The adjoint is thus the complex conjugate of the transposed matrix, and in the real case the two are the same. Observe that

$$(AB)^t = B^tA^t, \quad (AB)^* = B^*A^*, \quad A^{tt} = A, \quad A^{**} = A.$$

We may now state the following

Proposition 1.4. Let A be a self-adjoint matrix, that is, $A = A^*$. (If A is real, this means that A is symmetric.) Then the eigenvalues of A are all real.

Proof: Let λ be an eigenvalue with corresponding eigenvector (perhaps complex) v, so that $Av = \lambda v$. If we multiply by $v^* = \bar{v}^t$, we get

$$v^*Av = \lambda v^*v. \tag{1.11}$$

But $v^*v = \Sigma_{i=1}^m v_i\bar{v}_i = ||v||^2$, which is real and positive. Since v^*Av is a number and since $A = A^*$, we have

$$\overline{v^*Av} = (v^*Av)^* = v^*A^*v^{**} = v^*Av.$$

From this it follows that v^*Av is real, whence $\lambda = v^*Av\cdot(v^*v)^{-1}$ is real.

1.2. Linear Operators Between Banach Spaces

We have thus far considered matrices as "sets of vectors". Their norm was linked to the fact that we identified them with vectors of $\mathbb{R}^d$ where $d = mn$. Now we shall see that the matrices are best discussed from the more general point of view of linear operators between Banach spaces, and that it is possible to introduce another norm equivalent to the one we already have.

Let X and Y be two Banach spaces; a function $A: X \to Y$ is called a *linear operator* if, for every pair of elements x_1, x_2 in X, and for every pair of real (or complex, if the spaces are over the complex field) numbers λ_1, λ_2, it follows that

$$A(\lambda_1x_1 + \lambda_2x_2) = \lambda_1Ax_1 + \lambda_2Ax_2. \tag{1.12}$$

(We shall henceforth follow the tradition of omitting the parentheses with linear operators and shall write Ax for $A(x)$.) The continuity of the linear operators is equivalent to two other simpler conditions, as we see from the following:

Theorem 1.5. Let A be a linear operator between two Banach spaces X and Y. The following conditions are equivalent:

a. A is continuous at every point x of X.

b. A is continuous at one point x_0 of X.

c. There is a constant K such that for every x in X,

$$||Ax|| \leq K||x||. \tag{1.13}$$

Proof: It is clear that a. ⇒ b. If b. is true, then, given $\varepsilon > 0$, there exists $\delta > 0$ such that, if $||x|| \leq \delta$, we have

$$||A(x + x_0) - A(x_0)|| \leq \varepsilon.$$

The linearity of A implies that this is equivalent to $||Ax|| \leq \varepsilon$; for all x such that $||x|| = \delta$ it follows that

$$||Ax|| \leq \frac{\varepsilon}{\delta} ||x||.$$

Since every element x' of X can be expressed as λx, with $||x|| = \delta$, we have, again by linearity,

$$||Ax'|| = |\lambda| \; ||Ax|| \leq \frac{\varepsilon}{\delta} |\lambda| \; ||x|| = \frac{\varepsilon}{\delta} ||x'||$$

and so c. follows, with $K = \varepsilon/\delta$. If, finally, c. holds, then, because of linearity, a. clearly follows.

Theorem 1.6. The space $L(X,Y)$ of continuous linear operators from X to Y is a Banach space under the operations

$$(A + B)x = Ax + Bx \tag{1.14}$$

$$(\alpha A)x = \alpha(Ax) \tag{1.15}$$

and the norm

$$||A|| = \sup_{\substack{x \neq 0 \\ x \in X}} \frac{||Ax||}{||x||} = \sup_{||x||=1} ||Ax||. \tag{1.16}$$

Proof: All properties, save the triangle inequality and completeness, are quite obvious. To prove the former, note that

$$\begin{aligned} ||A+B|| &= \sup_{||x||=1} ||(A+B)x|| \leq \sup_{||x||=1} (||Ax||+||Bx||) \\ &\leq \sup_{||x||=1} ||Ax|| + \sup_{||x||=1} ||Bx|| = ||A|| + ||B||. \end{aligned}$$

The completeness of $L(X,Y)$ follows from that of Y. Let A_n be a Cauchy sequence. Then, for every $\varepsilon > 0$, there is an n_ε such that for $n,m > n_\varepsilon$ we have, for every x in X,

$$||(A_n - A_m)x|| \leq ||A_n - A_m|| \; ||x|| < \varepsilon||x||. \tag{1.17}$$

If, for every x, we put $A_n x = y_n$, it follows that $(y_n)_n$ is a Cauchy sequence, and therefore converges to an element y of Y. We define

$$Ax = y = \lim_{n\to\infty} A_n x. \tag{1.18}$$

The linearity of A is evident. We have

$$||A|| \leq ||A - A_n|| + ||A_n||$$

with $||A - A_n|| = \sup_{||x||=1} ||Ax - A_n x||$; if n is such that (1.17) holds, we have $||Ax - A_n x|| = \lim_{m\to\infty} ||A_m x - A_n x|| < \varepsilon ||x||$ (recall that the norm is a continuous function with respect to the metric which it induces), and so $||A - A_n|| < \varepsilon$. This implies that $A \in L(X,Y)$ and lim $\lim_{n\to\infty} ||A_n - A|| = 0$.

At this point, we introduce an *operator norm* on the space of $m \times n$ matrices. If A is an $m \times n$ matrix then the function that maps $x \in \mathbf{R}^n$ into Ax is a linear operator $\mathbf{R}^n \to \mathbf{R}^m$ which we will again denote by A. We transfer the operator norm of $L(\mathbf{R}^n, \mathbf{R}^m)$ on the space of $m \times n$ matrices by defining

$$||A|| = \sup_{\substack{x \neq 0 \\ x \in \mathbf{R}^n}} \frac{||Ax||}{||x||} = \sup_{||x||=1} ||Ax||. \tag{1.19}$$

We shall now show how this norm is related to the eigenvalues. We have

$$||A||^2 = \sup_{||x||=1} ||Ax||^2.$$

Since $||Ax||^2 = \sum_{i=1}^{m} \left(\sum_{j=1}^{n} a_{ij} x_j \right)^2$, we may, upon differentiating with respect to the x_s and noting the constraint $\sum_{j=1}^{n} x_j^2 = 1$, obtain by Lagrange multipliers

$$\sum_{i=1}^{m} \left(\sum_{j=1}^{n} a_{ij} x_j \right) a_{is} - \lambda x_s = 0, \qquad s = 1,2,\ldots,n$$

or

$$\sum_{j=1}^{n} \left(\sum_{i=1}^{m} a_{is} a_{ij} \right) x_j - \lambda x_s = 0, \qquad s = 1,2,\ldots,n.$$

Thus, since $A^t = A^*$ (both are real matrices),

$$A^*Ax - \lambda x = 0,$$

which has nonzero solutions when λ is a eigenvalue of A^*A.

Since $A^*A = (A^*A)^*$, such eigenvalues, by Proposition 1.4, are real. If we pick for these eigenvalues eigenvectors x such that $||x|| = 1$, we obtain

$$||Ax||^2 = (Ax)^*Ax = x^*(A^*Ax) = \lambda x^*x = \lambda||x||^2,$$

and so $||A||^2$ is the maximum eigenvalue of A^*A.

One could prove directly that if $A = A^*$, then $||A||$ can be calculated more simply, in that $||A|| = \max\{|\lambda_s|: \lambda_s$ is an eigenvalue of $A\}$. This will be an immediate consequence of Theorem 3.5.

The connection between the norms that we have introduced is made clear by the following:

<u>Theorem 1.7</u>. If $||\cdot||_1$ and $||\cdot||_2$ are the norms introduced in (1.1) and (1.16) respectively, we have, for every $m \times n$ matrix A,

$$||A||_1 \leq \sqrt{n}\ ||A||_2; \qquad ||A||_2 \leq ||A||_1. \tag{1.20}$$

<u>Proof</u>: Let $||A||_2 = M$. We put $B = A^*A$. The eigenvalues λ_i, $i = 1,2,\ldots,n$, of B are all such that $0 \leq \lambda_i = M^2$; the first coefficient of the characteristic equations will therefore satisfy the relation

$$I_B = \sum_{i=1}^{n} \lambda_i \leq n \cdot M^2.$$

On the other hand,

$$I_B = \sum_{i=1}^{n} b_{ii} = \sum_{i=1}^{n} \sum_{j=1}^{m} \bar{a}_{ji} a_{ji} = ||A||_1^2.$$

The estimate (1.20) with $\sqrt{n}$ cannot be improved in that if I is the $n \times n$ identity matrix, then

$$||I||_1 = \sqrt{n}\ ||I||_2.$$

Now let A be a square matrix and λ one of its eigenvalues. We have, then, for every index i,

$$\left|\sum_{j=1}^{n} a_{ij}u_j\right|^2 = |\lambda|^2|u_i|^2.$$

Even more, if we apply (1.3) to the $n \times 1$ matrix u_j and the $1 \times n$ matrix a_{ij} with i fixed, we get

$$|\lambda|^2|u_i|^2 = \left|\sum_{j=1}^{n} a_{ij}u_j\right|^2 \leq \sum_{j=1}^{n} |a_{ij}|^2 \cdot ||u||^2.$$

Adding with respect to i and dividing by $||u||^2$ produce

$$|\lambda|^2 \leq \sum_{i=1}^{n} \sum_{j=1}^{n} |a_{ij}|^2 = ||A||_1^2.$$

We therefore have, since A*A is square,

$$||A||_2^2 = (\text{the maximum eigenvalue of } A^*A) \leq ||A^*A||_1,$$

but, since $||A^*||_1 = ||A||_1$, it follows from (1.3) that

$$||A^*A||_1 \leq ||A^*||_1 ||A||_1 = ||A||_1^2,$$

and the theorem is proved. The bound thus obtained cannot be improved; it suffices to consider the case when $a_{11} = 1$ and $a_{ij} = 0$ for any other pair (i,j).

If we now consider L(X), the space of continuous linear operators from one Banach space X into itself, we may define the product AB of two operators by

$$(AB)x = A(Bx).$$

It follows from

$$||AB|| = \sup_{0 \neq x \in X} \frac{||ABx||}{||x||} \leq ||A|| \cdot \sup_{0 \neq x \in X} \frac{||Bx||}{||x||} = ||A|| \cdot ||B||,$$

that the norms satisfy

$$||AB|| \leq ||A|| \cdot ||B||. \tag{1.21}$$

This assures us that L(X) is a Banach algebra, that is, that it is a Banach space in which an associative (though, in general, noncommutative) product can be defined to satisfy the distributive laws, and to have an identity element (the identity operator $Ix = x$).

1.3. Canonical Form of Matrices

In this section, we shall present the principal theorems on the form to which matrices may be reduced by means of transformations of the type $T^{-1}AT$. We recall that if T is a nonsingular matrix, matrices of the form $T^{-1}AT$ are said to be similar to A. We shall provide proofs only of those theorems that are needed in what follows.

Theorem 1.8. Every square matrix A (with complex entries) is similar to an upper (or lower) triangular matrix. (A matrix is upper triangular if $a_{ij} = 0$ for $i > j$, lower triangular if $a_{ij} = 0$ for $i < j$.) If the matrix and its eigenvalues are real, then the triangular matrix is also real. The diagonal elements of the triangular matrix are the eigenvalues of A.

<u>Proof</u>: Let λ_1 be an eigenvalue of A, and let $x^{(1)}$ be a corresponding eigenvector. Construct a basis for $\mathbf{C}^n$ by adding $n-1$ other independent vectors $x^{(2)},\ldots,x^{(n)}$ that need not be eigenvectors. (We speak of $\mathbf{C}^n$ instead of $\mathbf{R}^n$ because the eigenvalues may be complex. If they are real, we need consider only $\mathbf{R}^n$.) Let P be the matrix defined by $p_{ij} = x_i^{(j)}$, that is, $P = (x^{(1)};x^{(2)};\ldots;x^{(n)})$. We denote the entries of P^{-1} by $\bar{p}_{ij}$ and observe that they satisfy the relation $\Sigma_{s=1}^{n} \bar{p}_{ij}p_{sj} = \delta_{ij}$. The matrix $A' = P^{-1}AP$ will have $(\lambda_1,0,\ldots,0)$ for its first column since

$$a'_{i1} = \sum_{s=1}^{n}\sum_{t=1}^{n} \bar{p}_{is}a_{st}x_t^{(1)} = \lambda_1 \sum_{s=1}^{n} \bar{p}_{is}x_s^{(1)} = \lambda_1 \sum_{s=1}^{n} \bar{p}_{is}p_{s1},$$

so $a'_{i1} = \lambda_1\delta_{1i}$. This depends on the fact that $x^{(1)}$ is an eigenvector associated with λ_1.

We now proceed by induction. For $n = 1$, the theorem is obvious. For the sake of brevity, we shall use block matrices. These have the property that if

$$A = \begin{pmatrix} A' & A'' \\ A''' & A^{IV} \end{pmatrix}, \qquad B = \begin{pmatrix} B' & B'' \\ B''' & B^{IV} \end{pmatrix},$$

then

$$AB = \begin{pmatrix} A'B' + A''B'' & A'B'' + A''B^{IV} \\ A'''B' + A^{IV}B''' & A'''B'' + A^{IV}B^{IV} \end{pmatrix}$$

Let A' be the matrix previously defined, and let A'_{11} be the one obtained by removing the first row and first column from A'. By the inductive hypothesis, there is an invertible $(n-1) \times (n-1)$ matrix Q' such that $(Q')^{-1}A'_{11}Q'$ is upper triangular. We set

$$Q = \begin{pmatrix} 1 & 0\ldots\ldots0 \\ 0 & \\ \vdots & Q' \\ 0 & \end{pmatrix}.$$

We see at once that $A'' = Q^{-1}A'Q$ is also upper triangular; in fact, A'' is given by

$$\begin{pmatrix} 1 & 0\ldots\ldots0 \\ 0 & \\ \vdots & (Q')^{-1} \\ 0 & \end{pmatrix}\begin{pmatrix} \lambda_1 & *\ldots\ldots* \\ 0 & \\ \vdots & A'_{11} \\ 0 & \end{pmatrix}\begin{pmatrix} 1 & 0\ldots\ldots0 \\ 0 & \\ \vdots & Q' \\ 0 & \end{pmatrix}$$

$$= \begin{pmatrix} \lambda_1 & *\ldots\ldots* \\ 0 & \\ \vdots & (Q')^{-1}A'_{11}Q' \\ 0 & \end{pmatrix} = \begin{pmatrix} \lambda_1 & * & & * \\ 0 & \lambda_2 & * & \\ \vdots & 0 & \ddots & * \\ 0 & & & \lambda_n \end{pmatrix}.$$

Observe, however, that although we have called $\lambda_1,\ldots,\lambda_n$ the eigenvalues, they need not necessarily be distinct.

<u>Theorem 1.9</u>. If the eigenvalues are all distinct, the square matrix A is similar to a diagonal matrix (whose diagonal entries are the eigenvalues of A).

<u>Proof</u>: Let $x^{(i)}$ be an eigenvector associated with the eigenvalue λ_i. We observe that the n vectors $x^{(i)}$ are linearly independent; in fact, if the first $k-1$ are linearly independent, and if

$$x^{(k)} = \sum_{i=1}^{k-1} \alpha_i x^{(i)}$$

we would have

$$\sum_{i=1}^{k-1} \alpha_i\lambda_i x^{(i)} = \sum_{i=1}^{k-1} \alpha_i Ax^{(i)} = Ax^{(k)} = \lambda_k x^{(k)} = \lambda_k \sum_{i=1}^{k-1} \alpha_i x^{(i)}$$

whence $\Sigma_{i=1}^{k-1} \alpha_i(\lambda_i - \lambda_k)x^{(i)} = 0$. Since $x^{(1)},\ldots,x^{(k-1)}$ are linearly independent and $\lambda_i - \lambda_k \neq 0$, we have $\alpha_i = 0$ for $i = 1,2,\ldots,k-1$, so that $x^{(k)} = 0$.

We now let P be the matrix whose entries are $p_{ij} = x_i^{(j)}$; let $P^{-1} = (\bar{p}_{ij})$. It follows that

$$A' = P^{-1}AP = \begin{pmatrix} \lambda_1 & & 0 \\ & \lambda_2 & \\ & & \ddots \\ 0 & & \lambda_n \end{pmatrix} = (\lambda_j\delta_{ij})$$

since

$$a'_{ij} = \sum_{s=1}^{n}\sum_{t=1}^{n} \bar{p}_{is}a_{st}x_t^{(j)} = \sum_{s=1}^{n} \bar{p}_{is}\cdot\lambda_j x_s^{(j)} = \lambda_j \sum_{s=1}^{n} \bar{p}_{is}p_{sj} = \lambda_j\delta_{ij}.$$

We now pose the more general problem of how close to a diagonal matrix the representation can be when the eigenvalues are not distinct. There is a case when the answer is easy, namely, when the original matrix is symmetric (self-adjoint if the entries are complex). In this case, not only is the matrix diagonizable and the eigenvalues real, but in addition, the matrix of the transformation T can be chosen so as to be orthogonal (unitary in the complex case). In the other cases, there is no fast algorithm to determine whether or not a matrix is diagonizable. We shall see in Lemma 1.10 that there is a type of matrix for which diagonalization is impossible if the eigenvalues are not distinct, but the proof is indirect and is postponed until further on (Lemma 3.6). What one can say in general is that every matrix is similar to a Jordan matrix, which we proceed to define step by step.

A matrix of the form

$$\begin{pmatrix} \lambda_1 & & & 0 \\ 1 & \lambda_1 & & \\ & 1 & \ddots & \\ & & \ddots & \ddots \\ 0 & & 1 & \lambda_1 \end{pmatrix} = J(\lambda_1,k) \tag{1.22}$$

is called a $k \times k$ *elementary Jordan matrix,* while one of the form

$$\begin{pmatrix} J(\lambda_1,k) & 0 & 0 & 0 \\ 0 & J(\lambda_1,k') & 0 & 0 \\ 0 & 0 & \ddots & 0 \\ 0 & 0 & 0 & J(\lambda_1,k^{(s)}) \end{pmatrix} = J(\lambda_1,k,k',\ldots,k^{(s)}) \tag{1.23}$$

is called a *Jordan block.* A Jordan matrix is a diagonal block matrix where every block on the diagonal is a Jordan block.

The representation theorem affirms that every matrix is similar to a Jordan matrix having a Jordan block for each distinct eigenvalue; the dimension of the block is equal to the multiplicity of the eigenvalue. What must be determined case by case is the manner in which Jordan blocks are subdivided into elementary Jordan matrices; in the two extreme cases, the Jordan block may be a diagonal matrix or a unique elementary Jordan matrix. For further details, the reader should consult a linear algebra text.

At this point, we insert two technical lemmas on particular types of matrices; we shall need these results in the sequel.

Lemma 1.10. Let A be a $k \times k$ elementary Jordan matrix with eigenvalue λ. Then the inverse of A is of the form

$$A^{-1} = \begin{pmatrix} \lambda^{-1} & & & & 0 \\ -\lambda^{-2} & \lambda^{-1} & & & \\ \lambda^{-3} & -\lambda^{-2} & \ddots & & \\ & \lambda^{-3} & \ddots & \ddots & \\ (-1)^{k+1}\lambda^{-k} & \cdots & \lambda^{-3} & -\lambda^{-2} & \lambda^{-1} \end{pmatrix} \tag{1.24}$$

Proof: We denote by J_s the square matrix of order k, $k \geq s$, in which $p_{ij} = 1$ for $i-j = s$ and 0 elsewhere. It follows that $J_s \cdot J_{s'}$ is $J_{s+s'}$ if $s + s' \leq k$ and 0 otherwise. Thus the elementary Jordan matrix may be written in the form $A = \lambda J_0 + J_1$. One may immediately verify that $A^{-1} = \lambda^{-1}J_0 - \lambda^{-2}J_1 + \ldots + (-1)^{k+1}\lambda^{-k}J_k$, which is the same as (1.24).

Lemma 1.11. Let A be the following $k \times k$ matrix.

$$A \equiv \begin{pmatrix} \lambda & -1 & & & & 0 \\ & \lambda & -1 & & & \\ 0 & & \lambda & -1 & & \\ & & & \ddots & \ddots & \\ & & & & \lambda & -1 \\ \alpha_0 & \alpha_1 & \alpha_2 & \cdots & \alpha_{k-2} & (\lambda+\alpha_{k-1}) \end{pmatrix}$$

Let $P(\lambda)$ be its characteristic polynomial $\lambda^k + \alpha_{k-1}\lambda^{k-1} + \ldots + \alpha_1\lambda + \alpha_0$. We set, for $s \leq k$, $P_s(\lambda) = \lambda^k + \alpha_{k-1}\lambda^{k-1} + \ldots + \alpha_s\lambda^s$ and $Q_s(\lambda) = P_0 - P_s$. Then A^{-1} can be written in the form

$$A^{-1} = \frac{1}{P_0(\lambda)}\begin{pmatrix} P_1(\lambda)\lambda^{-1} & P_2(\lambda)\lambda^{-2} & P_3(\lambda)\lambda^{-3} & \cdots & P_k(\lambda)\lambda^{-k} \\ -Q_1(\lambda) & P_2(\lambda)\lambda^{-1} & P_3(\lambda)\lambda^{-2} & \cdots & P_k(\lambda)\lambda^{-k+1} \\ -\lambda Q_1(\lambda) & -Q_2(\lambda) & & & \\ & & & & \\ -\lambda^{k-2}Q_1(\lambda) & -\lambda^{k-3}Q_2(\lambda) & & -Q_{k-1}(\lambda) & P_k(\lambda)\lambda^{-1} \end{pmatrix}.$$

Proof: Multiply A^{-1} by A and observe that $(P_s - P_{s-1}) = -\alpha_{s-1}\lambda^{s-1}$ and that $P_k = \lambda^k$. Note that if $\alpha_0 = \alpha_1 = \ldots = \alpha_{k-1} = 0$, then the case is similar to that of Lemma 1.10 above.

1.4. Spectrum and Eigenvalues of a Linear Operator

The resolvent operator plays a most important role in the theory of continuous linear operators from a Banach space into itself. If $\lambda \in \mathbf{C}$, and $A \in L(X)$, the resolvent of A is given by

$$R(\lambda,A) = (\lambda I - A)^{-1}, \quad \text{whenever} \quad (\lambda I - A)^{-1} \quad \text{exists.} \tag{1.25}$$

One can prove that, given A, there is always at least one value of λ for which the operator $R(\lambda,A)$ is not defined. We have already proved this fact for matrices since, if λ is an eigenvalue of the matrix A, then $\lambda I - A$ has nonzero kernel, and the operator $\lambda I - A$ is not invertible.

The set of numbers $\lambda \in \mathbf{C}$ for which $(\lambda I - A)^{-1}$ does not exist in $L(X)$ is called the spectrum of the operator A and is denoted by $\sigma(A)$.

The following two theorems are useful.

Theorem 1.12. Let X be a Banach space and $A \in L(X)$ a continuous linear operator. Then the spectrum of A is closed in $\mathbf{C}$.

Proof: It is sufficient to show that the complement of $\sigma(A)$ is open. Let λ_0 be such that $R(\lambda_0,A)$ exists. For every λ, we have $\lambda I - A = \lambda_0 I - A - (\lambda_0 - \ )I$; if we apply this operator side by side to the operator $R(\lambda_0,A)$, we get

$$(\lambda I - A)R(\lambda_0,A) = I - (\lambda_0 - \lambda)R(\lambda_0,A). \tag{1.26}$$

If $|\lambda_0 - \lambda|$ is such that $|\lambda_0 - \lambda|\ ||R(\lambda_0,A)|| < 1$, then the operator $(\lambda_0 - \lambda)R(\lambda_0,A)$ is a contraction. Then the operator on the right-hand side of (1.26) is invertible by the fixed point theorem for contractions of Sec. 4.1 of Chapter I. Therefore, for λ sufficiently near to λ_0, $R(\lambda_0,A)^{-1}\cdot R(\lambda,A)$ exists, so $R(\lambda,A)$ exists as well.

Theorem 1.13. (The spectral radius theorem). Let X be a Banach space and $A \in L(X)$. If $\lambda \in \sigma(A)$, then

$$|\lambda| \leq ||A||. \tag{1.27}$$

That is, the spectrum is contained in the closed ball with center at the origin and radius $||A||$.

Proof: Let $|\lambda| > ||A||$; the term $\frac{A}{\lambda}$ of the operator $\lambda^{-1}(\lambda I - A) = I - \frac{A}{\lambda}$ is a contraction, so the operator itself is invertible by the theorem cited above. As a result, $\lambda I - A$ is also invertible, so $\lambda \notin \sigma(A)$.

The relationship between eigenvalues and elements of the spectrum is trivial in one direction: every eigenvalue is an element of the spectrum. The converse is not in general true; nevertheless, if the space X has finite dimension k, the spectrum coincides with the set of eigenvalues. In fact, the linear operators in that case are given by $k \times k$ matrices (note that they are automatically continuous), and the matrix $\lambda I - A$ is invertible if and only if $\det(\lambda I - A) \neq 0$. On the other hand, the λ such that $\det(\lambda I - A) = 0$, are as we have seen, exactly the eigenvalues of A.

1.5. Limits of Operators

In this section, we introduce the concept of limit in the norm and of strong limit of a sequence of operators; we shall, however, easily see that in the case of matrices the two definitions give the same result. We shall then introduce the derivative and integral of operators depending on a parameter and shall discuss series of operators.

Let X be a Banach space and A_n a sequence of operators in $L(X)$. We shall say that A_n *converges in norm to the operator* $A \in L(X)$ *if*

$$\lim_{n\to\infty} ||A_n - A|| = 0. \tag{1.28}$$

If, for each element $x \in X$,

$$\lim_{n\to\infty} ||A_n x - Ax|| = 0, \tag{1.29}$$

we shall say that A_n *converges strongly to the operator* $A \in L(X)$. It is immediate that (1.28) implies (1.29). The converse is not in general true. However, in the case of a finite dimensional Banach space it is true, for if we put $x = e^{(i)}$, we have from (1.29)

$$\lim_{n\to\infty} ||(a_{i1}^{(n)}, a_{i2}^{(n)}, \ldots, a_{ik}^{(n)})|| = 0 \quad \text{for } i = 1,2,\ldots,k.$$

This insures that all the components tend to 0, and since there are only a finite number of them, the limit is uniform.

It is now possible to define the meaning of the symbol

$$\sum_{s=1}^{\infty} A_s;$$

we take it to be the limit in the norm of the partial sums $\Sigma_{s=1}^{N} A_s$, that are defined by the structure of $L(X)$. In the case of matrices, we obviously have

$$\sum_{s=1}^{\infty} A_s = A \tag{1.30}$$

if and only if, for every component,

$$\sum_{s=1}^{\infty} a_{ij}^{(s)} = a_{ij}. \tag{1.31}$$

We say that the series of operators (1.30) converges absolutely if the series

$$\sum_{s=1}^{\infty} ||A_s||. \tag{1.32}$$

converges. In the case of matrices, this happens if and only if, for every component,

$$\sum_{s=1}^{\infty} |a_{ij}^{(s)}| \tag{1.33}$$

converges.

We now let A be an operator depending on a real parameter t taking values in an interval $[a,b]$. Let t_0 be a point in that interval. If h is sufficiently small, the operator

$$\frac{A(t_0 + h) - A(t_0)}{h} \tag{1.34}$$

can be defined. If its limit as $h \to 0$ exists, we say that the operator A is differentiable at t with respect to t. The limiting operator is denoted by $\frac{d}{dt} A(t)$. We thus have

$$\lim_{h\to 0}\left|\left|\frac{A(t+h) - A(t)}{h} - \frac{dA}{dt}\right|\right| = 0. \tag{1.35}$$

In the case of matrices, the limit exists if and only if each component $a_{ij}(t)$ is differentiable, and we have

$$\frac{d}{dt} A(t) = \left(\frac{d}{dt} a_{ij}(t)\right).$$

In general, if A and B are two differentiable operators in $L(X)$, the following product rule is valid.

$$\frac{d}{dt}(AB) = \frac{dA}{dt} B + A \frac{dB}{dt} .$$

We shall say that a series of operators depending on a parameter converges uniformly if, for every $\varepsilon > 0$, there exists a ν_ε such that for every $t \in [a,b]$ and for every $\nu > \nu_\varepsilon$

$$\left\| \sum_{s=\nu}^{\infty} A_s(t) \right\| < \varepsilon.$$

If the operators $A_s(t)$ *are differentiable and the series*

$$\sum_{s=0}^{\infty} \frac{d}{dt} A_s(t)$$

converges uniformly, then also the operator $A = \Sigma_{s=0}^{\infty} A_s(t)$ *is differentiable, and we have*

$$\frac{dA}{dt} = \sum_{s=0}^{\infty} \frac{d}{dt} A_s.$$

For the time being, we shall limit our definition of the integral to the case of operators which can be represented by matrices; the more general case of operators that depend continuously on a parameter will be taken up later. Let $A(t)$ be a matrix depending on a parameter t; we suppose that the components $a_{ij}(t)$ are all integrable functions over the interval $[t_0,t]$. We shall say that the matrix

$$\left(\int_{t_0}^{t} a_{ij}(\tau)d\tau\right),$$

which we write as

$$\int_{t_0}^{t} A(\tau)d\tau,$$

is the *integral of the matrix* $A(t)$ between t_0 and t. It follows from the definition that if the functions a_{ij} are continuous at t, then

$$\frac{d}{dt}\int_{t_0}^{t} A(\tau)d\tau = A(t).$$

If, finally, A is an $m \times n$ matrix, and L is a number greater than or equal to the absolute values of $a_{ij}(t)$, then

$$\left\|\int_{t_0}^{t} A(\tau)d\tau\right\| \le \sqrt{m \cdot n} \cdot L|t-t_0|.$$

2. LINEAR SYSTEMS OF ORDINARY DIFFERENTIAL EQUATIONS

2.1. Formal Solution of Linear Systems

The linear systems that we shall consider in this section are those of the type

$$y_i' = \sum_{s=1}^{n} a_{is}(x)y_s + b_i(s), \qquad i = 1,2,\ldots,n. \tag{2.1}$$

We shall write

$$y(x) \equiv \begin{pmatrix} y_1(x) \\ \vdots \\ y_n(x) \end{pmatrix}$$

for an n-tuple of continuous functions and thereby obtain $n \times 1$ matrices depending on a parameter. We assign to $C([a,b],R^n)$ the following norm

$$||y(x)|| = \left\{ \sum_{i=1}^{n} \max_{a\le x\le b} |y_i(x)|^2 \right\}^{1/2}.$$

If $A(x)$ is the matrix of the coefficients of the system (2.1), it may, if we suppose $b_i(x) \equiv 0$, be expressed in the form

$$\frac{dy}{dx} = A(x)y(x) \tag{2.2}$$

which, if integrated from x_0 to x, yields

$$y(x) = y_0 + \int_{x_0}^{x} A(\tau)y(\tau)d\tau \quad \text{where} \quad y_0 = y(x_0). \tag{2.3}$$

(2.3) is the integral form of the system (2.1) in the homogeneous case, and $y(x)$ is the $n \times 1$ matrix whose elements are the functions $y_i(x)$ that satisfy the system (2.1) with the initial conditions $y(x_0) = y_0$.

We now consider the transformation

$$Ty(x) = y_0 + \int_{x_0}^{x} A(\tau)y(\tau)d\tau, \qquad x_0,x \in [a,b]. \tag{2.4}$$

It maps every element of the space of n-tuples of continuous functions on $C([a,b],R^n)$ into another such element. For every $x \in [a,b]$, we put $y_0(x) = y_0$, and try to find, by the method of successive approximation, the fixed point of the transformation in (2.4). We take $y_0(x)$ for our first approximation and produce

$$y_1(x) = Ty_0(x) = y_0(x) + \left(\int_{x_0}^{x} A(\tau)d\tau\right)y_0(x) = \left[I + \int_{x_0}^{x} A(\tau)d\tau\right]y_0,$$

$$y_2(x) = T^2y_0(x) = y_0(x) + \int_{x_0}^{x} A(\tau)y_1(\tau)d\tau$$

$$= \left[I + \int_{x_0}^{x} A(\tau)d\tau + \int_{x_0}^{x} d\tau \int_{x_0}^{\tau} A(\tau)A(\tau_1)d\tau_1\right]y_0,$$

$$y_3(x) = T^3y_0(x) = y_0(x) + \int_{x_0}^{x} A(\tau)y_2(\tau)d\tau$$

$$= \left[I + \int_{x_0}^{x} A(\tau)d\tau + \int_{x_0}^{x} d\tau \int_{x_0}^{\tau} A(\tau)A(\tau_1)d\tau_1 + \int_{x_0}^{x} d\tau \int_{x_0}^{\tau} d\tau_1 \int_{x_0}^{\tau_1} A(\tau)A(\tau_1)A(\tau_2)d\tau_2\right]y_0.$$

Proceeding in this way, we put $y(x) = \lim_{n\to\infty} y_n(x)$ for those x for which it exists. If we denote by $\overset{x}{\underset{x_0}{E}} A(\tau)$ the sum of the series

$$\overset{x}{\underset{x_0}{E}} A(\tau) = I + \int_{x_0}^{x} A(\tau)d\tau + \int_{x_0}^{x} d\tau \int_{x_0}^{\tau} A(\tau)A(\tau_1)d\tau_1 + \cdots \tag{2.5}$$

we get

$$y(x) = \overset{x}{\underset{x_0}{E}} A(\tau)y_0. \tag{2.6}$$

It is easy to show that the series (2.5) converges uniformly in the whole interval in which the coefficients of the system (2.1) are defined. To do so, let M be a number greater than the absolute values of the terms $a_{ij}(x)$ in $[a,b]$. Then the absolute values of the terms of $\int_{x_0}^{x} A(\tau)d\tau$ are less than $M|x-x_0|$, while those of the matrix $\int_{x_0}^{x} d\tau \int_{x_0}^{\tau} A(\tau)A(\tau_1)d\tau_1$ are bounded by $\frac{1}{2}\, nM^2(x-x_0)^2$. Furthermore, the general term of the series $\overset{x}{\underset{x_0}{E}} A(\tau)$ is a matrix the absolute values of whose terms are bounded by $\frac{n^{s-1}M^s}{s!}(x - x_0)^s$. We thus conclude that the series $\overset{x}{\underset{x_0}{E}} A(\tau)$ converges uniformly in every interval in which the coefficients of $A(x)$ are bounded. The matrix in (2.5) is called the *resolvent matrix* of the system (2.2). Observe that if the matrix A is constant, the terms of the series (2.6) are given by

$$\frac{A^s(x - x_0)^s}{s!} .$$

The series thus obtained is by definition $\exp[A(x-x_0)]$, as will be explained in Sec. 3.1; in Sec. 3.2 we shall give calculation formulas for this function.

If we now differentiate (2.5) term by term, we get

$$A(x) + A(x)\int_{x_0}^{x} A(\tau)d\tau + A(x)\int_{x_0}^{x} d\tau \int_{x_0}^{\tau} A(\tau)A(\tau_1)d\tau_1 + \cdots,$$

which is uniformly convergent in the same interval and has sum $A(x) \underset{x_0}{\overset{x}{E}} A(\tau)$. We thus deduce the relation

$$\frac{d}{dx} \underset{x_0}{\overset{x}{E}} A(\tau) = A(x) \underset{x_0}{\overset{x}{E}} A(\tau) \tag{2.7}$$

which allows us to verify that (2.6) is a solution of system (2.2).

2.2. Fundamental Systems of Solutions and Adjoint Systems

We denote by $\gamma_{ij}(x)$ the entries of the matrix $\underset{x_0}{\overset{x}{E}} A(\tau)$ of (2.5). From (2.6) and (2.7), we deduce that the n-tuple of functions

$$\gamma_{1j}(x),\ \gamma_{2j}(x),\ldots,\gamma_{nj}(x), \qquad (j = 1,2,\ldots,n)$$

is the particular solution of the ysystem (2.2) that satisfies the initial conditions

$$\gamma_{ij}(x_0) = 0 \quad \text{if } i \neq j, \quad \gamma_{jj}(x_0) = 1,$$

so that the matrix $\underset{x_0}{\overset{x}{E}} A(\tau)$ reduces to the identity when $x = x_0$. We now prove that the matrix $\underset{x_0}{\overset{x}{E}} A(\tau)$ is regular for every x. If $D(x)$ is its determinant, we have

$$\frac{d}{dx} D(x) = \sum_{i=1}^{n} \det \begin{pmatrix} \gamma_{11}(x) & \gamma_{12}(x)\cdots\cdots & \gamma_{1n}(x) \\ & \cdots\cdots\cdots\cdots & \\ \gamma_{i-1,1}(x) & \gamma_{i-1,2}(x)\cdots & \gamma_{i-1,n}(x) \\ \gamma'_{i,1}(x) & \gamma'_{i,2}(x)\cdots\cdots & \gamma'_{i,n}(x) \\ \gamma_{i+1,1}(x) & \gamma_{i+1,2}(x)\cdots & \gamma_{i+1,n}(x) \\ & \cdots\cdots\cdots\cdots & \\ \gamma_{n1}(x) & \gamma_{n2}(x) & \gamma_{nn}(x) \end{pmatrix}$$

and, noting that

$$Y'_{ij}(x) = \sum_{s=1}^{n} a_{is}(x)Y_{sj}(x),$$

we get

$$\frac{d}{dx} D(x) = D(x) \cdot \sum_{j=1}^{n} a_{jj}(x).$$

If we now integrate from x_0 to x, we produce the *formula of Jacobi*

$$D(x) = D(x_0) \cdot \exp\left(\int_{x_0}^{x} \sum_{j=1}^{n} a_{jj}(\tau)d\tau\right).$$

Since $D(x_0) = 1$, we see that $D(x)$ is never 0. It can be shown that one can derive a formula such as (2.5) even when beginning with any system of n linearly independent solutions, the general integral being a linear combination of these. In Sec. 3.2, we make use of the operational calculus to calculate the matrix of (2.5) for the case when the coefficients are constant.

Let us consider once again the system (2.2), $y' = A(x)y$. We consider a function $z(x)$ in $C([a,b],\mathbb{R}^n)$ that has derivatives. Let $Z^*(x) = (z_1(x), z_2(x), \ldots, z_n(x))$ be the transpose of the $n \times 1$ matrix $z(x)$. If we multiply (2.2) on the left by $z^*(x)$, we get

$$z^*y' = z^*Ay.$$

Since

$$z^*y' = (z^*y)' - (z^*)'y$$

we have

$$z^*y' - z^*Ay = (z^*y)' - (z^*)'y - z^*Ay.$$

This provides us with an identity valid for all z^*;

$$z^*(y' - Ay) + ((z^*)' + z^*A)y = (z^*y)'. \tag{2.8}$$

If (2.2) holds and

$$(z^*)' = -z^*A,$$

(which, in terms of the transpose, means

$$z' = -A^*z), \tag{2.9}$$

then we get from (2.8)

$$z^*(x)y(x) = \text{a constant}. \tag{2.10}$$

The system (2.9) of differential equations is called the *system of differential equations adjoint to* (2.2), and vice versa. The relation (2.10) holds between any solution of (2.2) and any of (2.9).

A system of differential equations that coincides with its adjoint is called *self-adjoint*. The matrix $A(x)$ must be antisymmetric, that is,

$$a_{ij}(x) = -a_{ji}(x) \quad \text{for } i \neq j, \quad a_{ii} = 0.$$

Observe that if matrix A is self-adjoint, it is not necessary that the system determined by the matrix A be self-adjoint. It is to avoid this incongruity that in the more advanced treatments of differential equations (with constant coefficients in general), one uses the operator $i\frac{d}{dx}$ instead of $\frac{d}{dx}$ and associates with it the corresponding matrix (with complex entries even if the original equation is over the real field).

If a system is self-adjoint, we have between any two of its general solutions $(y_1(x),\ldots,y_n(x))$ and $(z_1(x),\ldots,z_n(x))$ the relation

$$\sum_{i=1}^{n} y_i(x)\bar{z}_i(x) = \text{a constant}$$

or, in particular,

$$\sum_{i=1}^{n} |y_i(x)|^2 = \text{a constant.}$$

Given the resolvent matrix $\underset{x_0}{\overset{x}{E}} A(\tau)$ of the system (2.2), we shall now determine the resolvent matrix of the system (2.9) adjoint to it. Equivalently, we propose to determine a matrix $M(x)$ such that if $z(x_0) = z_0$, then

$$z^*(x) = z_0^*M^*(x).$$

Then, taking note of (2.10), we have

$$z_0^*M^*(x)\left(\underset{x_0}{\overset{x}{E}} A(\tau)\right)y_0 = \text{a constant,}$$

which, upon being differentiated, yields

$$z_0^* \frac{d}{dx}\left[M^*(x) \underset{x_0}{\overset{x}{E}} A(\tau)\right]y_0 = 0.$$

Since this relation must hold for every pair z_0^*, y_0, we must have

$$\frac{d}{dx}\left[M^*(x) \mathop{E}_{x_0}^{x} A(\tau)\right] = 0.$$

Because, finally, for $x = x_0$, $M(x_0) = \mathop{E}_{x_0}^{x_0} A(\tau) = I$, we have

$$M(x) = \left(\left[\mathop{E}_{x_0}^{x} A(\tau)\right]^{-1}\right)^* = \left[\mathop{E}_{x_0}^{x} A^*(\tau)\right]^{-1}$$

which gives the formula for the resolvent of the system (2.9),

$$z(x) = \left[\left(\mathop{E}_{x_0}^{x} A(\tau)\right)^{-1}\right]^* z_0. \tag{2.11}$$

2.3. Nonhomogeneous Systems

We now study nonhomogeneous sytems of linear differential equations; these may be written in the form

$$\frac{dy}{dx} = A(x)y + b(x). \tag{2.12}$$

From identity (2.8) we get

$$z^*(y' - A(x)y) + ((z^*)' + z^*A)y = (z^*y).$$

If we now take $z^* = z_0^*(\mathop{E}_{x_0}^{x} A(\tau))^{-1}$, a solution of the equation $z' = -A^*z$, and y a solution of (2.12), we have

$$z_0^*\left(\mathop{E}_{x_0}^{x} A(\tau)\right)^{-1} b = z_0^* \frac{d}{dx}\left\{\left(\mathop{E}_{x_0}^{x} A(\tau)\right)^{-1} y(x)\right\}$$

whence, because z_0^* is arbitrary, we get

$$\left[\mathop{E}_{x_0}^{x} A(\tau)\right]^{-1} b = \frac{d}{dx}\left\{\left(\mathop{E}_{x_0}^{x} A(\tau)\right)^{-1} y(x)\right\}.$$

If we integrate from x_0 to x, we produce

$$\left[\mathop{E}_{x_0}^{x} A(\tau)\right]^{-1} y(x) = \int_{x_0}^{x} \left[\mathop{E}_{x_0}^{t} A(\tau)\right]^{-1} b(t)dt + y(x_0)$$

which, upon being multiplied on the left by $\mathop{E}_{x_0}^{x} A(\tau)$, gives

$$y(x) = \mathop{E}_{x_0}^{x} A(\tau) \int_{x_0}^{x} \left[\mathop{E}_{x_0}^{t} A(\tau)\right]^{-1} b(t)dt + \mathop{E}_{x_0}^{x} A(\tau)y(x_0).$$

This relation furnishes a solution of the system (2.12). Since the difference of two solutions of Eq. (2.12) satisfies the associated homogeneous system $y' = Ay$, we deduce the expression for the resolvent of the system (2.12):

$$y(x) = \overset{x}{\underset{x_0}{E}} A(\tau)y_0 + \int_{x_0}^{x} \overset{x}{\underset{x_0}{E}} A(\tau)\left[\overset{t}{\underset{x_0}{E}} A(\tau)\right]^{-1} b(t)dt. \tag{2.13}$$

(Recall here the method of variation of arbitrary constants.)

In conclusion, we shall prove that if the matrix A has only constant entries we have the following equality:

$$\left[\overset{x}{\underset{x_0}{E}} A\right]\left[\overset{t}{\underset{x_0}{E}} A\right]^{-1} = \overset{x-t}{\underset{0}{E}} A = \exp[A\cdot(x - t)]. \tag{2.14}$$

In fact, if the entries are constant, all the terms of the two series

$$\overset{x_2}{\underset{x_1}{E}} A, \qquad \overset{x_2+\alpha}{\underset{x_1+\alpha}{E}} A;$$

coincide. On the other hand, we have in each case, since the series are the resolvents of the equation $y' = Ay$,

$$\overset{x_2}{\underset{x_1}{E}} A(\tau)\ \overset{x_1}{\underset{x_0}{E}} A(\tau) = \overset{x_2}{\underset{x_0}{E}} A(\tau),$$

and so

$$\overset{x-t}{\underset{0}{E}} A\ \overset{t}{\underset{x_0}{E}} A = \overset{x}{\underset{t}{E}} A\ \overset{t}{\underset{x_0}{E}} A = \overset{x}{\underset{x_0}{E}} A,$$

which is (2.14). Formula (2.13) can therefore be written in the form

$$y(x) = \left(\overset{x}{\underset{x_0}{E}} A\right)y_0 + \int_{x_0}^{x}\left(\overset{x-t}{\underset{0}{E}} A\right)b(t)dt. \tag{2.15}$$

3. OPERATIONAL CALCULUS

3.1. Analytic Functions of Operators

(The reader must understand the elementary notions of the theory of functions of a complex variable in order to follow this section.)

A function analytic in a neighborhood of a point λ_0 of $\mathbf{C}$ can be expanded in the series

$$f(\lambda) = \sum_{s=0}^{\infty} a_s(\lambda - \lambda_0)^s \tag{3.1}$$

which converges absolutely for $|\lambda - \lambda_0| < \rho$, where ρ is the radius of convergence given by $\rho^{-1} = \limsup_{s\to\infty} |a_s|^{1/s}$. As is well known, for every s, $a_s = \frac{1}{s!}[d^s f/d\lambda^s](\lambda_0)$; furthermore, in every neighborhood $|\lambda - \lambda_0| \leq \rho_0 < \rho$, the series converges uniformly and is differentiable and integrable term by term.

There is no difficulty in generalizing this concept by constructing series of operators in $L(X)$. Let A be an operator such that, given λ_0, $||A - \lambda_0 I|| < \rho$, where ρ is the radius of convergence of the series (3.1). The series

$$f(A) = \sum_{s=0}^{\infty} a_s (A - \lambda_0 I)^s \tag{3.2}$$

will then be well-defined, and we may take it to define an analytic function f of the operator A. It is actually more convenient to use another definition, which we shall give after having defined the integral of an operator; this definition will coincide with (3.2) when that one makes sense. It is important to note that analytic functions of operators do not in general have the same properties as analytic functions in the complex domain. For example, if A and B are operators in $L(X)$, it is not usually true that

$$\exp(A + B) = \exp(A) \exp(B)$$

or

$$\sin(A + B) = \sin(A) \cos(B) + \cos(A) \sin(B).$$

Such properties hold in some important special cases, e.g., if A and B commute.

We now pass to the definition of the integral of an operator over a path in the complex field. Let $A(\lambda)$ be an operator depending continuously on a complex parameter λ in the sense that

$$\lim_{\lambda-\lambda_0} ||A(\lambda) - A(\lambda_0)|| = 0. \tag{3.3}$$

Let Γ be an oriented curve of finite length in the complex plane, lying in the domain of $A(\lambda)$ and with continuous derivatives. We can then construct a Cauchy integral in the following way.

We suppose that the curve is a C^1 function from the real interval $[0,1]$ to the complex plain. We subdivide the interval $[0,1]$ into N subintervals $[x_i, x_{i+1}]$ such that $|x_{i+1} - x_i| < \varepsilon$. We choose an arbitrary point ξ_i on the corresponding arcs Γ_i with end points

$\lambda_i = \Gamma(x_i)$, $\lambda_{i+1} = \Gamma(x_{i+1})$ and set

$$I_\varepsilon(A;\Lambda_\varepsilon,\Xi_\varepsilon) = \sum_{i=0}^{N-1} (\lambda_{i+1} - \lambda_i)A(\xi_i). \tag{3.4}$$

For each sequence $\varepsilon_s \to 0$ we choose corresponding systems Λ_s, Ξ_s of subdivisions x_i and of points ξ_i respectively and consider the sequence of operators

$$I_{\varepsilon_s}(A;\Lambda_s,\Xi_s). \tag{3.5}$$

We now show that this sequence has a limit that does not depend on the sequence ε_s nor on the choice of Λ_s or Ξ_s. The limit of this sequence is called the integral of $A(\lambda)$ on the oriented curve Γ and will be denoted by

$$\int_\Gamma A(\lambda)d\lambda. \tag{3.6}$$

We first of all observe that on every compact set (and so, in particular, on the curve Γ), the function $A(\lambda)$ is, by Heine's theorem, uniformly continuous; thus, for every $\eta > 0$, there is a $\delta_\eta > 0$ such that if $|\lambda_1 - \lambda_2| < \delta_\eta$ then $||A(\lambda_1) - A(\lambda_2)|| < \eta$. Furthermore, since the curve Γ is of class C^1, there is a constant M such that if $x_1, x_2 \in [0,1]$, then the corresponding values λ_1 and λ_2 satisfy

$$|\lambda_1 - \lambda_2| \leq M|x_1 - x_2|.$$

We shall say that a pair (Λ',Ξ') is finer than a pair (Λ,Ξ) if $\Lambda' \supset \Lambda$ (i.e., the points of the subdivision Λ are among those of Λ') and $\Xi' \supset \Xi$ (the values of Ξ are among those of Ξ'). Then, for every $\eta > 0$, if $\varepsilon < \delta_\eta/M$ and (Λ',Ξ') is finer than $(\Lambda_\varepsilon,\Xi_\varepsilon)$, we must have

$$||I_\varepsilon(A;\Lambda_\varepsilon,\Xi_\varepsilon) - I_\varepsilon(A;\Lambda'_\varepsilon,\Xi'_\varepsilon)|| < M\eta. \tag{3.7}$$

This is an immediate consequence of the fact that if $|x_1 - x_2| < \frac{1}{M}\delta_\eta$, then the corresponding points of the curve satisfy the relation $|\lambda_1 - \lambda_2| < \delta_\eta$ and so $||A(\lambda_1) - A(\lambda_2)|| < \eta$.

One consequence of (3.7) is that if $\eta > 0$, if $\varepsilon < \frac{1}{M}\delta_\eta$, and if two pairs $(\Lambda_\varepsilon,\Xi_\varepsilon)$, $(\Lambda'_\varepsilon,\Xi'_\varepsilon)$ are given, then

$$||I_\varepsilon(A;\Lambda_\varepsilon,\Xi_\varepsilon) - I_\varepsilon(A;\Lambda'_\varepsilon,\Xi'_\varepsilon)|| < 2M\eta. \tag{3.8}$$

To see this, it is sufficient to introduce a subdivision finer than the two that are given.

Since a subdivision of order $\varepsilon_1 < \varepsilon$ is also a subdivision of order ε, (3.8) allows us to conclude that (3.5) is a Cauchy sequence and the limit depends neither on the sequence ε_s nor on (Λ_s, Ξ_s).

The properties of the integral (3.6) are the usual properties of a linear operator and are established by the usual methods. Observe that the inequality

$$\left|\left|\int_\Gamma A\, d\lambda\right|\right| \leq \left|\int_\Gamma ||A||\, d\lambda\right| \tag{3.9}$$

does not hold, since the integration is over the complex field. It is, however, valid if the integration is along a line segment (in the complex plane) or if the integration of the operator takes place in the real field. (The definition of the integral is in that case substantially the same.) We always, though, have

$$\left|\left|\int_\Gamma A\, d\lambda\right|\right| \leq (\max_\Gamma ||A||)\cdot(\text{length of } \Gamma). \tag{3.10}$$

We recall at this point that if a complex valued function of a complex variable is analytic in a simply connected set $\Omega \subset \mathbf{C}$, and if Γ is a regular closed curve contained in Ω, then

$$\int_\Gamma f(\lambda)d\lambda = 0. \tag{3.11}$$

This result has an important generalization for integrals of operators.

<u>Theorem 3.1</u>. Let $A \in L(X)$. Let Ω be a simply connected subset of $\mathbf{C}$ in which $R(\lambda,A)$, that is, $(\lambda I - A)^{-1}$ exists. Then $R(\lambda,A)$ is a continuous function of λ. Moreover, for every function f analytic in Ω and for every regular closed curve Γ contained in Ω, we have

$$\int_\Gamma f(\lambda)R(\lambda,A)d\lambda = 0. \tag{3.12}$$

<u>Proof</u>: Let us suppose that $R(\lambda,A) \in L(X)$ exists at λ. Then for every $\mu \in \mathbf{C}$ such that $|\lambda-\mu| \leq ||R(\lambda,A)||^{-1}$, we have

$$R(\mu,A) = R(\lambda,A)\cdot \sum_{s=0}^{\infty} [(\lambda-\mu)R(\lambda,A)]^s. \tag{3.13}$$

In fact, for fixed μ, the series on the right converges absolutely, and if we multiply both sides by $\mu I - A$, we get, if we recall that $\mu I - A = (\mu-\lambda)I + (\lambda I-A)$,

$$\sum_{s=0}^{\infty} [(\lambda I-A) - (\lambda-\mu)I](\lambda-\mu)^s R^{s+1}(\lambda,A) = I.$$

The continuity of the function $R(\lambda,A)$ with respect to λ follows from (3.13). In particular, if G is the compact subset whose boundary is Γ, there is an $M < \infty$ such that $M = \max_{\lambda\in G} ||R(\lambda,A)||$.

We observe that since the integral (3.6) is additive and oriented, we may integrate, instead of along the curve Γ, along two curves Γ_1 and Γ_2, both of which are closed and oriented in such a way as to coincide in every sense with Γ on Γ and to coincide with each other, but with opposite orientation, off Γ. The curves thus obtained will be

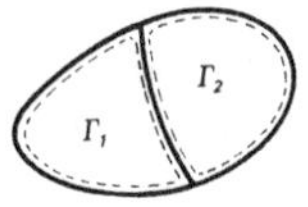

Figure 2.1

piecewise regular, but this is enough for the purposes of the integrals (3.6). This procedure can be repeated a sufficient number of times to decompose $\int_\Gamma \ldots d\lambda$ into a sum of integrals $\int_{\Gamma_i} \ldots d\lambda$ where the curves Γ_i have the property that $\text{diam}(\Gamma_i) < 1/M$. This subdivision is possible because G is a simply connected curve. We now choose an arbitrary point λ_i on each curve Γ_i. Because of the condition on the diameter of Γ_i, the expansion of (3.13) is valid for all points of Γ_i with total convergence as μ varies. The integral therefore becomes

$$\begin{aligned}\int_{\Gamma_i} f(\lambda)R(\lambda,A)d\lambda &= \sum_{s=0}^{\infty} \int_{\Gamma_i} f(\lambda)R(\lambda_i,A)[(\lambda_i-\lambda)R(\lambda_i,A)]^s d\lambda \\ &= \sum_{s=0}^{\infty} \left\{\left[\int_{\Gamma_i} f(\lambda)(\lambda_i-\lambda)^s d\lambda\right] R(\lambda_i,A)^{s+1}\right\}.\end{aligned}$$

Since $f(\lambda)$ is, by hypothesis, analytic in all of G, we have $\int_{\Gamma_i} f(\lambda)(\lambda_i-\lambda)^s d\lambda = 0$, whence the last part of the theorem follows.

We are now in a position to give the definition of an *analytic function of an operator* $A \in L(X)$. Let $\sigma(A)$ be the spectrum of A, and let $f(\lambda)$ be an analytic function in an open neighborhood $\Omega \supset \sigma(A)$. Let Γ be any closed regular curve contained in Ω such that its interior G contains $\sigma(A)$. The orientation of Γ is positive so as to leave

G on the left. We define

$$f(A) = \frac{1}{2\pi i} \int_\Gamma f(\lambda)R(\lambda,A)d\lambda. \tag{3.14}$$

Because of Theorem 3.1, the integral does not depend on the curve chosen.

Operational calculus is dominated by the following fundamental theorem.

<u>Theorem 3.2</u>. Let $f(\lambda)$ and $g(\lambda)$ be two functions analytic in an open neighborhood Ω of $\sigma(A)$. Then

$$(f\cdot g)(A) = f(A)g(A) = g(A)f(A), \tag{3.15}$$

that is, if Γ_1 and Γ_2 are two closed curves in Ω, oriented in the positive sense containing $\sigma(A)$ in their interior, then

$$\frac{1}{2\pi i}\int_{\Gamma_2} f(\mu)g(\mu)R(\mu,A)d\mu = -\frac{1}{4\pi^2}\int_{\Gamma_1} f(\lambda)R(\lambda,A)d\lambda\cdot\int_{\Gamma_2} g(\mu)R(\mu,A)d\mu. \tag{3.16}$$

<u>Proof</u>: First of all, if we multiply $I = \frac{1}{\mu-\lambda}[(\mu I - A) - (\lambda I - A)]$ on the left by $R(\lambda,A)$ and on the right by $R(\mu,A)$, (respectively on the left by $R(\mu,A)$ and on the right by $R(\lambda,A)$), we have

$$R(\lambda,A)R(\mu,A) = \frac{1}{\mu-\lambda}[R(\lambda,A) - R(\mu,A)] = R(\mu,A)R(\lambda,A). \tag{3.17}$$

Note that $R(\lambda,A)$ commutes with A, which implies, by Definition (3.14), that $f(A)$ commutes with A. From (3.17) we then have that $f(A)$ commutes with $g(A)$, whence follows the second equality of (3.15). To prove (3.16), we choose the curve Γ_2 in such a way that it is in the interior of Γ_1. (We can do this by virtue of Theorem 3.1.) We then have, by (3.17),

$$\begin{aligned}
&\left(\frac{1}{2\pi i}\right)^2\int_{\Gamma_1} d\lambda \int_{\Gamma_2} f(\lambda)g(\mu)R(\lambda,A)R(\mu,A)d\mu \\
&\quad = \left(\frac{1}{2\pi i}\right)^2\int_{\Gamma_1} d\lambda \int_{\Gamma_2} f(\lambda)\frac{g(\mu)}{\mu-\lambda}R(\lambda,A)d\mu \\
&\qquad\qquad - \left(\frac{1}{2\pi i}\right)^2\int_{\Gamma_2} d\mu \int_{\Gamma_1} \frac{f(\lambda)}{\mu-\lambda}\, g(\mu)R(\mu,A)d\lambda \\
&\quad = 0 + \frac{1}{2\pi i}\int_{\Gamma_2} f(\mu)g(\mu)R(\mu,A)d\mu,
\end{aligned}$$

since, in the first integral, $g(\mu)/\mu-\lambda$ is a function analytic in the interior of Γ_2 for $\lambda \in \Gamma_1$, while, in the second integral,

$$\frac{1}{2\pi i}\int_{\Gamma_1}\frac{f(\lambda)}{\mu-\lambda}\,d\lambda = -f(\mu) \quad \text{for} \quad \mu \in \Gamma_2.$$

The following theorem offers a relationship between the definition given in (3.14) and that given by the series expansion in (3.2).

Theorem 3.3. Let $f(\lambda)$ be a function analytic for $|\lambda| < \rho_0$ with the series expansion

$$f(\lambda) = \sum_{s=0}^{\infty} a_s\lambda^s. \tag{3.18}$$

If A is an operator in $L(X)$ with $||A|| < \rho_0$, then

$$f(A) = \frac{1}{2\pi i}\int_{\Gamma} f(\lambda)R(\lambda,A)\,d\lambda = \sum_{s=0}^{\infty} a_s A^s. \tag{3.19}$$

Proof: By Theorem 3.1 the curve Γ can be chosen arbitrarily so long as it contains $\sigma(A)$ in its interior; since $\sigma(A)$ is contained in $\{\lambda\colon |\lambda| \le ||A||\}$, we may choose Γ to be in the open set $\{\lambda\colon ||A|| < |\lambda| < \rho_0\}$. The series (3.18) converges uniformly there, as does

$$R(\lambda,A) = \frac{1}{\lambda}\sum_{s=0}^{\infty} (A/\lambda)^s.$$

We may therefore write

$$\begin{aligned}\frac{1}{2\pi i}\int_{\Gamma} f(\lambda)R(\lambda,A)\,d\lambda &= \sum_{r=0}^{\infty}\sum_{s=0}^{\infty}\frac{1}{2\pi i}\int_{\Gamma} a_r\frac{\lambda^r}{\lambda^{s+1}}A^s\,d\lambda\\ &= \sum_{r=0}^{\infty}\sum_{s=0}^{\infty}\frac{1}{2\pi i}\left(\int_{\Gamma}\lambda^{r-s-1}\,d\lambda\right)a_r A^s.\end{aligned}$$

Since $\frac{1}{2\pi i}\int_{\Gamma}\lambda^{r-s-1}\,d\lambda = 0$ for $r \neq s$ and equal to 1 for $r = s$, (3.19) follows.

Corollary 3.4. If $f(\lambda) = \lambda$, then $f(A) = A$. Moreover, the spectrum of an operator is not empty.

Proof: The first part follows clearly from (3.19). If the spectrum of A were empty, then $\frac{1}{2\pi i}\int_{\Gamma}\lambda R(\lambda,A)\,d\lambda = 0$ for every closed curve Γ, which is absurd since there must be curves for which this integral is A.

We will use the following theorem for operators between finite dimensional spaces. It is a particular case of a theorem for more general operators.

Theorem 3.5. Let A be an $n \times n$ matrix, let $\lambda_1,\ldots,\lambda_n$ be its eigenvalues, and let $v_1,\ldots,v_m$ $(m \leq n)$ be the eigenvectors associated with them. Then, if f is an analytic function, $f(\lambda_i)$ are the eigenvalues of $f(A)$ and $v_1,\ldots,v_m$ are the eigenvectors associated with them.

Proof: We observe that the spectrum of $f(A)$ is contained in $f(\sigma(A))$, since if $\lambda_0 \notin f(\sigma(A))$, the function $(\lambda_0 - f(\lambda))^{-1}$ is analytic in an open neighborhood of $\sigma(A)$, whence

$$\frac{1}{2\pi i}\int_\Gamma (\lambda_0 - f(\lambda))^{-1} R(\lambda,A)\,d\lambda$$

exists and by Theorem 3.2 is actually $[\lambda_0 I - f(A)]^{-1}$. Now let λ_1 be an eigenvalue of A and u_1 an eigenvector associated with it. We have $Au_1 = \lambda_1 u_1$; so, for every λ, $(A - \lambda I)u_1 = (\lambda-\lambda_1)u_1$. Let λ be an element of the resolvent of A; then we may apply $R(\lambda,A)$ on the left and get

$$Iu_1 = (\lambda - \lambda_1)R(\lambda,A)u_1,$$

so that u_1 is an eigenvector of $R(\lambda,A)$ relative to the eigenvalue $(\lambda - \lambda_1)^{-1}$. If we multiply by $f(\lambda)$ and integrate, we produce

$$\frac{1}{2\pi i}\int_\Gamma \frac{f(\lambda)}{\lambda-\lambda_1}\,d\lambda\cdot u_1 = \frac{1}{2\pi i}\left[\int_\Gamma f(\lambda)R(\lambda,A)\,d\lambda\right]u_1.$$

Since $\lambda_1 \in \sigma(A)$ and the path of integration contains $\sigma(A)$ in its interior, we have

$$f(\lambda_1)\cdot u_1 = f(A)u_1.$$

Since, in the case of matrices, the spectrum consists entirely of eigenvalues, we have, finally, $f(\sigma(A)) = \sigma(f(A))$.

3.2. Linear Systems with Constant Coefficients

Consider the system of linear equations with initial conditions

$$\begin{cases} y' = Ay \\ y(x_0) = y_0. \end{cases} \tag{3.20}$$

The solution is given by

$$y(x) = \left[\sum_{s=0}^{\infty} A^s \frac{(x-x_0)^s}{s!}\right] y_0.$$

Because of Theorem 3.3, we have

$$y(x) = \exp[A(x - x_0)]y_0. \tag{3.21}$$

We now want to calculate $\exp[A(x-x_0)]$. In order to understand better the structure of $\exp[A(x-x_0)]$, we shall use the canonical form of matrices; we observe that if M is a nonsingular matrix, we can put $y = Mz$. We shall then have $y' = Mz'$, and Eq. (3.20) then becomes

$$\begin{cases} Mz' = AMz \\ z(x_0) = M^{-1}y_0 \end{cases}$$

which, if multiplied on the left by M^{-1}, yields (writing B for the matrix $M^{-1}AM$)

$$\begin{cases} z' = Bz \\ z(x_0) = M^{-1}y_0. \end{cases} \tag{3.22}$$

The solution is thus not only expressed by (3.21), but also by

$$y(x) = M\exp[B(x - x_0)]M^{-1}y_0. \tag{3.23}$$

The dependence on the variable x occurs via the exponential function only, so if the eigenvalues of A are all distinct, it is possible, by Theorem 1.2, to choose M so that the matrix B is diagonal. We then have, for $\lambda \notin \sigma(B)$,

$$R(\lambda,B) = (\lambda I-B)^{-1} = \begin{pmatrix} (\lambda-\lambda_1)^{-1} & & & 0 \\ & (\lambda-\lambda_2)^{-1} & & \\ 0 & & \ddots & \\ & & & (\lambda-\lambda_n)^{-1} \end{pmatrix}.$$

Integrating component by component leads to

$$\frac{1}{2\pi i}\int_\Gamma e^{(x-x_0)}R(\lambda,B)d\lambda = \begin{pmatrix} e^{\lambda_1(x-x_0)} & & & \\ & e^{\lambda_2(x-x_0)} & & 0 \\ & & \ddots & \\ 0 & & & e^{\lambda_n(x-x_0)} \end{pmatrix}.$$

Therefore, all the terms of the matrix $\exp[A(x-x_0)] = M\exp[B(x-x_0)]M^{-1}$ can be expressed as linear combinations of $\exp[\lambda_1(x-x_0)]$, $\exp[\lambda_2(x-x_0)]$, $\ldots,\exp[\lambda_n(x-x_0)]$.

We now suppose that the eigenvalues are not all distinct; more precisely, let the eigenvalues $\lambda_1,\ldots,\lambda_p$ have multiplicities $s_1,\ldots,s_p$. By Theorem 3.1, we may choose the matrix M so that $B = M^{-1}AM$ is triangular. $(\lambda I - B)$ will then be of the form

$$s_1\left\{\begin{array}{c}\\ \\ \\ \end{array}\right.\begin{pmatrix} \lambda-\lambda_1 & & & & & & & * \\ & \lambda-\lambda_1 & * & & & & & \\ & & \ddots & & & & & \\ & & & \lambda-\lambda_1 & & & & \\ & & & & \ddots & & & \\ & 0 & & & & \lambda-\lambda_p & * & \\ & & & & & & \ddots\ \lambda-\lambda_p & \\ & & & & & & & \lambda-\lambda_p \end{pmatrix}\left.\begin{array}{c}\\ \\ \\ \end{array}\right\} s_p$$

Let $P = P(\lambda) = \prod_{i=1}^{p} (\lambda-\lambda_1)^{s_i}$ be the characteristic polynomial of $\lambda I - B$; we shall then have

$$(\lambda I - B)^{-1}\begin{pmatrix} (\lambda-\lambda_1)^{-1} & & & & & \\ & (\lambda-\lambda_1)^{-1} & & & & \\ & & \ddots\ (\lambda-\lambda_1)^{-1} & & r_{ij} & \\ & & & \ddots & & \\ & & & & (\lambda-\lambda_p)^{-1} & \\ & & & & \ddots & \\ & & & & & (\lambda-\lambda_p)^{-1} \end{pmatrix},$$

where the terms r_{ij} are of the form $Q_{ij}(\lambda)/P(\lambda)$. Since the terms Q_{ij} are, except perhaps for their sign, the determinants of the adjoint of the term $(\lambda I - B)_{ji}$, they are polynomials in λ of degree not higher than $n - 1$. It can happen that the greatest common divisor of all the polynomials $Q_{ij}(\lambda)$ is not the identity and is a divisor of $P(\lambda)$. In this case, we may divide all the $Q_{ij}(\lambda)$ and $P(\lambda)$ by this greatest common divisor and observe that the r_{ij} take on the form

$$\frac{R_{ij}(\lambda)}{P_1(\lambda)}$$

with $P_1(\lambda) = (\lambda-\lambda_1)^{t_1}(\lambda-\lambda_2)^{t_2}\cdots(\lambda-\lambda_p)^{t_p}$, with $t_\ell \le s_\ell$ and $\sum_{t=1}^{p} t_\ell \le n.$

We now consider the integral $\frac{1}{2\pi i}\int_\Gamma \exp[\lambda(x-x_0)]R(\lambda,B)d\lambda$ and integrate term by term. For the entries on the diagonal we have poles of order 1, so

$$\frac{1}{2\pi i}\int_\Gamma \exp[\lambda(x-x_0)](\lambda-\lambda_\ell)^{-1}d\lambda = \exp[\lambda_\ell(x-x_0)].$$

As for the other nonzero entries, there may be poles of order $q_\ell \le t_\ell$ at the points λ_ℓ; we therefore have

$$r_{ij} = \frac{1}{2\pi i}\int_\Gamma \exp[\lambda(x-x_0)]\frac{R_{ij}(\lambda)}{P_1(\lambda)}d\lambda$$

$$= \sum_{\ell=1}^{p} c_\ell^{(ij)}(x-x_0)^{q_\ell-1}\exp[\lambda_\ell(x-x_0)].$$

On the other hand, since there are no longer any common divisors of all the $R_{ij}(\lambda)$ and $P(\lambda)$, for every ℓ there will be at least one element r_{ij} in which the term $(x-x_0)^{t_\ell-1}\exp[\lambda_\ell(x-x_0)]$ appears. Therefore, the solutions of the system (3.20) will be given by linear combinations of elements of the form

$$(x-x_0)^{t_\ell-1}\exp[\lambda_\ell(x-x_0)],\ (x-x_0)^{t_\ell-2}\exp[\lambda_\ell(x-x_0)],\ldots,$$

$$(x-x_0)\exp[\lambda_\ell(x-x_0)],\ \exp[\lambda_\ell(x-x_0)].$$

We have seen that there are solutions in which $\exp[\lambda_\ell(x-x_0)]$ and $(x-x_0)^{t_\ell-1}\exp[\lambda_\ell(x-x_0)]]$ actually appear. In fact, there are solutions in which all the intermediate powers also appear; it is enough to note that for a linear system with constant coefficients, even the derivatives of a solution are themselves a solution. Thus, if the term $(x-x_0)^{t_\ell-1}\exp[\lambda_\ell(x-x_0)]$ appears in a solution, the terms of lower degree will appear in the derivatives.

If we had assumed that we knew the transformation that maps every matrix to some Jordan matrix (see Sec. 1.3), it would then have been sufficient to invert $k \times k$ matrices of the type

$$\begin{pmatrix} \lambda-\lambda_1 & & & & \\ 1 & \cdot & & 0 & \\ & 1 & \cdot & & \\ & & \cdot & \cdot & \\ & & & 1 & \lambda-\lambda_1 \end{pmatrix}$$

which appear in Lemma 1.10; we would have obtained the same result directly. So we have established the following

Lemma 3.6. If λ_1 is an eigenvalue of multiplicity $p_1 > 1$ for the matrix A, then A is similar to a Jordan matrix whose elementary Jordan matrices relative to λ_1 are of maximum dimension t_1, where $p_1 - t_1$ is the exponent of $(\lambda-\lambda_1)$ in the greatest common divisor of the terms of the matrix $P(\lambda)[\lambda I - A]^{-1}$.

We shall now consider a particular type of system for which it is possible to give information directly on the structure of the solutions even when there are multiple eigenvalues. If we take the ordinary equation

$$v^{(k)} + \alpha_{k-1}v^{(k-1)} + \alpha_{k-2}v^{(k-2)} + \dots + \alpha_1 v' + \alpha_0 v = 0$$

with initial conditions

$$v(0) = v_0; \quad v'(0) = v_1; \dots; v^{(k-1)}(0) = v_{k-1}$$

we may associate with it the system $y' = Ay$ with $y = (v, v', v'', \dots, v^{(k-1)})$ and

$$A = \begin{pmatrix} 0 & 1 & & & \\ & 0 & 1 & & 0 \\ 0 & & 0 & 1. & \\ & & & 0.\ddots & 1 \\ -\alpha_0 & -\alpha_1 & \dots & \dots & -\alpha_{k-1} \end{pmatrix}.$$

To get $(\lambda I - A)^{-1}$, we make use of Lemma 1.11; it follows that $(\lambda I - A)^{-1} = 1/P_0(\lambda)$, so that if there are eigenvalues of multiplicity s_ℓ, the 1,k component of $\exp A(x-x_0)$ will accordingly be

$$\frac{1}{2\pi i}\int \exp[\lambda(x-x_0)][P_0(\lambda)]^{-1}d\lambda = \sum_{\ell=1}^{p} c_\ell(x-x_0)^{s_\ell-1}\exp[\lambda_\ell(x-x_0)].$$

In view of what we have established above, we may conclude with the following two theorems.

Theorem 3.7. Let the system $y' = Ay$ be given, and let $\lambda_1,\dots,\lambda_n$ be the eigenvalues of A. If these values are all distinct, then the solutions are linear combinations of $\exp(\lambda_1 x), \exp(\lambda_2 x), \dots, \exp(\lambda_n x)$. If they are not all distinct, and s_ℓ are their respective multiplicities, then the solutions are linear combinations of

$P_{s_1}\exp(\lambda_1 x)$, $P_{s_2}\exp(\lambda_2 x),\ldots,P_{s_p}\exp(\lambda_p x)$, where P_ℓ is a polynomial of degree less than or equal to $s_\ell-1$. In each case, a fundamental system of solutions is given by the columns of the matrix $\exp(Ax)$.

Theorem 3.8. Let the equation

$$y^{(n)} + \alpha_{n-1}y^{(n-1)} + \alpha_{n-2}y^{(n-2)} + \ldots + \alpha_0 y = 0. \tag{3.24}$$

be given. If the characteristic equation

$$\lambda^n + \alpha_{n-1}\lambda^{(n-1)} + \ldots + \alpha_1\lambda + \alpha_0 = 0 \tag{3.25}$$

admits distinct roots λ_ℓ, then a fundamental system of solutions is given by $y_\ell = \exp(\lambda_\ell x)$; if the roots are not all distinct, and s_ℓ are the respective multiplicities, then a fundamental system is given by

$$\begin{cases} \exp(\lambda_1 x),\ x\exp(\lambda_1 x),\ldots,x^{s_1-1}\exp(\lambda_1 x); \\ \exp(\lambda_2 x),\ x\exp(\lambda_2 x),\ldots,x^{s_2-1}\exp(\lambda_2 x); \\ \quad\ldots\ldots \\ \exp(\lambda_p x),\ x\exp(\lambda_p x),\ldots,x^{s_p-1}\exp(\lambda_p x). \end{cases} \tag{3.26}$$

Proof: There are exactly n linearly independent solutions; since in the associated system the dependence on the variable x is exactly via the n functions (3.26), these must be linearly independent and therefore themselves constitute a fundamental system of solutions.

Note, however, that even if the coefficients are real, the system (3.26) may consist of complex functions, while the fundamental system given by the columns of $\exp(Ax)$ automatically consists of real functions.

As one learns in elementary courses, the two preceding theorems may be proved directly; furthermore, the matrix $\exp(Ax)$ can be obtained from the fundamental system of solutions by finding one after the other the solutions of the n initial value problems

$$\begin{cases} y(0) = 0 \\ y'(0) = 0 \\ \cdots\cdots\cdots \\ y^{(i-1)}(0) = 0 \\ y^{(i)}(0) = 1 \\ y^{(i+1)}(0) = 0 \\ \cdots\cdots\cdots \\ y^{(n-1)}(0) = 0. \end{cases}$$

Such solutions constitute the columns of the matrix $\exp(Ax)$.

For an example, we consider the case when A is a 2×2 matrix. We set

$$A = \begin{pmatrix} a & b \\ c & d \end{pmatrix}; \quad (\lambda I - A)^{-1} = \frac{1}{\lambda^2-(a+d)\lambda+ad-bc} \begin{pmatrix} \lambda-d & b \\ c & \lambda-a \end{pmatrix}. \tag{3.27}$$

The eigenvalues of A are distinct provided that $(a-d)^2 + 4bc \neq 0$; if not, the matrix assumes the form

$$\begin{pmatrix} \lambda_1 \pm \sqrt{-bc} & b \\ c & \lambda_1 \pm \sqrt{-bc} \end{pmatrix}$$

with $bc < 0$, and λ_1 is an eigenvalue. Suppose that the eigenvalues λ_1 and λ_2 are distinct and real. It follows from the calculation of the components of

$$\exp(Ax) = \frac{1}{2\pi i} \int_\Gamma \exp(\lambda x) R(\lambda, A)\, d\lambda$$

that

$$\frac{1}{2\pi i} \int_\Gamma \frac{\exp(\lambda x)}{(\lambda-\lambda_1)(\lambda-\lambda_2)} \lambda\, d\lambda = \frac{1}{\lambda_1-\lambda_2} [\lambda_1 \exp(\lambda_1 x) - \lambda_2 \exp(\lambda_2 x)]$$

$$\frac{1}{2\pi i} \int_\Gamma \frac{\exp(\lambda x)}{(\lambda-\lambda_1)(\lambda-\lambda_2)}\, d\lambda = \frac{1}{\lambda_1-\lambda_2} [\exp(\lambda_1 x) - \exp(\lambda_2 x)]$$

so that

$$\exp(Ax) = \frac{1}{\lambda_1-\lambda_2} \begin{pmatrix} (\lambda_1-d)\exp(\lambda_1 x)-(\lambda_2-d)\exp(\lambda_2 x) & b[\exp(\lambda_1 x)-\exp(\lambda_2 x] \\ c[\exp(\lambda_1 x)-\exp(\lambda_2 x)] & (\lambda_1-a)\exp(\lambda_1 x)-(\lambda_2-a)\exp(\lambda_2 x) \end{pmatrix} \tag{3.28}$$

In the particular case of the second order equation

$$y'' + \alpha y' + \beta y = 0, \tag{3.29}$$

A becomes $\begin{pmatrix} 0 & 1 \\ -\beta & -\alpha \end{pmatrix}$. Therefore, if $\alpha^2 - 4\beta > 0$ and if λ_1, λ_2 are the two roots of $\lambda^2 + \alpha\lambda + \beta = 0$, we have

$$\exp(Ax) = \frac{1}{\lambda_1 - \lambda_2}\begin{pmatrix} (\lambda_1+\alpha)\exp(\lambda_1 x) - (\lambda_2+\alpha)\exp(\lambda_2 x) & \exp(\lambda_1 x) - \exp(\lambda_2 x) \\ -\beta[\exp(\lambda_1 x) - \exp(\lambda_2 x)] & \lambda_1 \exp(\lambda_1 x) - \lambda_2 \exp(\lambda_2 x) \end{pmatrix}$$

and, so, noting that $\lambda_1\lambda_2 = \beta, \lambda_1 + \lambda_2 = \alpha$, we see that

$$\exp(Ax) = \frac{1}{\lambda_1 - \lambda_2}\begin{pmatrix} -\lambda_2 \exp(\lambda_1 x) + \lambda_1 \exp(\lambda_2 x) & \exp(\lambda_1 x) - \exp(\lambda_2 x) \\ -\lambda_1\lambda_2[\exp(\lambda_1 x) - \exp(\lambda_2 x)] & \lambda_1 \exp(\lambda_1 x) - \lambda_2 \exp(\lambda_2 x) \end{pmatrix}. \tag{3.30}$$

If the eigenvalues are complex conjugates, we obtain a more expressive form than that in (3.28) by setting $a + d = p$, $\sqrt{-bc - (a-d)^2/4} = \omega$; the roots will then be given by

$$\lambda_1 = \frac{p}{2} + i\omega, \quad \lambda_2 = \frac{p}{2} - i\omega.$$

If we apply Euler's formula, (3.28) becomes

$$\exp(Ax) = \begin{pmatrix} \exp(\frac{p}{2}x)\left[\frac{a-d}{2\omega}\sin \omega x + \cos \omega x\right] & \exp(\frac{p}{2}x)\frac{b}{\omega}\sin \omega x \\ \exp(\frac{p}{2}x)\frac{c}{\omega}\sin \omega x & \exp(\frac{p}{2}x)\left[\frac{d-a}{2\omega}\sin \omega x + \cos \omega x\right] \end{pmatrix}. \tag{3.31}$$

In the special case of Eq. (3.29), this becomes, upon putting $\omega = \sqrt{\beta - (\alpha^2/4)}$,

$$\exp(Ax) = \begin{pmatrix} \exp(-\frac{\alpha}{2}x)\left[\frac{\alpha}{2\omega}\sin \omega x + \cos \omega x\right] & \exp(-\frac{\alpha}{2}x)\left[\frac{1}{\omega}\sin \omega x\right] \\ \exp(-\frac{\alpha}{2}x)\left[-\frac{\beta}{\omega}\sin \omega x\right] & \exp(-\frac{\alpha}{2}x)\left[-\frac{\alpha}{2\omega}\sin \omega x + \cos \omega x\right] \end{pmatrix}. \tag{3.32}$$

Let us now suppose that the two eigenvalues are equal. If $b = c = 0$, we have the trivial situation in which the matrix A equals $\lambda_1 I$; in this case we have

$$\exp(Ax) = \begin{pmatrix} \exp(\lambda_1 x) & 0 \\ 0 & \exp(\lambda_1 x) \end{pmatrix}. \tag{3.33}$$

If we suppose that b and c are not both 0, the matrix is nondiagonizable and has the form

$$A = \begin{pmatrix} \lambda_1 - \alpha\beta & \alpha^2 \\ -\beta^2 & \lambda_1 + \alpha\beta \end{pmatrix},$$

where α and β are both real or both imaginary. We have

$$(\lambda I - A)^{-1} = \frac{1}{(\lambda-\lambda_1)^2} \begin{pmatrix} \lambda - \lambda_1 - \alpha\beta & \alpha^2 \\ -\beta^2 & \lambda - \lambda_1 + \alpha\beta \end{pmatrix}$$

and so, since

$$\frac{1}{2\pi i}\int_\Gamma \frac{\exp(\lambda x)}{\lambda - \lambda_1}\, d\lambda = \exp(\lambda_1 x), \qquad \frac{1}{2\pi i}\int_\Gamma \frac{\exp(\lambda x)}{(\lambda-\lambda_1)^2}\, d\lambda = x \exp(\lambda_1 x),$$

we have

$$\exp(Ax) = \exp(\lambda_1 x) \begin{pmatrix} 1 - \alpha\beta x & \alpha^2 x \\ -\beta^2 x & 1 + \alpha\beta x \end{pmatrix}.$$

4. LINEAR FINITE DIFFERENCES EQUATIONS

4.1. Homogeneous Linear Finite Differences Equations

In many problems, especially in economics and demography, we must consider functions whose values are known only for special values of time; we must also consider decisions which at a particular moment, involve a discontinuity in the functions which represent the state of the system. In these cases, the method of classical differential equations could become artificial; it is more convenient to adopt the techniques of equations with finite differences. These have the form of sequences defined by recurrence on certain of the preceding values, for example,

$$y_n = f(y_{n-1}, y_{n-2}, \ldots, y_{n-k}, n). \tag{4.1}$$

The number of elements on which they depend (k in the preceding example) is called the order of the equation. It is beyond the scope of this section to make a general study of Eq. (4.1) when the dependence of f on $y_{n-1}, \ldots, y_{n-k}$ is not linear. We shall therefore consider only the case of the homogeneous linear equation

$$y_n = a_k(n)y_{n-1} + a_{k-1}(n)y_{n-2} + \ldots + a_1(n)y_{n-k}. \tag{4.2}$$

The typical problem is an initial value problem in which $y_0, y_1, \ldots, y_{k-1}$ are taken as the initial data. Given the definition of recurrence, it is clear that the solution exists and is unique.

The problem is to find, if possible, a direct formula for calculating y_n entirely from the initial data, without using the recurrence formula. As we did in the case of differential equations, we shall study a more general problem, that of linear systems with finite differences, which have the form

$$y_n = A(n)y_{n-1}, \tag{4.3}$$

where $(y_n)_{n=1}^{\infty}$ is a sequence of vectors in $\mathbb{R}^m$ and $(A(n))_{n=1}^{\infty}$ is a sequence of $m \times m$ matrices. We have by recurrence that the solution of the initial value problem with initial datum y_0 is given by

$$y_n = A(n)A(n-1)\ldots A(2)A(1)y_0 \equiv P(n)y_0. \tag{4.4}$$

If, in particular, $A(n) = A$ for all n, then

$$y_n = A^n y_0. \tag{4.5}$$

To calculate A^n, we use the usual formula

$$A^n = \frac{1}{2\pi i}\int_{\Gamma} \lambda^n R(\lambda, A)\,d\lambda. \tag{4.6}$$

Results hold similar to those used in the preceding section to calculate $\exp(Ax)$. In particular, if all the eigenvalues $\lambda_1, \lambda_2, \ldots, \lambda_k$ of A are distinct, then A^n has the form

$$A^n = T^{-1}\begin{pmatrix} \lambda_1^n & & & 0 \\ & \lambda_2^n & & \\ & & \ddots & \\ 0 & & & \lambda_k^n \end{pmatrix} T,$$

where T is a suitable nonsingular matrix. If there are eigenvalues of multiplicity greater than 1, it is necessary to examine the structure of the matrix as we did in the preceding section; in particular, there will be terms of the type $n(n-1)\cdots(n-k)\lambda_s^{n-k-1}$.

We now analyze in greater depth the system associated with Eq. (4.2); we set

$$y_n = u_n^{(k)}$$
$$y_{n-1} = u_n^{(k-1)}$$
$$\cdots\cdots\cdots\cdots$$
$$y_{n-k+1} = u_n^{(1)}.$$

In this case, (4.2) becomes

$$u_n = A(n)u_{n-1}$$

with

$$A(n) = \begin{pmatrix} 0 & 1 & & 0 \\ & \cdot & \cdot & \\ & & \cdot & \cdot \\ & & & 1 \\ 0 & & 0 & \\ a_1(n) & a_2(n) & & a_k(n) \end{pmatrix} \tag{4.7}$$

The eigenvalues of this matrix are the roots of the algebraic equation obtained by substituting λ^s in place of y_s in (4.2) and dividing by λ^{n-k}. In the case in which the coefficients are constants, we have that λ^n is a solution of (4.2) if λ is an eigenvalue of A. To calculate $A^n = P(n)$ in this case, we may use (4.6), with $R(\lambda,A)$ given by Lemma 1.11. In particular, this allows us to conclude that if λ_s is an eigenvalue of multiplicity $p_s > 1$, then A^n has terms of the type $n(n-1)\cdots(n-k+2)\lambda_s^{n-k-1}$ for $k = 2,3,\ldots,p_s$.

To given an example, we study the case of 2×2 matrices. As in the preceding section, we have

$$A = \begin{pmatrix} a & b \\ c & d \end{pmatrix}, \quad (\lambda I - A)^{-1} = \frac{1}{\lambda^2-(a+d)\lambda+ad-bc} \begin{pmatrix} \lambda-d & b \\ c & \lambda-a \end{pmatrix}. \tag{4.8}$$

If we suppose that the eigenvalues λ_1 and λ_2 are distinct, we get

$$A^n = \frac{1}{2\pi i}\int_\Gamma \lambda^n R(\lambda,A)\,d\lambda = \frac{\lambda_1^{n+1}-\lambda_2^{n+1}}{\lambda_1-\lambda_2} I + \frac{\lambda_1^n-\lambda_2^n}{\lambda_1-\lambda_2}\begin{pmatrix} -d & b \\ c & -a \end{pmatrix}. \tag{4.9}$$

If the eigenvalues are complex conjugates, we put $\lambda_1 = \rho(\cos\theta + i\sin\theta)$, and $\lambda_2 = \rho(\cos\theta - i\sin\theta)$, and produce

$$A^n = \rho^n \frac{\sin[(n+1)\theta]}{\sin\theta} I + \rho^{n-1} \frac{\sin(n\theta)}{\sin\theta} \begin{pmatrix} -d & b \\ c & -a \end{pmatrix}. \tag{4.10}$$

If the eigenvalues coincide, the matrix A has the form

$$A = \begin{pmatrix} \lambda_1 - \alpha\beta & \alpha^2 \\ \beta^2 & \lambda_1 + \alpha\beta \end{pmatrix},$$

where α and β are both real or both imaginary, whence

$$A^n = \lambda_1^n I + n\lambda_1^{n-1} \begin{pmatrix} -\alpha\beta & \alpha^2 \\ \beta^2 & \alpha\beta \end{pmatrix}. \tag{4.11}$$

We observe now that the columns of the matrix $P(n)$ are the solutions associated with the initial values e_i; we may ask whether they also form a fundamental system of solutions and whether, in particular, a linear combination of them allows us to solve the initial value problem with an initial datum assigned for $n_0 \neq 0$. The answer is yes so long as all the matrices $A(n)$ are nonsingular; in that case, $\det P(n) \neq 0$. In general, if the matrices $A(n)$ are nonsingular, and if we take any k-tuple of solutions with Wronskian $W(n)$ at some instant n, we have

$$W(n + s) = \left(\prod_{i=1}^{s} \det A(n + i)\right) W(n),$$

so that it is enough to verify that the Wronskian is not 0 for one value of n to know that it is not 0 for all the others.

There is quite a strong analogy with linear ordinary differential equations except when the matrix $A(n)$ is singular for some n. We shall now explain the reasons for this similarity.

For simplicity, we consider only the case when the matrix A is constant. Suppose that A can be expressed as $\exp M$, where M is a matrix with real entries. We then have $A^n = \exp(Mn)$, so that the solution of the initial value problem is

$$y_n = A^n y_0 = \exp(Mn) y_0, \tag{4.12}$$

hence it is the solution of the system of differential equations

$$\begin{cases} y' = My \\ y(0) = y_0 \end{cases} \tag{4.13}$$

calculated at the points $1,2,\ldots,n,\ldots$.

It is now clear that from any equation of the type (4.13), we may pass to an equation with finite differences of the type (4.3). We shall now see when the opposite is true. To do this, we must calculate $\ell g\ A$; since the logarithm is a meromorphic function, it is necessary, if one wants to apply the theory of Section 3, to make a cut in the complex plane in order to get a holomorphic function. Such a cut may be arbitrary but must terminate at 0; it is therefore necessary, in order to be able to define

$$\ell g\ A = \int_{\Gamma} \ell g\ \lambda\ R(\lambda,A)\,d\lambda,$$

that 0 not be an eigenvalue of A. This condition is also sufficient, since $\sigma(A)$ consists of a finite number of points, so that it is possible to take for our cut a half line originating at 0 and not passing through any point of the spectrum and to take for Γ a path that does not touch the cut and which contains the spectrum in its interior. We shall choose in every case the principal branch of the logarithm with the cut on the negative real semiaxis, unless there are negative real numbers among the eigenvalues of A. To complete our analysis of this problem, we shall need the following lemma.

Lemma 4.1. Let $P(\lambda)$ be a polynomial with real coefficients. Let λ_1 and $\overline{\lambda}_1$ be two simple complex conjugate roots of the equation $P(\lambda) = 0$. If we put $P^+(\lambda) = (\lambda-\lambda_1)^{-1}P(\lambda)$, $P^-(\lambda) = (\lambda-\overline{\lambda}_1)^{-1}P(\lambda)$, we have

$$P^+(\lambda_1)P^-(\overline{\lambda}_1) \text{ is real,} \tag{4.14}$$

$$P^+(\lambda_1) + P^-(\overline{\lambda}_1) \text{ is real, and} \tag{4.15}$$

$$P^+(\lambda_1) - P^-(\overline{\lambda}_1) \text{ is pure imaginary.} \tag{4.16}$$

Proof: Let $p_1 \pm iq_1,\ldots,p_s \pm iq$ be the pairs of complex conjugate roots, and let $\xi_1,\ldots,\xi_r$ be the real roots (possibly multiple). We have

$$P^+(\lambda_1) = 2iq_1\cdot[(p_1-p_2+iq_1)^2 + q_2^2]\ldots[(p_1-p_s+iq_1)^2 + q_s^2]\cdot$$
$$\cdot[p_1-\xi_1 + iq_1]\ldots[p_1-\xi_r + iq_1],$$
$$P^-(\overline{\lambda}_1) = -2iq_1[(p_1-p_2-iq_1)^2 + q_2^2]\ldots[(p_1-p_s-iq_1)^2 + q_s^2]\cdot$$
$$\cdot[p_1-\xi_1-iq_1]\ldots[p_1-\xi_r-iq_1]$$

whence it immediately follows that $P^+(\lambda_1) = \overline{P^-(\overline{\lambda}_1)}$, and the lemma is proved.

We now have the following theorem.

Theorem 4.2. Let A be a matrix with real coefficients. If we suppose that there is no zero or real negative eigenvalue, then $\ell g\ A$ is defined and is a matrix with real entries.

Proof: For every component of $\ell g\ A$, we must calculate an integral of the form

$$\frac{1}{2\pi i} \int_{\Gamma} \frac{Q(\lambda)}{P(\lambda)}\, \ell g\ \lambda\, d\lambda. \tag{4.17}$$

Let us first of all suppose that the poles of $\frac{Q(\lambda)}{P(\lambda)}\, \ell g\ \lambda$ are all simple; to calculate the integral (4.17), we must add up the residues. At the poles on the positive real axis, the value of the residue is pure imaginary, whereas at poles which are complex conjugates λ_j and $\overline{\lambda}_j$, we calculate $\mathrm{Res}(\lambda_j) + \mathrm{Res}(\overline{\lambda}_j)$. Note that $Q(\lambda_j) = Q(\overline{\lambda}_j)$ and $\ell g(\lambda_j) = \ell g(\overline{\lambda}_j)$ (this last relation follows from our choice of the principal branch of the logarithm with a cut on the negative real semiaxis). We have

$$\frac{1}{2\pi i}\left[\mathrm{Res}(\lambda_j) + \mathrm{Res}(\overline{\lambda}_j)\right] = \frac{Q(\lambda_j)}{P^+(\lambda_j)}\, \ell g(\lambda_j) + \frac{Q(\overline{\lambda}_j)}{P^-(\overline{\lambda}_j)}\, \ell g(\overline{\lambda}_j).$$

Because of the preceding lemma, it follows that

$$\frac{1}{2\pi i}\left[\mathrm{Res}(\lambda_j) + \mathrm{Res}(\overline{\lambda}_j)\right] = \frac{Q(\lambda_j)}{P^+(\lambda_j)}\, \ell g(\lambda_j) + \frac{\overline{Q(\lambda_j)\,\ell g(\lambda_j)}}{\overline{P^+(\lambda_j)}},$$

which is therefore a real number. Upon passing to the limit, we get the same result in the case of multiple complex conjugate roots.

The theorem is not valid if there are negative real eigenvalues, not even if double. Such a case cannot be obtained by passing to the limit under the conditions in the preceding theorem, since the path of integration Γ would have to leave the domain in which $\ell g\ \lambda$ is holomorphic, while changing the position of the cut would mean that we would no longer have, in general, $\ell g\ \lambda = \ell g(\overline{\lambda})$.

The case in which there are negative real eigenvalues is not reducible to a differential equation with real coefficients; it is, however, always reducible to a differential equation, so that, even in this case, the analogues of the theorems on fundamental solutions are valid. However, the case in which 0 is an eigenvalue does not correspond to any differential equation; in effect, there are no fundamental systems of solutions in this case, and the problem can only be solved for increasing n, while in all the other cases it is possible to study even the problem with delay.

4.2. Nonhomogeneous Linear Finite Differences Equations

For the benefit of the reader, we present here a frequently used resolvent formula for the equations studied in Sec. 4.1 when they are not homogeneous. We observe that all the results obtained for differential equations in Sec. 2.3 are valid, mutatis mutandis, for these equations. In particular, we note that, given the equation

$$y_{n+1} - A(n)y_n = b_n, \tag{4.18}$$

and given two of its solutions $y_n^{(1)}, y_n^{(2)}$, the sequence $z_n = y_n^{(1)} - y_n^{(2)}$ satisfies the associated homogeneous equation

$$z_{n+1} - A(n)z_n = 0. \tag{4.19}$$

The standard integral methods for finding a particular solution of (4.18) starting from a fundamental system of solutions of (4.19) are in this case of little utility, since they imply a determination through recurrence equivalent to the solution through recurrence of (4.18). Nevertheless, certain types of b_n permit explicit solutions when the equations have constant coefficients. These types are given by

$$b_n = \begin{pmatrix} P_1(n)\exp(\lambda_1 n) \\ \vdots \\ P_k(n)\exp(\lambda_k n) \end{pmatrix}$$

where $P_j(n)$ are polynomials in n and λ_j are real or complex. In such a case, a particular solution of the system is still of the same form, with the exception that if $\exp(\lambda_j)$ coincides with some eigenvalue of A, the corresponding polynomial will have its degree increased by an amount equal to the multiplicity of the eigenvalue.

5. EXAMPLES

The reader will find the standard kind of exercises in the preceding chapter. More complex than those are cases of systems of equations of higher order. They occur in various important problems in the applied sciences, such as paired electrical circuits. A common situation leads to equations of the type

$$y'' = ay' + bz' + cy + dz$$

$$z'' = \alpha z' + \beta y' + \gamma z + \delta y.$$

If we set $u = y'$, $v = z'$, we are led to a system of four equations

$$\begin{pmatrix} y \\ u \\ z \\ v \end{pmatrix}' = \begin{pmatrix} 0 & 1 & 0 & 0 \\ c & a & d & b \\ 0 & 0 & 0 & 1 \\ \delta & \beta & \gamma & \alpha \end{pmatrix} \begin{pmatrix} y \\ u \\ z \\ v \end{pmatrix}; \quad x' = Ax$$

to which we may apply the theoretical treatment described above. In the particular case in which the two circuits are symmetric, with $a = \alpha$, $b = \beta$, $c = \gamma$, $d = \delta$, it is simpler to work with the pair of equations

$$(y+z)'' = (a+b)(y+z)' + (c+d)(y+z)$$

$$(y-z)'' = (a-b)(y-z)' + (c-d)(y-z).$$

In this case, if $E_1(t)$ is the resolvent matrix of the first equation and $E_2(t)$ that of the second, we have

$$\exp(At) = \begin{pmatrix} E_1(t) + E_2(t) & E_1(t) - E_2(t) \\ E_1(t) - E_2(t) & E_1(t) + E_2(t) \end{pmatrix}.$$

Observe that in the study of systems with constant coefficients, it is often more convenient, rather than to look for the general solution, to study directly, via the Laplace or Fourier transforms, the particular solution one wants; the consideration of such transforms, however, is beyond the objectives of this text.

Simple Examples of G-Convergence. This section, based on examples, is an introduction to G-convergence, treated at the end of Chapter III. We saw in Chapter I that the theorem of Kamke affirms that if the function $f_\lambda(x,y)$ in the equation $y' = f_\lambda(x,y)$ converges uniformly to a function $g(x,y)$, then the solutions of the initial value problem converge uniformly on a suitable small interval to the solution of the equation $y' = g(x,y)$ with the same initial data. We shall meet other cases of this sort below; they are more or less characterized by the fact that if the coefficients of the equation converge in some norm, then the solutions with the same initial datum also converge, and in a stronger norm. We now take the opposite point of view and start from the convergence of the solutions. We shall study the problem for homogeneous linear systems with variable coefficients.

We make the following definition. Given the sequence $y'_k = A_k(x)y_k$ of linear systems on the interval $[a,b]$, we say that it G-converges to the system $y' = Ay$ if, for every Cauchy problem with initial data $y_k(x_0) = y_0$, the respective solutions y_k converge uniformly in $[a,b]$ to the solution of the problem $y' = Ay$, $y(x_0) = y_0$. Other possible definitions of G-convergence are equivalent.

We now give a few examples of this convergence. For simplicity, we begin with linear systems whose coefficients are piecewise constant; in this case we must clarify what we mean by a solution to the system since the coefficients are discontinuous. We set

$$y'_k = A_k y_k \quad \text{for} \quad x_k < x < x_{k+1}.$$

By a solution of $y' = Ay$, $y(x_0) = y_0$, we mean the solution obtained by imposing continuity at the points x_k, that is,

$$y(x) = y_k(x) \quad \text{in} \quad x_k < x < x_{k+1},$$

with the initial conditions defined by recurrence,

$$y(x_0) = y_0, \quad y_k(x_k) = y_{k-1}(x_k) \quad \text{for} \quad k \geq 1.$$

We now suppose that we have a piecewise constant equation with periodic coefficients of period T. Let $x_0 = 0$, $x_0 < x_1 < x_2 < \dots < x_N = T$, and set

$$A(x) = A_k \quad \text{for} \quad sT + x_k < x < sT + x_{k+1}.$$

We now set $A_\lambda(x) = A(\lambda x)$ and show that the G-limit of $A_\lambda(x)$ as $\lambda \to \infty$ is given by $\overline{A}$ where

$$\overline{A} = \frac{1}{T}\int_0^T A(x)dx = \frac{1}{T}\sum_{i=0}^{N-1} A_i(x_{i+1} - x_i).$$

To see this, observe that the matrix $A_\lambda(x)$ is periodic with period $T_\lambda = T/\lambda$. We consider

$$E = \mathop{E}_0^{T_\lambda} A_\lambda = \exp\left[A_{N-1}\frac{x_N - x_{N-1}}{\lambda}\right]\exp\left[A_{N-2}\frac{x_{N-1}-x_{N-2}}{\lambda}\right]\dots\exp\left[A_0\frac{x_1-x_0}{\lambda}\right].$$

We then see that the solution of the problem

$$\begin{cases} y' = A_\lambda(x)y \\ y(0) = y_0 \end{cases}$$

at the points sT/λ satisfies the finite differences equation

$$y\left(\frac{sT}{\lambda}\right) = y_s = E_\lambda y_{s-1},$$

so that $y_s = E_\lambda^s y_0$. From what we saw in Sec. 4.1, it is now evident that y_s is a discretization of the solution of the problem with constant coefficients

$$\begin{cases} y' = \left[\ell g(E_\lambda)\cdot \frac{\lambda}{T}\right]y \\ y(0) = y_0. \end{cases}$$

We now proceed to show that $\lim\limits_{\lambda\to+\infty}\left[\ell g(E_\lambda)\cdot \frac{\lambda}{T}\right] = \bar{A}$. We use Theorem 3.3, which allows us to expand $\exp M$ and $\ell g(I + M)$ in series. We get

$$E_\lambda = \left(I + A_{N-1}\frac{x_N - x_{N-1}}{\lambda} + \frac{A_{N-1}^2(x_N - x_{N-1})^2}{2!\lambda^2} + \dots\right)\cdot\dots\cdot$$

$$\cdot\left(I + A_0\frac{x_1 - x_0}{\lambda} + \frac{A_0^2(x_1 - x_0)^2}{2!\lambda^2} + \dots\right)$$

$$= I + \frac{1}{\lambda}\bar{A}T + g(A_0, A_1, \dots, A_{N-1}),$$

where $||g(A_0,A_1,\dots,A_{N-1})|| \le (\text{a constant})\,\frac{1}{\lambda^2}$. For $\lambda > ||\bar{A}T + \lambda g(\cdots)||$, we may expand the logarithm and get

$$\frac{\lambda}{T}\,\ell g\, E_\lambda = \frac{\lambda}{T}\cdot\left(\frac{\bar{A}T}{\lambda} + g\right) - \frac{\lambda}{2T}\left(\frac{\bar{A}T}{\lambda} + g\right)^2 + \dots .$$

Passing to the limit as $\lambda \to \infty$ this produces $\lim\limits_{\lambda\to\infty}\frac{\lambda}{T}\,\ell g\, E_\lambda = \bar{A}$ in norm. We therefore have, over every finite interval $[a,b]$,

$$\lim_{\lambda\to\infty}||(E_\lambda)^{\lambda t/T} - \exp \bar{A}t|| = 0 \tag{5.1}$$

uniformly as t varies in $[a,b]$.

We now consider the convergence of a solution $y_\lambda(x)$ with initial datum y_0. It is sufficient to show that given $\varepsilon > 0$, there is a $\bar{\lambda}$ such that for $\lambda > \bar{\lambda}$ we have

$$|y_\lambda(t) - E_\lambda^{\lambda t/T}y_0|| < \varepsilon \quad \text{for every } t \in [a,b].$$

By construction, $|y_\lambda(t) - E_\lambda^{\lambda t/T}y_0| = 0$ for the values $t = sT/\lambda$. Let $y_0, y_1, \dots, y_s, \dots$ be such values. We observe first of all that in $[a,b]$ we have $|y_\lambda(t)| \le K|y_0|$ with

$$K = \exp\left\{\left[\frac{b-a}{T} + 1\right]\int_0^T ||A(t)||dt\right\}. \tag{5.2}$$

To see this, note that

$$(|y_\lambda|^2)' = 2\langle y_\lambda, y_\lambda'\rangle = 2\langle A_\lambda y_\lambda, y_\lambda\rangle \le 2||A_\lambda||\ |y_\lambda|^2$$

so

$$|y_\lambda(t)| \le |y_0| \exp \int_a^b ||A_\lambda(t)||dt,$$

which implies the conclusion. Therefore, there is a uniform bound on $y_\lambda(x)$ in $[a,b]$ independent of λ and x. For $\frac{sT}{\lambda} < t < \frac{s+1}{\lambda} T$, we have

$$|y_\lambda(t)-y_s| = \left|\int_{sT/\lambda}^t y_\lambda' d\tau\right| \le \int_{sT/\lambda}^t ||A_\lambda||\ |y_\lambda|d\tau$$

$$\le \frac{1}{\lambda}\left(\int_0^T ||A(\tau)||d\tau\right)\cdot \max|y_\lambda| \le \frac{K_1 K}{\lambda}|y_0|. \tag{5.3}$$

The same type of bound holds for

$$|E^{\lambda t/T} y_0 - y_s| = \left|\left(E_\lambda^{\frac{\lambda}{T}(t-\frac{sT}{\lambda})} y_s - y_s\right)\right|,$$

which is a solution of the equation with constant coefficients $y' = (\frac{\lambda}{T} \ell g\ E_\lambda)y$. For λ sufficiently large, we have

$$||\frac{\lambda}{T} \ell g\ E_\lambda|| \le 2||\bar{A}|| \le \frac{2}{T}\int_0^T ||A(t)||dt.$$

From this it follows that for sufficiently large λ,

$$|y_\lambda(t) - (E_\lambda)^{\lambda t/T} y_0| < \frac{C}{\lambda} \quad \text{for} \quad t \in [a,b].$$

This result, together with (5.1), completes the proof. Essentially, the technique is to substitute over each period an equation with constant coefficients that does not alter the transition between the extremities of the period, to control the error along the period, and to prove that the approximations with constant coefficients G-converge.

This result, which we have proved for equations with piecewise constant coefficients, can be extended without difficulty to the case of continuous coefficients; it is sufficient to take piecewise constant functions A_{1k} that agree with $A(x)$ at $s\,\frac{b-a}{2^k}$, to consider the sequence $A_{nk} = A_{1k}(nt)$, and examine the sequence A_{nn}. The result is formally the same; the G-limit is $\bar{A} = \frac{1}{T}\int_0^T A(t)dt$. This formula suggests that the result holds even if $A(x)$ is not continuous, provided that

$$\int_0^T ||A(\tau)||d\tau < +\infty, \tag{5.4}$$

a condition which is satisfied if $\int_0^T |a_{ij}(\tau)|d\tau < +\infty$. (Note that if $A(x)$ is not continuous, it is necessary to define the meaning not only of solutions, which are taken to be absolutely continuous functions, but also of the equation; this is beyond the scope of this text.) In effect, we consider in such a case approximating equations with constant coefficients

$$A_{1k}(t) = \int_{s\cdot 2^{-k}T}^{(s+1)2^{-k}T} A(\tau)d\tau \quad \text{for} \quad s2^{-k}T < t < (s+1)2^{-k}T$$

and the contractions $A_{nk} = A_{1k}(nt)$. The sequence A_{nn} G-converges to

$$\overline{A} = \frac{1}{T}\int_0^T A(\tau)d\tau.$$

(Note that (5.2) and (5.3) do not depend on the regularity of the coefficients but only on (5.4)).

We give an application of the previous results by considering equations of the type

$$D_\lambda y = (a_\lambda y')' + b_\lambda y' + (c_\lambda y)' + d_\lambda y = 0,$$

where a, b, c, and d are periodic functions with period T, $a_\lambda(t) = a(\lambda t)$, etc. (When the coefficients are not sufficiently regular to insure the existence of solutions in the classical sense, we take y and $(a_\lambda y' + c_\lambda y)$ to be absolutely continuous and understand the equation in the weak sense.) We are then able to write down the associated system by putting

$$a_\lambda y' + c_\lambda y = u.$$

We then have

$$y' = -\frac{c_\lambda}{a_\lambda}y - \frac{1}{a_\lambda}u$$

$$u' = \left[-d_\lambda + \frac{b_\lambda c_\lambda}{a_\lambda}\right] y + \frac{b_\lambda}{a_\lambda} u,$$

and it follows that

$$\int_0^T \left|\frac{c_\lambda}{a_\lambda} dt\right| < +\infty, \quad \int_0^T \left|\frac{1}{a_\lambda}\right| dt < +\infty, \quad \int_0^T \left|\frac{b_\lambda c_\lambda - a_\lambda d_\lambda}{a_\lambda}\right| < +\infty, \quad \int_0^T \left|\frac{b_\lambda}{a_\lambda}\right| dt < +\infty.$$

The equation $D_\lambda y = 0$ G-converges to the equation $Dy = ay'' + (b+c)y' + dy = 0$ with

$$a = \left[\frac{1}{T}\int_0^T \frac{1}{a(t)}\, dt\right]^{-1}, \quad b = a \cdot \frac{1}{T}\int_0^T \frac{b(t)}{a(t)}\, dt,$$

$$c = a \cdot \frac{1}{T}\int_0^T \frac{c(t)}{a(t)}\, dt, \quad d = \frac{bc}{a} + \frac{1}{T}\int_0^T \left[d(t) - \frac{b(t)c(t)}{a(t)}\right] dt.$$

These are the known homogenezation formulas for equations of the second order.

Note that the G-convergence of a system $P_s[y_s] = y_s' - A_s y_s = 0$ implies the G-convergence of the nonhomogeneous problems $P_s[y_s] = f$. To prove this, one uses the results of Sec. 2.3, in particular, (2.13).

In conclusion, we now take up the compactness theorem. Because we shall merely give examples in this section, we state it under somewhat simplified hypotheses. Let the equations

$$y_n' = A_n y_n \quad \text{on} \quad [a,b]$$

be given with $||A_n(x)|| \leq K < \infty$ for every x in $[a,b]$; then it is possible to extract a subsequence which G-converges to an equation $y' = Ay$, with $||A(x)|| \leq K$. We have imposed quite strong hypotheses on the coefficients. These can be weakened, but not so much that they have the "natural" condition

$$\int_a^b ||A_n||\, dt \leq K < +\infty.$$

To see this, it is enough to consider the sequence

$$y' = a_n y \quad \text{with} \quad a_n = \begin{cases} n & \text{in } [0,1/n] \\ 0 & \text{elsewhere.} \end{cases}$$

This is a result of the fact that we want a very strong sort of convergence of the solutions. Indeed, if we want to remain in the territory of ordinary differential equations, we cannot weaken further the type of convergence that we ask for, since if we did, the limits (see the example above) might no longer be solutions of differential equations.

To prove the stated compactness theorem, consider the sequence of constant operators

$$A_n^{(0)} = \frac{1}{b-a}\int_a^b A_n(t)dt,$$

and extract a convergent subsequence using the Bolzano-Weierstrass theorem. We consider the corresponding operators $A_{n_k}(t)$ which, for simplicity, we continue to call $A_n(t)$. We divide the interval $[a,b]$ into two parts, and put

$$A_n^{(1)}(t) = \frac{2}{b-a}\int_a^{(a+b)/2} A_n(t)dt, \quad a < t < \frac{a+b}{2}$$

$$A_n^{(1)}(t) = \frac{2}{b-a}\int_{(a+b)/2}^{b} A_n(t)dt, \quad \frac{a+b}{2} < t < b.$$

From the sequence of piecewise constant operators, we extract a convergent subsequence and consider the corresponding operators $A_{n_k}(t)$. Proceeding in this way, we obtain sequences $A_n^{(s)}$ of operators with piecewise constant coefficients corresponding to 2^s intervals. The diagonal sequence $A_n^{(n)}$ converges in solution to the operator $A(t)$, as we see upon taking note of the fact that

$$\lim_{t\to 0} ||\exp(A+B)t - \exp Bt \exp At|| = 0$$

uniformly with respect to $||A||, ||B|| \leq M$. To complete the proof, use the usual a priori bounds (Gronwall's Lemma). We do not persist any longer with these examples since we shall treat G-convergence in a more general form, for nonlinear equations, at the end of Chapter III.

It is, however, important to note that, quite differently from what will be the case in Chapter III, we were, in these examples, in a position to solve the homogenizing problems *explicitly*, whereas in general there are no explicit formulas for that problem.

6. BIBLIOGRAPHY

See the bibliography of Chapter I.

Chapter III
Existence and Uniqueness for the Cauchy Problem Under the Condition of Continuity

In Chapter I, we studied the Cauchy problem (or initial value problem) for systems of normal type assuming strong regularity, i.e., the Lipschitz condition, for the functions on the right-hand side of the equation. In this chapter, we shall study the same problem but shall only assume that those functions are continuous. The biggest difference between the two cases will be that the uniqueness theorem no longer holds, as we see from the following classic example.

We consider the first order differential equation

$$y' = \sqrt{|y|},$$

whose second member is defined for all points (x,y) in the plane and is continuous there, but does not satisfy a Lipschitz condition in neighborhoods of the points $(x,0)$. As a matter of fact, the increment of the function on the second side, as one proceeds from $(x_0,0)$ to (x_0,y), is infinitesimal of order $\frac{1}{2}$ in $|y|$. The zero function clearly satisfies the equation with the initial condition $y(x_0) = 0$, but the function

$$y(x) = \frac{1}{4}(x-x_0)^2 \quad \text{for} \quad x \geq x_0,$$

$$y(x) = -\frac{1}{4}(x-x_0)^2 \quad \text{for} \quad x \leq x_0$$

also satisfies the given equation with the same initial condition $y(x_0) = 0$. There are infinitely many other solutions to the same initial value problem; if a and b are two numbers such that $a \leq x_0 \leq b$, then

$$y(x) = -\frac{1}{4}(x - a)^2 \quad \text{for} \quad x \le a$$

$$y(x) = 0 \quad \text{for} \quad a \le x \le b$$

$$y(x) = \frac{1}{4}(x - b)^2 \quad \text{for} \quad x \ge b$$

is also a solution.

Since the initial value problem may have infinitely many solutions, we must study not only the existence of solutions for a given Cauchy problem, but also the properties of the set of such solutions, that is, its structure, and determine under what conditions we have uniqueness. The present chapter is devoted to this question. In particular, we shall see that the example we just gave is not the only one with infinitely many solutions; any initial value problem without uniqueness will turn out to have infinitely many solutions. We shall, in fact, show that there are two possibilities: either there is a unique solution, or there is a continuum of solutions.

1. EXISTENCE THEOREM

In this section, we shall show the existence of at least one solution for a given Cauchy problem for a system whose right-hand side consists of functions that are merely assumed to be continuous. We cannot, in this case, apply the procedure of successive approximations (see the bibliographical note in Chapter I) and must therefore resort to other methods for the proofs. To this end, we first consider a problem concerning sets of continuous functions, the characterization of compactness in the space $C(X,Y)$ of continuous functions.

1.1. Characterization of Compact Sets of Continuous Functions: Ascoli's Theorem

We consider a sequence $(f_n)_{n=1}^{\infty}$ of real-valued continuous functions defined on the same interval $[a,b]$ and suppose that the sequence converges uniformly there. As is known, the limiting function f is also continuous on $[a,b]$. If, therefore, we fix $\varepsilon > 0$, there is an index ν_ε such that for all x in $[a,b]$ and for all $n > \nu_\varepsilon$, we have

$$|f_n(x) - f(x)| < \varepsilon$$

and, furthermore, corresponding to the same ε, we may find a number $\delta_\varepsilon > 0$ such that for all x' and x'' in $[a,b]$ with $|x''-x'| < \delta_\varepsilon$, we have

$$|f(x'') - f(x')| < \varepsilon.$$

It then follows that, for all x' and x'' in $[a,b]$ such that

$$|x'' - x'| < \delta_\varepsilon$$

we have

$$|f_n(x'')-f_n(x')| \leq |f_n(x'')-f(x'')| + |f(x'')-f(x')| + |f(x')-f_n(x')| < 3\varepsilon \tag{1.0}$$

so long as $n > \nu_\varepsilon$. This relation expresses not only the uniform continuity of the functions f_n of the sequence (which we have already supposed), but also the fact that the difference

$$|f_n(x'') - f_n(x')|$$

may be made less than any preassigned positive number provided x'' and x' satisfy a relation of the type

$$|x'' - x'| < \delta$$

where δ may be taken *independent of* n. This follows from (1.0) for those functions f_n with $n > \nu_\varepsilon$ and from the consideration that the remaining functions $f_1, f_2, \ldots, f_{\nu_\varepsilon}$ are finite in number and uniformly continuous.

The functions of a sequence $(f_n)_{n=1}^{\infty}$ are said to be *equicontinuous* in an interval $[a,b]$ if, for $\varepsilon > 0$, there is a $\delta > 0$ such that for every pair x', x'' in $[a,b]$ satisfying

$$|x'' - x'| < \delta$$

we have

$$|f_n(x'') - f_n(x')| < \varepsilon \quad \text{for every } n.$$

This is equivalent to saying that the oscillation of *every* function of the sequence is less than ε in any interval of length less than δ. We have thus shown that *a uniformly convergent sequence of continuous functions on* $[a,b]$ *consists of equicontinuous functions.*

If, now, we put K equal to the maximum absolute value of f in $[a,b]$, it follows from

$$|f_n(x)| \leq |f_n(x) - f(x)| + |f(x)|$$

that, for every $n > \nu_\varepsilon$,

$$|f_n(x)| < K + \varepsilon.$$

If then $\overline{K}$ is some number greater than the absolute values of the finitely many functions $f_1, f_2, \ldots, f_{\nu_\varepsilon}$ on $[a,b]$ and greater than $K + \varepsilon$ as well, then we have, for each n,

$$|f_n(x)| \leq \overline{K}.$$

This means that the functions of the given convergent sequence are all less, in absolute value, than a suitable number. The functions of a sequence that have this property are called *uniformly bounded* or *equibounded*. We may now complete the statement made above by saying that *functions, continuous on* $[a,b]$ *and forming a uniformly convergent sequence, are both equicontinuous and uniformly bounded.*

We observe that if the functions of a sequence $(f_n)_{n=1}^{\infty}$ are equicontinuous, and if it is possible to find a number M independent of n such that every function of the sequence has absolute value less than M at at least one point (which may vary with n), then the functions of the sequence are uniformly bounded. To see this, note that since the functions of the sequence are equicontinuous, it is possible to find a bound, independent of n, on the oscillation of each function on $[a,b]$; it then follows that the absolute value of every function in the sequence is bounded by the sum of M and the bound just determined.

We now consider a sequence of functions which are defined, equicontinuous, and equibounded on $[a,b]$. It is evident that these hypotheses are not sufficient to conclude that the given sequence converges, that is, the converse of the preceding result is false. It is enough, indeed, to consider the example of the sequence $f_n(x) = (-1)^n$. Nevertheless, such sequences possess at least one subsequence which converges uniformly, a fact that follows from the theorem below, which characterizes compact sets of continuous functions. (Recall that, in accordance with what was said in Sec. 4.1 of Chapter I, a subset A of a metric space is compact if every sequence in A has at least one convergent subsequence.)

<u>Ascoli's Theorem</u>. Let X be a compact metric space, Y a Banach space, and $C(X,Y)$ the Banach space of continuous functions from X to Y with the sup norm. A subset H of $C(X,Y)$ has compact closure in $C(X,Y)$ if and only if

(i) H is equicontinuous, and

(ii) for every x in X, the set $H(x) = \{f(x) | f \in H\}$ has compact closure in Y.

If we call a set with compact closure *relatively compact,* then Ascoli's theorem may be reformulated as follows: H *is relatively compact in* $C(X,Y)$ *if and only if* H *is equicontinuous and pointwise relatively compact.*

The definition of an equicontinuous and equibounded set of functions from a metric space to a Banach space is similar to that given for sequences of real functions; $H \subseteq C(X,Y)$ is *equicontinuous* if, for every $\varepsilon > 0$, there is a $\delta > 0$ such that, for each $f \in H$, $d(x,y) < \delta$ implies

$$||f(x) - f(y)|| \leq \varepsilon,$$

where d is the metric in X. H is *uniformly bounded* or *equibounded* if there exists $M \in R^+$ such that

$$||f(x)|| \leq M \qquad (x \in X;\ f \in H).$$

If we recall that the theorem of Bolzano and Weierstrass states that the compact sets of R^n are the closed and bounded ones, Ascoli's theorem for R^n can be expressed in the following way. To be perfectly rigorous, we should say "pointwise bounded" instead of uniformly bounded, but in our situation the two concepts are equivalent. (See Exercise 2.)

Theorem of Ascoli for R^n. Let X be a compact metric space and $H \subseteq C(X,R^n)$. The set H is relatively compact in $C(X,R^n)$ if and only if H is equicontinuous and uniformly bounded.

This is the statement of Ascoli's theorem that we shall always use below, and it is from it that we may affirm that an equicontinuous and uniformly bounded sequence has at least one uniformly convergent subsequence.

Proof of Ascoli's Theorem: We shall first prove necessity, that is, that a relatively compact set in $C(X,Y)$ must necessarily have properties (i) and (ii). (ii) follows from the fact that the function $\phi_x\colon C(X,Y) \to Y$ defined by $\phi_x(f) = f(x)$ is continuous for every $x \in X$ and that the continuous image of a compact set is compact. As for (i), fix $\varepsilon > 0$. Since $\overline{H}$ is compact, there are finitely many balls $B(f_i,\varepsilon/3)$ that cover H. For each i, since f_i is continuous on a compact set and therefore uniformly continuous there, there is a $\delta_i > 0$ such that

$$||f_i(x) - f_i(y)|| \leq \frac{\varepsilon}{3} \qquad (d(x,y) \leq \delta_i)$$

where d is the metric of X. Let $\delta = \min \delta_i$. Then, for $d(x,y) \leq \delta$ and f in H we have, for i such that $f \in B(f_i,\varepsilon/3)$,

$$\|f(x)-f(y)\| \le \|f(x)-f_i(x)\| + \|f_i(x)-f_i(y)\| + \|f_i(y)-f(x)\|$$
$$\le \frac{\varepsilon}{3} + \frac{\varepsilon}{3} + \frac{\varepsilon}{3} = \varepsilon$$

and so (i) holds.

We now prove sufficiency, that is, that (i) and (ii) imply that H is relatively compact in $C(X,Y)$. Suppose that H satisfies (i) and (ii). To prove that $\overline{H}$ is compact, it is enough to prove that every sequence in H has a subsequence that converges in $C(X,Y)$. To do this, we prove the following proposition.

(*) Let $(f_n)_{n=1}^{\infty}$ be a sequence in H. For every $\varepsilon > 0$ there is a subsequence $(f_{\varepsilon n})_{n=1}^{\infty}$ such that

$$\|f_{\varepsilon n} - f_{\varepsilon m}\| \le \varepsilon \qquad (n,m \ge 1).$$

To see this, fix $\varepsilon > 0$. Equicontinuity implies that every point x in X has an open neighborhood U_x such that

$$\|f_n(u) - f_m(v)\| \le \frac{\varepsilon}{3} \qquad (u,v \in U_x). \tag{1.1}$$

The compactness of X implies the existence of finitely many points $x_1,\dots,x_k$ such that $X = U_{x_1} \cup \dots \cup U_{x_k}$. Since each $H(x_i)$ is relatively compact in Y, there is a subsequence $(f_{n_k})_{k=1}^{\infty}$ of the given sequence such that $(f_{n_k}(x_i))_{k=1}^{\infty}$ converges for each i; one constructs this subsequence by first taking a subsequence converging at x_1, then a subsequence of that converging at x_2, and so on until one arrives at x_k. (Recall that compactness is equivalent to the existence of a convergent subsequence for any given sequence.) From the convergence of $(f_{n_k})_{k=1}^{\infty}$ at the points x_i there follows the existence of k_ε such that

$$\|f_{n_k}(x_i) - f_{n_\ell}(x_i)\| \le \frac{\varepsilon}{3} \qquad (k,\ell \ge k_\varepsilon). \tag{1.2}$$

Now let $u \in X$ and i_u be such that $u \in U_{i_u}$. For every $k,\ell \ge k_\varepsilon$, we have

$$\|f_{n_k}(u) - f_{n_\ell}(u)\| \le \|f_{n_k}(u) - f_{n_k}(x_{i_u})\|$$
$$+ \|f_{n_k}(x_{i_u}) - f_{n_\ell}(x_{i_u})\|$$
$$+ \|f_{n_\ell}(x_{i_u}) - f_{n_\ell}(u)\| \le \frac{\varepsilon}{3} + \frac{\varepsilon}{3} + \frac{\varepsilon}{3}$$
$$\text{(by (1.1) and (1.2))}$$
$$= \varepsilon.$$

With $(f_{\varepsilon n})_{n=1}^{\infty}$ equal to $(f_{n_k})_{k \geq k_\varepsilon}$, proposition (*) is proved.

We now consider any sequence $(f_n)_{n=1}^{\infty}$ in H and construct a convergent subsequence in C(X,Y). We apply proposition (*) to $(f_n)_{n=1}^{\infty}$ with $\varepsilon = 1$, thereby obtaining a subsequence $(f_{1n})_{n=1}^{\infty}$ for which proposition (*) holds. We then apply proposition (*) to $(f_{1n})_{n=1}^{\infty}$ with $\varepsilon = \frac{1}{2}$ and obtain a subsequence $(f_{\frac{1}{2}n})_n$ for which proposition (*) holds. We repeat this with $\varepsilon = 1/3$ and obtain a subsequence $(f_{\frac{1}{3}n})_{n=1}^{\infty}$. Proceeding in this way for each $\varepsilon = 1/k$, we obtain a sequence $(f_{\frac{1}{k}n})_{n=1}^{\infty}$ for which proposition (*) holds and which is a subsequence of $(f_{\frac{1}{k-1}n})_{n=1}^{\infty}$. We now define a subsequence $(f_{n_k})_{k=1}^{\infty}$ of $(f_n)_{n=1}^{\infty}$ in the following way:

$$f_{n_k} = f_{\frac{1}{k}k}, \qquad k \geq 1.$$

It is then clear that $(f_{n_k})_{k=1}^{\infty}$ is a subsequence of the given sequence $(f_n)_{n=1}^{\infty}$ and that

$$||f_{n_k} - f_{n_\ell}|| \leq 1/k \qquad (\ell \geq k).$$

It follows from this relation that $(f_{n_k})_{k=1}^{\infty}$ is a Cauchy sequence in C(X,Y) and therefore converges in the Banach space C(X,Y). This completes our proof of the theorem of Ascoli.

In the case of differentiable functions, conditions (i) and (ii) in Ascoli's theorem are often easy to verify, as is clear from the following.

Theorem of Arzelà. Let $H \subseteq C([a,b],\mathbb{R}^n)$ be a set of functions whose incremental ratios are uniformly bounded. Then H is equicontinuous. If there is an x_0 in $[a,b]$ such that $H(x_0) = \{f(x_0) \mid f \in H\}$ is bounded, then H is equibounded.

In fact, the hypothesis of the theorem means that there is a positive constant M such that

$$\left|\left|\frac{f(x) - f(y)}{x - y}\right|\right| \leq M \qquad (x,y \in [a,b]).$$

From this it follows that

$$||f(x) - f(y)|| \leq M|x-y|,$$

so that the quantity on the left can be made less than ε so long as x and y satisfy the following condition, which is independent of $f \in H$:

$|x - y| < \varepsilon/M.$

The last part of the theorem can be proved upon observing that

$$||f(x)|| - ||f(x_0)|| \leq ||f(x) - f(x_0)|| \leq M|x - x_0|$$

implies that

$$||f(x)|| \leq ||f(x_0)|| + M|x - x_0| \leq N + M(b - a)$$

if

$$||f(x_0)|| \leq N \qquad (f \in H).$$

Exercise 1. What we have called an equicontinuous set is known in the literature as an "equiuniformly continuous set"; an equicontinuous set is actually a set $H \subseteq C(X,Y)$ such that for every x in X and every $\varepsilon > 0$, there is a neighborhood U of x such that for every f in H and y in Y,

$$||f(x) - f(y)|| \leq \varepsilon.$$

It so happens that we have done nothing worse than abuse the terminology, since when X is compact, then $H \subseteq C(X,Y)$ is equiuniformly continuous if and only if it is equicontinuous. Prove this.

Exercise 2. Prove that if X is compact and $H \subseteq C(X,Y)$ is equicontinuous and pointwise bounded, then H is uniformly bounded. Is this true even if H is pointwise bounded in a dense set in X?

Exercise 3. Ascoli's theorem is true if (ii) is assumed to be valid only for those x belonging to a dense subset of X and if Y is a metric space. What must be changed in the proof?

Exercise 4. Prove that if $(f_n)_{n=1}^{\infty}$ is equicontinuous (that is, $\{f_n | n \geq 1\}$ is equicontinuous), then $\lim_{n\to\infty} f_n = f$ uniformly implies that f is continuous.

Exercise 5. Prove that if $(f_n)_{n=1}^{\infty}$ is equicontinuous, then $(f_n)_{n=1}^{\infty}$ converges pointwise in a dense subset of X if and only if it converges uniformly, and the limit is the same.

Exercise 6. Use Exercises 4 and 5 to produce a different proof for Ascoli's theorem for $H \subseteq C([a,b],R^n)$ by means of the diagonal process of Cantor. Hint: Let $\{q_n \mid n \geq 1\}$ be the set of rational numbers in $[a,b]$. Given $(f_n)_{n=1}^{\infty}$ in H, take a subsequence $(f_{1n})_{n=1}^{\infty}$ that converges at q_1. Take a subsequence $(f_{2n})_{n=1}^{\infty}$ of this that converges at q_2, and so on. Then consider the diagonal sequence $(f_{nn})_{n=1}^{\infty}$.

Exercise 7. Ascoli's theorem can be extended in the following way to sequences of functions that are not necessarily continuous. Let $(f_n)_{n=1}^{\infty}$ be a sequence of arbitrary functions from a compact metric space X into a Banach space Y. We shall say that $(f_n)_{n=1}^{\infty}$ is *pseudoequicontinuous* if, for every $\varepsilon > 0$ and each x in X, there is an n_0 in N and a neighborhood U_x of x such that

$$||f_n(x) - f_n(y)|| < \varepsilon \qquad (n \geq n_0;\ y \in U_x).$$

Prove the following propositions for a pseudoequicontinuous sequence $(f_n)_{n=1}^{\infty}$.

(a) $(f_n)_{n=1}^{\infty}$ is equicontinuous if and only if each f_n is continuous.

(b) $f = \lim_{n\to\infty} f_n$ uniformly implies f is continuous.

(c) $f = \lim_{n\to\infty} f_n$ uniformly if and only if $f = \lim_{n\to\infty} f_n$ pointwise on a dense subset of X.

(d) If $\{f_n | n \geq 1\}$ is pointwise relatively compact in a dense subset of X, then there exists a subsequence that converges uniformly to a continuous function. This generalizes Ascoli's theorem; prove that it implies Dini's theorem on increasing sequences.

Exercise 8. A sequence $(f_n)_{n=1}^{\infty}$ of $C(X,R^m)$ is equicontinuous if and only if the sequence $(f_{ni})_{n=1}^{\infty}$ of coordinate functions is equicontinuous for every $i = 1,\ldots,m$.

1.2. Local Existence

We now return to the starting problem of the existence of integrals for a system of differential equations whose second member is continuous. We begin, for the sake of simplicity, with the case of one differential equation

$$y' = f(x,y) \tag{1.3}$$

where f is continuous in a strip $S = \{a \leq x \leq b,\ |y| < \infty\}$. Equation (1.3) thus requires any possible integral $y = y(x)$ through (x,y) to have a fixed slope equal to $f(x,y)$. This observation suggests that an integral relative to the initial value problem $y(x_0) = y_0$ can be approximated by polygonal lines passing through (x_0,y_0) whose sides have slopes equal to a value of the function f in a neighborhood of (x_0,y_0). For example, one may construct the polygonal line beginning at (x_0,y_0) with slope equal to $f(x_0,y_0)$ right up to a point (x_1,y_1); from there proceed with a side of slope $f(x_1,y_1)$ and so on. One may predict that such polygonal lines will converge to an integral if successive vertices are infinitely close to each other.

We may make a similar analysis even in the case of a system where the integrals can be represented as curves with equation $x = x$, $y_1 = y_1(x),\dots,y_n = y_n(x)$ in $(n+1)$- dimensional space with general point $(x,y_1,\dots,y_n)$. Such curves are then graphs with respect to the x-axis in the sense that every hyperplane $x = \text{constant}$ meets them in only one point. Even in this case, the system determines the tangent of every possible integral curve at every point, and so permits us to construct, in a manner similar to the one above, polygonal lines that may be expected to converge to integrals.

The preceding considerations of an intuitive nature lie at the base of the Cauchy-Lipschitz method of proving the existence of integrals. We now propose to use this method to prove the following:

Lemma. Let f_i, $i = 1,\dots,n$, be functions continuous on the strip $S = [a,b] \times \mathbb{R}^n$, and let there exist a constant M such that $|f(x,y)| \le M$ in S. If (x_0,y^0) is any point whatsoever of S, then there is at least one family of differentiable functions $y_1,\dots,y_n$ defined on $[a,b]$ and satisfying there

$$y_i'(x) = f_i(x,y(x)), \quad y_i(x_0) = y_i^0, \qquad i = 1,\dots,n. \tag{1.4}$$

We shall prove the existence of at least one family of functions $y_1,\dots,y_n$ satisfying (1.4) on the interval $[x_0,b]$; in a similar manner, one may show the existence of a family on the interval $[a,x_0]$, completing the proof of the lemma.

If $b = x_0 + \delta$, we divide the interval $[x_0,x_0+\delta]$ into m equal parts with end points

$$x_{mj} = x_0 + j\,\frac{\delta}{m} \qquad (j = 0,1,\dots,m)$$

and consider the function ϕ_m defined on $[x_0,x_0+\delta]$ in the following way (we use vector notation):

$$\phi_m(x_0) = y^0$$

$$\phi_m(x) = y^0 + (x-x_0)f(x_0,y^0) \qquad \text{for} \quad x_0 < x \le x_{m1}$$

$$\phi_m(x) = \phi_m(x_{mj}) + (x-x_{mj})f(x_{mj},\phi_m(x_{mj})) \qquad \text{for} \quad x_{mj} \le x \le x_{mj+1}.$$

We now introduce the function ψ_n defined on $[x_0,x_0+\delta]$ in the following manner:

$$\psi_m(x) = f(x_{mj},\phi_m(x_{mj})) \quad \text{for} \quad x_{mj} \le x \le x_{mj+1}.$$

One immediately sees that the coordinate functions ϕ_{mi} satisfy the relation

$$\phi_{mi}(x) = y^0 + \int_{x_0}^{x} \psi_{mi}(t)dt, \tag{1.5}$$

for if x belongs to the interval $[x_0,x_{m1}]$, (1.5) follows at once, whereas if x belongs to the next interval $[x_{m1},x_{m2}]$, we have

$$\phi_{mi}(x) = y^0 + \int_{x_0}^{x_{m1}} \psi_{mi}(t)dt + \int_{x_{m1}}^{x} \psi_{mi}(t)dt = y^0 + \int_{x_0}^{x} \psi_{mi}(t)dt.$$

Proceeding in this way, we establish (1.5) in general. From (1.5) it follows that if x' and x'' are arbitrary points of $[x_0,x_0+\delta]$, then the coordinate functions ϕ_{mi} of the ϕ_m satisfy

$$|\phi_{mi}(x'') - \phi_{mi}(x')| = \left|\int_{x'}^{x''} \psi_{mi}(t)dt\right| \le \left|\int_{x'}^{x''} |\psi_{mi}(t)|dt\right| \le M|x'' - x'| \tag{1.6}$$

since the maximum of the absolute value of the function ψ_m on $[x_0,x_0+\delta]$ is not greater than that of the function f on S.

We have thus proved that the incremental ratios of the functions ϕ_{mi}, and therefore also of ϕ_m, are equibounded, and so, by the theorem of Arzelà, the ϕ_m are equicontinuous. But when $x = x_0$, these all assume the value y^0, and so, again by the theorem of Arzelà, they are also equibounded.

By Ascoli's theorem, we now can extract from $(\phi_m)_{m=1}^{\infty}$ a subsequence $(\phi_{m_k})_{k=1}^{\infty}$ that converges uniformly to a continuous function on $[x_0,x_0+\delta]$ that we shall call y. To complete the proof of the lemma, it will suffice to show that this last function satisfies the relation

$$y_i(x) = y^0 + \int_{x_0}^{x} f_i(t,y(t))dt.$$

But since

$$\phi_{m_k i}(x) = y^0 + \int_{x_0}^{x} \psi_{m_k i}(t)dt, \tag{1.7}$$

it will be enough to be sure that the sequence $(\psi_{m_k})_{k=1}^{\infty}$ converges uniformly in $[x_0,x_0+\delta]$ to the function $f(x,y(x))$ and then pass to the limit under the integral sign in (1.7).

We first of all note that, because of (1.6), we have

$$|\phi_{mi}(x) - y^0| \le M\delta \quad \text{for all } i \text{ and } x.$$

Given $\varepsilon > 0$, we determine a $\sigma > 0$ by the uniform continuity of f so that we have

$$|f_i(x'',y'') - f_i(x',y')| < \varepsilon$$

for every i and every pair of points (x',y'), (x'',y'') in the rectangle $[x_0,x_0+\delta] \times [y^0-M\delta,\ y^0+M\delta]$ that satisfy the conditions

$$|x' - x''| < \sigma, \quad |y_i' - y_i''| < \sigma.$$

Corresponding to this σ we determine an $\eta > 0$ less than σ so that we have for every i

$$|y_i(x') - y_i(x'')| < \frac{\sigma}{2}, \qquad (|x'-x''| < \eta). \tag{1.8}$$

We now consider an arbitrary point $x \in [x_0,x_0+\delta]$. Let x_{m_kj} be the point of the m_k-th subdivision that satisfies the relation $x_{m_kj} \le x \le x_{m_kj+1}$. We then have

$$\begin{aligned}|f_i(x,y(x))-\psi_{m_ki}(x)| &= |f_i(x,y(x))-f_i(x_{m_kj},\phi_{m_k}(x_{m_kj}))| \\ &\le |f_i(x,y(x))-f_i(x,\phi_{m_k}(x_{m_kj}))| + |f_i(x,\phi_{m_k}(x_{m_kj})) \\ &\qquad - f_i(x_{m_kj},\phi_{m_k}(x_{m_kj}))|.\end{aligned} \tag{1.9}$$

We determine an index ν such that for $m_k > \nu$, we have

$$\frac{\delta}{m_k} < \eta < \sigma$$

so that, since

$$|x - x_{m_kj}| \le |x_{m_kj+1} - x_{m_kj}| = \frac{\delta}{m_k} < \sigma,$$

the last term in (1.9) is less than ε. We now take ν still larger in order to have

$$|y_i(x) - \phi_{m_ki}(x)| < \sigma/2, \qquad (x \in [x_0,x_0+\delta];\ m_k > \nu). \tag{1.10}$$

Then, from

$$|y_i(x)-\phi_{m_k i}(x_{m_k j})| \leq |y_i(x)-y_i(x_{m_k j})| + |y_i(x_{m_k j})-\phi_{m_k i}(x_{m_k j})|,$$

we deduce by means of (1.8) and (1.10) that even the next to last term in (1.9) is less than ε. We may therefore conclude that we have

$$\lim_{k\to\infty} \psi_{m_k}(x) = f(x,y(x))$$

uniformly in $[x_0,x_0+\delta]$, and the lemma is completely proved.

The following theorem is an immediate consequence of the lemma.

Peano's Local Existence Theorem. Let $f_1,\ldots,f_n$ be continuous functions in the rectangle R defined by $x_0-a \leq x \leq x_0+a$, $y_i^0-a_i \leq y_i \leq y_i^0+a_i$. Then there is at least one n-tuple $y_1,\ldots,y_n$ of functions continuous and differentiable in an interval $[x_0-\delta,x_0+\delta]$ with $0 < \delta \leq a$ satisfying there the initial value problem

$$y_i' = f_i(x,y(x)), \quad y_i(x_0) = y_i^0, \qquad (i = 1,\ldots,n).$$

In fact, considering the auxiliary functions $r_i\colon \mathsf{R} \to [y_i^0-a_i, y_i^0+a_i]$ defined by

$$r_i(t) = \begin{cases} y_i^0 + a_i & \text{if } t \geq y_i^0 + a_i \\ t & \text{if } y_i^0 - a_i \leq t \leq y_i^0 + a_i \\ y_i^0 - a_i & \text{if } t \leq y_i^0 - a_i, \end{cases}$$

we may define functions $g_i\colon [x_0-a,x_0+a] \times \mathsf{R}^n \to \mathsf{R}$ with

$$g_i(x,y_1,\ldots,y_n) = f_i(x,r_1(y_1),\ldots,r_n(y_n))$$

in order to be in a position to apply the lemma. The g_i are continuous and bounded. It thus follows from the lemma that the initial value problem

$$y_i' = g_i(x,y), \qquad y_i(x_0) = y_i^0$$

has at least one solution $y_1,\ldots,y_n$ in $[x_0-a,x_0+a]$. From the continuity of the y_i, we know that there is a $0 < \delta \leq a$ such that

$$|y_i(x) - y_i^0| \leq a_i \qquad (|x - x_0| \leq \delta). \tag{1.11}$$

Since g_i and f_i agree in R for each i, it follows from (1.11) that the restrictions of the y_i to $[x_0-\delta,x_0+\delta]$ constitute a solution of the initial value problem

$$y_i' = f_i(x,y), \qquad y_i(x_0) = y_i^0,$$

and the theorem is proved.

Remark. The idea behind the proof of the lemma plays a role in a great many procedures of numerical analysis. By defining the functions ϕ_m, we associate a sequence of equations with finite differences of non-linear type with the given system. We shall return to this topic in Chapter V.

Exercise 1. The polygonal line ϕ_m can be defined in a different way from that used in the proof of the lemma so that even the derivatives converge. To simplify the notation, let us suppose that $n = 1$. For every integer $k \geq 1$, we define a polygonal line ϕ_k on a finite number of intervals $[t_i^k, t_{i+1}^k]$ in the following manner. Let

$$\psi_1(x) = y^0 + (x-x_0)f(x_0,y^0).$$

This function is defined for $x \geq x_0$. Let

$$t_2^k = \sup\{x \in [x_0,b] \mid |\psi_1'(x) - f(x,\psi_1(x))| < \tfrac{1}{k}\}.$$

If we set $t_1^k = x_0$, we may define $\phi_k = \psi_1$ on $[t_1^k, t_2^k]$. If $t_2^k = b$, we stop at this point. Otherwise we define

$$\psi_2(x) = \psi_1(t_2^k) + (x-t_2^k)f(t_2^k,\phi_1(t_2^k))$$

$$t_3^k = \sup\{x \in [t_2^k,b] \mid |\psi_2'(x) - f(x,\psi_2(x))| < \tfrac{1}{k}\}$$

and set $\phi_k = \psi_2$ on $[t_2^k, t_3^k]$. If $t_3^k = b$, we stop. Otherwise we proceed *as long as is possible*. Prove that the points t_i^k are finite in number and that the last is b. We thus have a polygonal line ϕ_m on $[x_0,b]$ such that

$$|\phi_k'(x) - f(x,\phi_k(x))| < \frac{1}{k}$$

for all x in $[x_0,b]$ except for those $x = t_i^k$. It is now enough to use the theorem of Ascoli and to pass to the limit in the preceding inequality in order to finish the proof of the theorem. This proof is much simpler than that given in the text but is not as constructive because of the definition of the points t_i^k.

Exercise 2. Use the results of Chapter I to prove the local existence theorem. Hint: Approximate the functions f_{mi} uniformly in R with a sequence $(f_{mi})_{m=1}^{\infty}$ of C^1 functions (for example, polynomials).

Then use the results of Chapter I to show that if m is sufficiently large, the initial value problems

$$y'_{mi} = f_{mi}(x,y_{mi}), \qquad y_{mi}(x_0) = y_i^0$$

all have solutions y_{mi}, $i = 1,\ldots,n$, defined on the same interval $[x_0-\delta, x_0+\delta]$. Apply Ascoli's theorem to $(y_{mi})_{m=1}^{\infty}$ to complete the argument

1.3. Global Existence

We consider the initial value problem

$$y_i' = f_i(x,y), \quad y_i(a) = y_i^0, \qquad (i = 1,\ldots,n), \tag{1.12}$$

where the f_i are defined and continuous in the cylinder $S = [a,b] \times \mathbb{R}^n$. The theorem of the preceding section does not allow us to affirm the existence of a solution over the whole interval $[a,b]$, that is, a global solution or a solution in the large. It is, however, important to know about global solutions for the purpose of applications. We shall now study certain sufficient conditions for the existence of global solutions and resume our discussion below, in Section 2.4.

We begin with a criterion which will be used frequently.

Theorem on A Priori Bounds. If a solution to (1.12) is bounded in its maximal interval of existence, then it exists in the whole interval [

This means that if the solution in question is not yet defined in all of $[a,b]$, it can be extended to the whole interval so as to be a solution of (1.12) there. Because of local existence, it follows from the theorem just stated that it is enough to know that the *possible* solutions in the large are bounded to conclude that they exist.

To prove the theorem, we consider a solution $y_1,\ldots,y_n$ of (1.12) and a constant $K > 0$ such that

$$|y_i(x)| \leq K$$

for every i and every x in the domain of y_i. We suppose that the y_i are defined in a proper subset of $[a,b]$ since there is otherwise nothing to prove. It is enough to show that the assumption that the y_i cannot be extended as solutions of (1.12) to all of $[a,b]$ leads to a contradiction. To do this, let A be the set of t, $a \leq t \leq b$ such that the y_i are defined or can be extended so as to be solutions of (1.12) on $[a,t]$. Put $\alpha = \sup A$. We suppose that $\alpha < b$ and produce a contradiction. Let M be the maximum of the values of

$|f_i(x,u)|$ in the rectangle $a \le x \le b$, $-K \le u_i \le K$. From the integral representation of the solutions, we get, for $s \le t < \alpha$,

$$|y_i(t) - y_i(s)| \le \int_s^t |f_i(\xi,y(\xi))|d\xi$$
$$\le \int_s^t M\, d\xi$$
$$= M|t - s|.$$

This proves that

$$\ell_i = \lim_{x\uparrow\alpha} y_i(x)$$

exists, since the Cauchy criterion is satisfied as $M|t-s|$ becomes infinitely small as $|t-s|$ does. If we apply the local existence theorem, we obtain a solution of

$$z_i' = f_i(x,z), \quad z_i(\alpha) = \ell_i, \qquad (i = 1,\dots,n)$$

in an interval $[\alpha,\alpha+\delta] \subseteq [a,b]$. Then the functions

$$u_i(x) = \begin{cases} y_i(x) & \text{for} \quad a \le x \le \alpha \\ z_i(x) & \text{for} \quad \alpha \le x \le \alpha+\delta \end{cases}$$

are solutions of (1.12) in $[a,\alpha+\delta]$ (see Exercise 1 of Sec. 1.7, Chapter I) that extend the y_i. This contradicts the definition of α, and the theorem is proved.

From this theorem, we get a result similar to one established in Chapter I for functions satisfying a Lipschitz condition.

Corollary 1. Every solution of (1.12) has its graph contained in that of a nonextendable solution of (1.12).

This means that a given solution of (1.12) can be extended into a maximal interval I as a solution of (1.12), and that either this interval coincides with $[a,b]$, or the solution cannot be extended to sup I or to the right of sup I. Regarding the extension to the single point sup I, see Exercise 1 of Sec. 1.7 of Chapter I.

Proof of Corollary 1: Let $y_1,\dots,y_n$ be a solution of (1.12), and let A be the set of all t, $a \le t \le b$ such that the y_i are defined or can be extended into $[a,t]$ as solutions of (1.12). If $\sup A = \alpha < b$, then, by the preceding theorem, one of the y_i must be unbounded, so it cannot be extended beyond $[a,\alpha[$. If $\alpha = b$, there are two cases to be examined: either all the y are bounded in $[a,b[$, or at least one is unbounded. In the second case, it is not possible to extend the solution to b, while in the first it is, if one proceeds as in Exercise 1 of Sec. 1.7 of Chapter I. The corollary is thus completely proved.

We now give the following

Corollary 2. Let $f: [a,b] \times \mathbf{R} \to \mathbf{R}$ be a continuous function, and let $\alpha,\beta: [a,b] \to \mathbf{R}$ be two continuous and differentiable functions such that $\alpha < \beta$ and

$$\alpha'(x) \leq f(x,\alpha(x)), \quad \beta'(x) \geq f(x,\beta(x)).$$

Every solution y of

$$y' = f(x,y), \quad y(a) = y^0$$

with $\alpha(a) < y^0 < \beta(a)$ satisfies $\alpha \leq y \leq \beta$ and exists in $[a,b]$.

There are cases in which functions satisfying the hypotheses of this corollary are easily found. For example, if there are two constants $c_1 < c_2$ such that

$$0 \leq f(x,c_1), \quad 0 \geq f(x,c_2),$$

then the constant functions $\alpha \equiv c_1$ and $\beta \equiv c_2$ satisfy the conditions of Corollary 2. When $f(x,y) = g(x,y) + h(x,y)$ with $|h(x,y)| \leq M$, and there exist constants $c_1 < c_2$ such that

$$0 \leq f(x,c_1) - M, \quad 0 \geq f(x,c_2) + M,$$

then the constant functions $\alpha = c_1$ and $\beta = c_2$ satisfy the hypotheses of Corollary 2. In connection with this example, see Exercise 1.

We note that the lemma of Sec. 1.2, which, in substance, is a criterion for global existence, is contained in Corollary 2 when $n = 1$. One sees this by setting

$$\alpha(x) = y^0 - M(x-x_0), \quad \beta(x) = y^0 + M(x-x_0).$$

Proof of Corollary 2: We consider the auxiliary functions

$$f_n(x,y) = f(x,y) + \frac{-2(y-\alpha(x))+\beta(x)-\alpha(x)}{n}$$

defined for every integer n. For every n, we have

$$\alpha'(x) < f_n(x,\alpha(x)), \quad \beta'(x) > f_n(x,\beta(x)). \tag{1.13}$$

For each n, the local existence theorem guarantees that the initial value problem

$$y_n' = f_n(x,y_n), \quad y_n(a) = y^0$$

has at least one local solution. Let us prove that

$$\alpha \le y_n \le \beta. \tag{1.14}$$

To do so, note that if

$$y_n(x_0) < \alpha(x_0)$$

were the case at x_0, then $x_1 < x_0$ would exist such that

$$y_n(x) < \alpha(x) \quad \text{for} \quad x_1 < x \le x_0 \quad \text{and} \quad y_n(x_1) = \alpha(x_1).$$

If we write down the incremental ratio of y and α at x_1 and take the right-hand limit, we get

$$y_n'(x_1) \le \alpha'(x_1).$$

But this contradicts what we obtain from (1.13):

$$\alpha'(x_1) < f_n(x_1,\alpha(x_1)) = f_n(x_1,y_n(x_1)) = y_n'(x_1).$$

We therefore have $\alpha \le y_n$. We similarly establish that $y_n \le \beta$, and so (1.14) is true. From the theorem on a priori bounds (α and β are bounded because they are continuous), it follows that y_n exists in $[a,b]$. If we let N be the maximum of $|f(x,y)|$ for $a \le x \le b$, $\alpha(x) \le y \le \beta(x)$, we have

$$\begin{aligned} |y_n'(x)| = |f_n(x,y_n(x))| &\le N + \max_x \frac{|\beta(x) - \alpha(x)|}{n} \qquad \text{(by (1.14))} \\ &\le N + \max_x |\beta(x) - \alpha(x)|. \end{aligned}$$

The theorem of Arzelà therefore implies that $(y_n)_{n=1}^{\infty}$ is an equicontinuous and equibounded sequence. This in turn implies that there is a subsequence $(y_{n_k})_{k=1}^{\infty}$ that converges uniformly in $[a,b]$ to a continuous function y. If we pass to the limit in

$$y_{n_k}(x) = y^0 + \int_a^x \left(f(t,y_{n_k}(t)) - \frac{2(y_{n_k}(t)-\alpha(t))+\beta(t)-\alpha(t)}{n_k} \right) dt$$

we get

$$y(x) = y^0 + \int_a^x f(t,y(t))dt,$$

so y is a solution of the given initial value problem. From (1.14) it follows that $\alpha \le y \le \beta$, which completes the proof of the corollary.

Another strategy for obtaining a priori bounds on the solutions, and therefore global existence theorems, is that of considering the behavior of the functions on the right-hand side of the given system. In general, one does this by comparing the given system with another, the properties of whose solutions are known. The situation is illustrated by the following

Comparison Theorem. Besides (1.12), let us consider the system

$$y_i' = F_i(x,y), \qquad (i = 1,\dots,n)$$

where the F_i are defined and continuous in the cylinder S, nonnegative, and increasing in each of the variables $y_1,\dots,y_n$. If $Y_1,\dots,Y_n$ is a solution in $[a,b]$ of the initial value problem

$$Y_i' = F_i(x,Y), \qquad Y_i(a) = Y_i^0,$$

then

$$|f_i(x,y_1,\dots,y_n)| \leq F_i(x,|y_1|,\dots,|y_n|) \tag{1.15}$$

and

$$|y_i^0| \leq Y_i^0, \qquad (i = 1,\dots,n)$$

imply the existence of at least one solution $y_1,\dots,y_n$ of (1.12) such that

$$|y_i(x)| \leq Y_i(x), \qquad (i = 1,\dots,n)$$

in $[a,b]$.

In the proof, we shall use a method of Tonelli that is at least formally simpler than that of Cauchy and Lipschitz; it furnishes a priori bounds as well as existence of solutions.

To prove the theorem, we observe that, since the functions $F_i(x,Y(x))$ are defined and continuous in the interval $[a,b]$, there is a positive number N such that

$$F(x,Y(x)) \leq N \tag{1.16}$$

in the given interval. We now consider, for each integer m, the functions over the interval $[a,b]$ defined by

$$\begin{aligned} y_i^{(m)}(x) &= y_i^0 && \text{for } x_0 \leq x \leq x_0 + \frac{\delta}{m}, \\ y_i^{(m)}(x) &= y_i^0 + \int_{x_0}^{x-\frac{\delta}{m}} f_i(t,y^{(m)}(t))dt && \text{for } x > x_0 + \frac{\delta}{m} \end{aligned} \tag{1.17}$$

where $\delta = b - a$. The definition of these functions, which are called *Tonelli approximations,* depends on the fact that the second relation in (1.17) determines the values of the $y_i^{(m)}(x)$ in the interval $[a,a+h]$ when the values in the interval $[a,a+h-\frac{\delta}{m}]$ are already known. Therefore, since, by the first relation of (1.17), the values in the interval $[a,a+\frac{\delta}{m}]$ are known, those in the interval $[a+\frac{\delta}{m}, a+2\frac{\delta}{m}]$ are defined; since these are defined, so are those in the interval $[a+2\frac{\delta}{m}, a+3\frac{\delta}{m}]$, and so on.

We now observe that, by $F_i \geq 0$, the functions Y_i, $i = 1,\ldots,n$, are nondecreasing, and therefore we have

$$|y_i^{(m)}(x)| = |y_i^0| \leq Y_i(x_0) \leq Y_i(x)$$

in $[a, a+\frac{\delta}{m}]$. In $[a+\frac{\delta}{m}, a+2\frac{\delta}{m}]$, we also have, because of this last fact and of (1.15),

$$\begin{aligned} |y^{(m)}(x)| &\leq |y_i^0| + \int_a^{x-\frac{\delta}{m}} F_i(t,|y_1^{(m)}|,|y_2^{(m)}|,\ldots,|y_n^{(m)}|)dt \\ &\leq |y_i^0| + \int_a^{x-\frac{\delta}{m}} F_i(t,Y_1(t),\ldots,Y_n(t))dt \\ &= Y_i(x-\tfrac{\delta}{m}) \leq Y_i(x). \end{aligned}$$

If we continue to reason in this way, we find that the bounds

$$|y_i^{(m)}(x)| \leq Y_i(x) \tag{1.18}$$

hold in all of $[a,a+\delta]$. If, now, x' and x'' are two points in $[a,b]$ with $x' \leq x''$, we get

$$\begin{aligned} |y_i^{(m)}(x'')-y_i^{(m)}(x')| &\leq \int_{x'-\frac{\delta}{m}}^{x''-\frac{\delta}{m}} |f_i(t,y^{(m)}(t)|dt \\ &\leq \int_{x'-\frac{\delta}{m}}^{x''-\frac{\delta}{m}} F_i(t,|y_1^{(m)}(t)|,\ldots,|y_n^{(m)}(t)|)dt \qquad \text{(by (1.15))} \\ &\leq \int_{x'-\frac{\delta}{m}}^{x''-\frac{\delta}{m}} F_i(t,Y_1(t),\ldots,Y_n(t))dt \end{aligned}$$

(by (1.18) and the fact that the F_i are increasing with respect to the variables y_k)

where the bounds of integration are replaced by a whenever they are less than a. In each case, we conclude, by (1.16),

$$|y_i^{(m)}(x'') - y_i^{(m)}(x')| \leq N|x'' - x'|.$$

The functions $y_i^{(m)}$, and therefore $y^{(m)}$, are thus seen to be equicontinuous, and, by (1.18), equibounded. Ascoli's theorem now implies the existence of a subsequence $(y^{(m_k)})_{k=1}^{\infty}$ of $(y^{(m)})_{m=1}^{\infty}$ that converges uniformly in $[a,b]$ to a limit y which is continuous there:

$$\lim_{k\to\infty} y_i^{(m_k)}(x) = y_i(x), \qquad (i = 1,2,\ldots,n). \tag{1.19}$$

It remains to prove that the functions y_i, $i = 1,2,\ldots,n$, constitute a solution of (1.12). To show this, it is enough to demonstrate that we have, uniformly in $[a,b]$,

$$\lim_{k\to\infty} f_i(x,y^{(m_k)}(x)) = f_i(x,y(x)). \tag{1.20}$$

This is a consequence of the uniform continuity of the $f_i(x,y)$ in the bounded domain $x_0 \leq x \leq x_0+\delta$; $|y_i| \leq Y_i(x)$, $i = 1,2,\ldots,n$, and the uniform convergence established by (1.19). After this, we observe that if we take the limits in (1.17) using the continuity of the integral, we get

$$y_i(x) = y_i^0 + \int_a^x f_i(t,y(t))dt.$$

This completes the proof of the theorem.

The following is an interesting case of the preceding theorem. Let us consider

$$|f_i(x,y_1,\ldots,y_n)| \leq \sum_{j=1}^{n} a_{ij}|y_j| + b_i$$

in the cylinder S, where a_{ij}, b_i are nonnegative constants. Then the integrals of the system (1.12) are defined in the whole interval $[a,b]$, and we have thereby generalized the lemma of the preceding section. To see this, it is enough to recall that the integrals of the linear system

$$Y_i' = \sum_{j=1}^{n} a_{ij}Y_j + b_i$$

are defined in all of $[a,b]$.

The preceding comparison theorem can be more useful in determing the domain of definition of the integrals of a system of differential equations than the standard existence theorems. Thus, for example, if we suppose that we have

$$|f_i(x,y_1,\ldots,y_n)| \leq A|y_i|^{\alpha}, \quad (\alpha > 1)$$

in S, we can affirm the existence of the integrals in $[a,a+\delta]$ with $\delta = 1/A(\alpha-1)\eta^{1-\alpha}$ where $\eta = \max |y_i^0|$, because the integrals of the system $Y_i' = AY_i$ are defined there.

We can easily deduce from the theorem above another formulation of Gronwall's lemma:

If a function $y(x)$ defined on $[a,b]$ is positive there, and if its derivative satisfies the bound

$$|y'(x)| \leq My(x) + N$$

where M and N are positive constants, then we have

$$0 \leq y(x) \leq (y(a) + \frac{N}{M})e^{M(b-a)}.$$

It is enough to observe that the function y is an integral of the differential equation

$$Y' = y'(x)\,\frac{MY + N}{My(x) + N}\,.$$

Since the absolute value of the second side is less than $M|Y| + N$, the absolute value of y is less than the integral of the equation

$$Y' = MY + N$$

that satisfies the initial condition $Y(x_0) = y(x_0)$, that is,

$$0 \leq y(x) \leq y(x_0)e^{M(x-x_0)} + \frac{N}{M}\left(e^{M(x-x_0)} - 1\right).$$

The desired inequality easily follows from this.

We shall resume the question of global existence in the next section after first introducing the notions of maximal and minimal solutions.

Exercise 1. With reference to the second example given in illustration of Corollary 2, consider the function

$$f(x,y) = \mu x^{\nu} + h(x,y),$$

where h is continuous and bounded. Determine those values of μ and ν for which there exist C^1-functions α and β such that $\alpha < \beta$, $\alpha' \leq f(x,\alpha)$, $\beta' \geq f(x,\beta)$.

Exercise 2. Prove the following generalization of Gronwall's lemma: Let $\omega: [a,b] \times [\alpha,\beta] \to R$ be continuous and increasing in the second variable. Then

$$v(t) \leq c + \int_a^t \omega(s,v(s))ds$$

implies that $v \leq u$ in $[a,a+\delta]$ for at least one solution u of $u' = \omega(t,u)$, $u(a) = c$. Hint: Use Tonelli's approximations with a suitable restriction of the domain of ω.

Exercise 3. Prove that if

$$|f_i(x,y)| \leq \omega(t,||y||)$$

with ω nonnegative, continuous, and increasing in the second variable, then, for every solution u of

$$u' = \omega(t,u), \qquad u(a) = u_0,$$

there is a solution $y_1,\dots,y_n$ of

$$y_i' = f_i(x,y), \qquad y_i(a) = y_i^0$$

where $|y_i^0| \leq u_0$, such that $|y_i(x)| \leq u(x)$.

Exercise 4. Generalize the existence theorem established in Chapter I for the Lipschitz case by proving that if

$$|f_i(x,y)| \leq h(x)g(||y||),$$

where g is increasing, then

$$\int_a^b h(s)ds \leq \int_a^b \frac{ds}{g(s)}$$

implies global existence for the system $y_i' = f_i(x,y)$, while for $b = \infty$, the condition

$$\int_a^{+\infty} h(s)ds < \int_a^{+\infty} \frac{ds}{g(s)}$$

implies that all of the solutions of $y_i' = f_i(x,y)$, $(i = 1,\dots,n)$ are bounded in $[a,\infty[$ and have a limit as $x \to \infty$.

Exercise 5. The comparison theorem is valid if the F_i satisfy the following weaker type of monotonicity: for every $(x,y_1,\dots,y_n)$ and $(x,\overline{y}_i,\dots,\overline{y}_n)$, the condition $y_j \leq \overline{y}_j$ for $j \neq i$ and $y_i = \overline{y}_i$ implies

$$F_i(x,y_1,\ldots,y_n) \le F_i(x,\overline{y}_1,\ldots,\overline{y}_n).$$

Prove it by considering the solutions $Y_{n,i}$ of the Cauchy problem

$$Y'_{n,i} = F_i(x,Y_n) + \frac{1}{n}, \qquad Y_{n,i}(a) = y_i^0$$

and by showing that $y_i \le Y_{n,i}$ for every n and i (see the proof of Corollary 2) and by using Ascoli's theorem. It follows that the comparison theorem holds for $n = 1$ without assuming that the F_i are increasing.

Exercise 6. Let $U \subseteq R^n$ be open, $y^0 \in U$, and $f_1,\ldots,f_n$ be defined and continuous on $[a,b] \times U$. Prove that every solution of

$$y_i' = f_i(x,y), \qquad y_i(a) = y_i^0$$

has the property that its graph is contained in that of a solution which cannot be extended. If α is the least upper bound of the domain of definition of a solution that cannot be extended to the right of α, then there are two possible cases; either at least one of the limits

$$\ell_i = \lim_{x\uparrow\alpha} y_i(x)$$

does not exist, or all exist but $(\ell_1,\ldots,\ell_n) \in \partial U$. Prove this result.

Exercise 7. Show that all the initial value problems $y_i' = f_i(x,y)$ have global solutions if

$$||f(x,y)|| \le A||y|| + B.$$

2. THE PEANO PHENOMENON

We shall now study the set of all solutions of a given initial value problem. We shall begin by examining the case of an equation of the first order defined on the real line. We shall refine the method of Cauchy and Lipschitz and show that every solution can be approximated by the new method; the procedure we use leads in a natural way to the consideration of the existence of two special solutions, one greater and one smaller than all the others. The question then arises whether the set of solutions intersects in an interval every line parallel to the y-axis. This does happen, and is called Peano phenomenon. After having generalized the Peano phenomenon to systems, we shall apply the theorems on the existence of maximal solutions to differential inequalities and to the problem of global existence.

2.1. Approximation of all Solutions of a Given Cauchy Problem

To prove the lemma of Sec. 1.2, we used a procedure suited to the approximation of at least one of the solutions. It is natural to wonder if it is possible to find some more refined methods that would permit us to approximate every solution of the given equation. We shall describe one for equations of order 1 on the real line; it is a modification of that of Cauchy and Lipschitz and was inspired by the Riemann sums that are used to approximate integrals.

Consider the initial value problem

$$y' = f(x,y), \qquad y(x_0) = y_0, \tag{2.1}$$

where f is a real-valued continuous function bounded in absolute value by M in the strip $[a,b] \times \mathbb{R}$. We shall limit ourselves to considering solutions of (2.1) only to the right of x_0, in the interval $[x_0,b]$. Set $\delta = b - x_0$, and define R to be the rectangle determined by the conditions

$$a \leq x \leq b, \qquad y_0 - M\delta \leq y \leq y_0 + M\delta.$$

From the integral representation of the solutions, it follows that every integral curve of (2.1) is contained in R.

We denote by $D^{(h)}$ the subdivision of the rectangle R obtained by dividing the interval $[a,b]$ at the points

$$a \equiv x_1^{(h)} < x_2^{(h)} < \dots < x_n^{(h)} \equiv b$$

and the interval $[y_0-M\delta, y_0+M\delta]$ at the points

$$y_0 - M\delta \equiv y_1^{(h)} < y_2^{(h)} < \dots < y_m^{(h)} \equiv y_0 + M\delta$$

and drawing from these points the lines parallel to the axes.

We indicate by $R_{ij}^{(h)}$ the rectangle defined by

$$x_{i-1}^{(h)} \leq x \leq x_i^{(h)}$$

$$y_{j-1}^{(h)} \leq y \leq y_j^{(h)}$$

and suppose that the maximum diagonal δ_h of the rectangles $R_{ij}^{(h)}$ tends to 0 as $h \to \infty$. For every subdivision $D^{(h)}$ of R into rectangles $R_{ij}^{(h)}$, we shall call $R_{ij}^{(h)}$ the *rectangle associated with the point* $P = (x,y)$ of R if

$$x_{i-1}^{(h)} \leq x < x_i^{(h)}, \quad y_{j-1}^{(h)} < y < y_j^{(h)},$$

or, alternatively,

$$R_{ij}^{(h)} \cup R_{ij+1}^{(h)} \quad \text{if} \quad x_{i-1}^{(h)} \leq x < x_i^{(h)}, \quad y = y_j^{(h)}.$$

We shall indicate the rectangle associated with P by $R^{(h)}(P)$. This rectangle is therefore composed of only one of the $R_{ij}^{(h)}$ if P does not belong to a horizontal line of the subdivision $D^{(h)}$; if P does belong to such a horizontal line, the rectangle associated with P consists of two $R_{ij}^{(h)}$ with a horizontal side in common.

We denote by $m^{(h)}(P)$ and $M^{(h)}(P)$ respectively the minimum and maximum of f in the rectangle $R^{(h)}(P)$ associated with the point P and by $\mu^{(h)}(P)$ a number satisfying the inequality

$$m^{(h)}(P) \leq \mu^{(h)}(P) \leq M^{(h)}(P)$$

in such a way that we have $|\mu^{(h)}(P)| \leq M$ in R. Given a subdivision $D^{(h)}$, we trace in the plane the half-ray passing through $P_0 = (x_0, y_0)$ with slope $\mu^{(h)}(P_0)$, and let $P_{1,h}$ be the point common to this half-ray and to the boundary of $R^{(h)}(P_0)$; we draw through $P_{1,h}$ the half-ray with slope $\mu^{(h)}(P_{1,h})$, and let $P_{2,h}$ be the point (distinct from $P_{1,h}$) which is common to this half-ray and to the boundary of $R^{(h)}(P_{1,h})$. We proceed in this way until we have drawn through a certain point $P_{r-1,h}$ the half-ray with slope $\mu^{(h)}(P_{r-1,h})$ that intersects the line $x = b$ at a point $P_{r,h}$. We have thus produced a polygonal line $P_0P_{1,h}\ldots P_{r,h}$; let $y = y_h(x)$ be its equation. For every value of h, we introduce an auxiliary function $f_h(x,y)$ obtained from $f(x,y)$ by changing its definition only at the points of the polygonal line $y = y_h(x)$; we define it there, at each point different from the vertices, to be equal to the value of the slope of the side to which the point belongs; at the vertices, we assign one of the two possible values.

It is clear that

$$y_h(x) = y_0 + \int_{x_0}^{x} f_h(t, y_h(t))dt. \tag{2.2}$$

By the theorem on uniform continuity, if $\varepsilon > 0$, we may determine $\bar{h}$ in such a way that, for $h > \bar{h}$,

$$|f_h(x,y) - f(x,y)| < \varepsilon,$$

that is

$$\lim_{h\to\infty} f_h(x,y) = f(x,y)$$

uniformly in R.

Since the polygonal lines $y = y_h(x)$ pass through P_0 and have equibounded slopes, we deduce that the functions $y_h(x)$ are equibounded and equicontinuous, so that is is possible, by virtue of the theorem of Arzelà and Ascoli, to extract from $(y_h)_h$ a subsequence uniformly convergent in $[x_0,b]$ that we, for simplicity, still denote by $(y_h)_h$. Let $y = \lim_{h\to\infty} y_h$.

We now deduce that $\lim_{h\to\infty} f_h(x,y_h(x)) = f(x,y(x))$ uniformly in $[x_0,b]$; this and (2.2) imply that

$$y(x) = y_0 + \int_{x_0}^{x} f(x,y(x))dx.$$

We have thus shown that the constructive procedure that we have adopted permits us to find at least one integral of (2.1). We now propose to show that *if we fix a sequence* $D^{(h)}$ *of subdivisions, then we can, by varying the values of* $\mu^{(h)}(P)$, *approximate any solution of* (2.1) *by the procedure just described.* To see this, given the subdivision $D^{(h)}$, and given an integral γ of (2.1) passing through P_0 with equation $y = \gamma(x)$, let $Q_{1,h}$ be the first point (with x coordinate greater than x_0) common to both γ and the boundary of $R^{(h)}(P_0)$; let $Q_{2,h}$ be the first point after $Q_{1,h}$ common to both γ and the boundary of $R^{(h)}(Q_{1,h})$ and so on. Continue this process until we reach the right endpoint of γ, which we denote by $Q_{i,h}$. If now, $y = \gamma_h(x)$ is the equation of the polygonal line $P_0Q_{1,h}\ldots Q_{i,h}$, the mean value theorem implies that the arcs of γ with end points P_0 and $Q_{1,h}$, $Q_{1,h}$ and $Q_{2,h},\ldots,Q_{i-1,h}$ and $Q_{i,h}$, have the same slopes as the tangent lines at points interior to the portions of the curve γ with endpoints P_0 and $Q_{1,h}$, $Q_{1,h}$ and $Q_{2,h},\ldots,Q_{i-1,h}$ and $Q_{i,h}$ respectively, and so equal values assumed by f in the respective rectangles associated with the first endpoints. Since it is clear that

$$\lim_{h\to\infty} \gamma_h(x) = \gamma(x)$$

the proof of the claim is complete.

It will be useful in the sequel to note now that we have also proved the following

<u>Proposition</u>. If a sequence $D^{(h)}$ of subdivisions is fixed, each integral of (2.1) can be approximated by polygonal lines such that the slope of each side is a value assumed by f on the rectangle $R_{ij}^{(h)}$ of $D^{(h)}$ to which that side belongs.

We observe that the procedure of Cauchy and Lipschitz that was described in Sec. 1.2 does not allow us to approximate every solution, as we can verify by applying it to the integrals of $y' = \sqrt{|y|}$ originating from $(0,0)$.

2.2. Maximal and Minimal Solutions. The Peano Phenomenon

If we consider particular cases of the approximating procedure described in the preceding section, we obtain special properties of the set of solutions. For example, if we take for the slope of the sides of the polygonal lines the minimum or the maximum values of f in each rectangle, we obtain solutions smaller or greater respectively than all the others. We prove this in the following

<u>Theorem on the Existence of Maximal and Minimal Solutions</u>. There are two special integrals $y = G(x)$ and $y = g(x)$ of (2.1) such that if $y = y(x)$ is an arbitrary solution of (2.1), then

$$g(x) \le y(x) \le G(x).$$

The two integrals $y = g(x)$ and $y = G(x)$ are called the *inferior and superior integrals* respectively originating from the point P_0; they are also known as the *minimal and maximal solutions* of the given initial value problem.

Let us prove, as an example, the existence of the superior integral. To do this, having chosen a subdivision $D^{(h)}$, we construct, according to the described procedure, the polygonal line with equation $y = G_h(x)$ by taking

$$\mu^{(h)}(P) = M^{(h)}(P),$$

that is, by drawing the half-rays that make up the polygonal lines with slopes equal to the maximum permitted. Such a polygonal line has the property that the slope of each of its sides is greater or equal to the maximum value assumed by f in the rectangle $R_{ij}^{(h)}$ to which that side belongs.

On the other hand, every integral y of (2.1) can be approximated by polygonal lines with equation $y = y_h(x)$ that satisfy the conditions stated in the preceding section. From this, we easily deduce that

$$G_h(x) \geq y_h(x). \tag{2.3}$$

As we saw in the preceding section, there is a sequence $(G_{h_k})_{k=1}^{\infty}$ extracted from the sequence $(G_h)_{h=1}^{\infty}$ which converges uniformly to a continuous function G. This is a solution of (2.1); moreover, we have by (2.3) that

$$G(x) \geq y(x). \tag{2.4}$$

If we are able to extract from $(G_h)_{h=1}^{\infty}$ another sequence converging to another function $\overline{G}$, this too would be a solution of (2.1), and we would have, by (2.4),

$$G(x) \geq \overline{G}(x).$$

But from (2.3) we know that

$$\overline{G}(x) \geq G(x),$$

and we conclude that the sequence $(G_h)_{h=1}^{\infty}$ converges to a function $G(x)$ that satisfies (2.4) for every integral $y = y(x)$ originating from the point P_0.

As for the inferior integral, we proceed in the same way, by choosing

$$\mu^{(h)}(P) = m^{(h)}(P).$$

The following proposition, which follows immediately from the one just proved, states another property of the set of integrals of a differential equation that pass through a given point.

Theorem (Peano Phenomenon). If $y = g(x)$ and $y = G(x)$ are the inferior and superior integrals of (2.1), then through every point of the region $R(P_0)$ given by

$$x_0 \leq x \leq x_0 + \delta$$
$$g(x) \leq y \leq G(x)$$

there passes an integral of (2.1) originating from P_0.

Proof: Let Q be a point of the region $R(P_0)$. The integral going to the left from Q can be extended in such a way that if it does not arrive at P_0, it will meet either the inferior or the superior integral at some point and coincide with it after that point. In either case, we conclude that there is a solution originating from P_0 passing through Q.

In other words, we may say that if the set of integrals of a first order differential equation does not consist of a unique curve, then it constitutes a region like $R(P_0)$, so that the intersections of this region with lines parallel to the y axis are intervals (perhaps reduced to a point). We express this result by saying that the differential equations exhibit the *Peano phenomenon* and that the set of integral curves is *brush of Peano*.

We now make use of this procedure and prove the following comparison theorem; in the particular case of a single differential equation it is a generalization of the theorem proved in Sec. 2.3.

Theorem (Comparison of Extremal Solutions). Let f_1 and f_2 be functions continuous and bounded in S, and suppose that

$$f_1 \geq f_2.$$

If G_1 and G_2 are the superior integrals and g_1 and g_2 the inferior integrals relative to $P_0 = (x_0, y_0)$ of the differential equations

$$y' = f_1(x,y), \qquad y' = f_2(x,y),$$

then we have

$$G_1 \geq G_2, \qquad g_1 \geq g_2.$$

The proof is immediate. It is enough to observe that from $f_1 \geq f_2$ we have

$$G_h^{(1)} \geq G_h^{(2)},$$

where we have denoted by $y = G_h^{(i)}(x)$ the polygonal line that approximates G_i in the subdivision $D^{(h)}$ according to the proof of the theorem on the existence of maximal and minimal solutions. We take the limit as $h \to \infty$ and obtain

$$G_1 \geq G_2.$$

Similarly, one shows $g_1 > g_2$.

We shall end this section by observing that if the sequence of subdivisions $D^{(h)}$ is obtained by adding new points of subdivision to those we already have at each state, and if a uniqueness theorem holds, we have the following formula:

$$|y(x) - y_h(x)| \leq G_h(x) - g_h(x).$$

This gives a bound on the error committed by substituting one of the approximations for the integral $y = y(x)$. We leave the proof to the reader.

Exercise 1. Use the following procedure to prove the existence of the maximal solution. (A similar one works for the minimal solution.) Prove that *every* solution of (2.1) is less than or equal to every solution of

$$y_n' = f(x,y_n) + \frac{1}{n}, \quad y_n(a) = y^0.$$

Conclude by applying the theorem of Ascoli to $(y_n)_{n=1}^{\infty}$.

Exercise 2. Use the following procedure to prove the existence of the maximal solution. (A similar one works for the minimal solution.) Prove that the set H of solutions has a countable dense subset $\{u_n : n \geq 1\}$. Define the sequence $(v_n)_{n=1}^{\infty}$ by induction:

$$v_1 = u_1, \quad v_{n+1} = \max\{u_n, v_n\},$$

Then prove that each $v_n \in H$ and that $(v_n)_{n=1}^{\infty}$ converges to the maximal solution.

Exercise 3. Let $f_1,\ldots,f_n$ be functions continuous and bounded in the strip $[a,b] \times \mathbb{R}^n$ and satisfying there the following monotonicity condition. If $y_i \leq \bar{y}_j$ for $j \neq i$, and $y_i = \bar{y}_i$ then $f_i(x,y_1,\ldots,y_n) \leq f_i(x,\bar{y}_1,\ldots,\bar{y}_n)$. Prove that the initial value problem

$$y_i' = f_i(x,y), \quad y_i(a) = y_i^0$$

has a maximal solution $\bar{\bar{y}}_i$ and a minimal solution $\underline{\bar{y}}_i$ in the sense that if y_i is any other solution, then $\bar{y}_i \leq y_i \leq \bar{\bar{y}}_i$ for each i.

2.3. The Peano Phenomenon for Systems

In this section, we study the Peano phenomenon for systems whose second members are bounded continuous functions. *We shall prove that the set obtained by intersecting the integral curves originating at one given point with a hyperplane* $x = c$ *is a continuum,* that is, a nonempty, compact, connected set (which may, however, consist of one point only). In the case of a single equation, this result reduces to that proved in the preceding section, since the only continua on the line are the closed and bounded intervals. In general, the intersection with the hyperplane $x = c$ is not a simply connected set as is the case for a single equation. We shall see that this is so in Exercise 2.

We shall deduce the Peano phenomenon from the following

<u>Theorem</u>. Let A be a closed subset of a Banach space X, T a continuous proper function from A into a Banach space Y, and $(T_n)_{n=1}^{\infty}$ a sequence of injective functions from A into Y such that T_n^{-1}: $T_n(A) \to A$ is continuous for every n. If $\lim_{n\to\infty} T_n = T$ uniformly, and if y is an interior point of $\bigcap_{n=1}^{\infty} T_n(A)$, then $T^{-1}(y)$ is a continuum.

As we said above, a *continuum* is a compact, connected, nonempty set. A function T is called *proper* if the inverse image of a compact set is compact, in other words, $T^{-1}(K)$ is compact if K is. Let T be a mapping between metric spaces (as in our case). To say that T is proper is equivalent to saying that if $(T(x_n))_{n=1}^{\infty}$ converges, then the set $\{x_n: n \geq 1\}$ is relatively compact.

We give a simple example before proceeding to the proof of the theorem. Let f be a function defined on $[a,b]$ which can be uniformly approximated there by a sequence $(f_n)_{n=1}^{\infty}$ of continuous, invertible functions. Since each f_n is strictly monotone it follows that f is monotone on the interval $[a,b]$. The transformation $y = f(x)$ shows the Peano phenomenon at every point y in the interval whose endpoints are the minimum and maximum values of f on $[a,b]$. ($f^{-1}(y)$ is an interval.)

We now pass on to the proof of the theorem. Let y be as in the hypothesis; $T^{-1}(y)$ is compact because T is proper.

To complete the proof, let us suppose that $T^{-1}(y)$ can be decomposed into the disjoint union of two closed sets X_1 and X_2, and let x_1 and x_2 be points of X_1 and X_2 respectively. We then set

$$y_n^{(1)} = T_n(x_1); \quad y_n^{(2)} = T_n(x_2).$$

By the hypothesis, there is a neighborhood U of y contained in each of the sets $T_n(A)$, and moreover

$$\lim_{n\to\infty} y_n^{(1)} = \lim_{n\to\infty} y_n^{(2)} = y.$$

It follows that it is possible to determine an index ν such that for $n > \nu$, both points $y_n^{(1)}$ and $y_n^{(2)}$ are contained in U, as well as the points $y_n^{(\lambda)}$ on the line segment S_n joining these two points:

$$S_n\colon y_n^{(\lambda)} = (1 - \lambda)y_n^{(1)} + \lambda y_n^{(2)}.$$

We therefore have

$$\lim_{n\to\infty} y_n^{(\lambda)} = y$$

uniformly as λ varies in $[0,1]$.

We denote by Γ_n the curve connecting x_1 with x_2 and formed by al the points of A that are mapped by T_n into points of S_n. Consider the set H of accumulation points of the curves Γ_n, that is, the set of points of A with the property that every neighborhood of one of them contains points of infinitely many curves Γ_n.

H contains x_1 and x_2 and is clearly a closed set. We can furthermore easily see that H is contained in $T^{-1}(y)$, for if x_0 is a point of H, and if $(x_{n_k})_{k=1}^{\infty}$ is a sequence of points converging to x_0 with x_{n_k} on Γ_{n_k}, then the point $T_{n_k}(x_{n_k})$ belongs to the segment S_{n_k}, so

$$\lim_{k\to\infty} T_{n_k}(x_{n_k}) = y.$$

On the other hand, it follows from $||T_{n_k}(x_{n_k})-T(x_0)|| \leq ||T_{n_k}(x_{n_k})-T(x_{n_k})|| + ||T(x_{n_k})-T(x_0)||$ that

$$\lim_{k\to\infty} T_{n_k}(x_{n_k}) = T(x_0).$$

whence $T(x_0) = y$, so x_0 belongs to $T^{-1}(y)$.

As a consequence of this and of our assumption about the set $T^{-1}(y)$, it is possible to decompose H into the sum of two closed disjoint sets H_1 and H_2 containing x_1 and x_2 respectively. The sets H_1 and H_2 are compact because they are closed subsets of the compact set $T^{-1}(y)$.

We now denote by H_ε the set formed by the points of A whose distance from H is less than ε, and we prove that the curves Γ_n are, from a certain n on, contained in H_ε. Let us suppose that infinitely many curves have points outside of H_ε, and let us form a sequence $(x_{n_k})_{k=1}^{\infty}$ of such points. Since the point $T_{n_k}(x_{n_k})$ belongs to the segment S_{n_k}, the sequence $(T_{n_k}(x_{n_k}))_{k=1}^{\infty}$ converges to y, while on account of the uniform convergence the sequence $(T(x_{n_k}) - T_{n_k}(x_{n_k}))_{k=1}^{\infty}$ converges to the origin of the space Y. Since

$$T(x_{n_k}) = T_{n_k}(x_{n_k}) + \left[T(x_{n_k}) - T_{n_k}(x_{n_k})\right],$$

it follows that the sequence $(T(x_{n_k}))_{k=1}^{\infty}$ converges, and so, since T is proper, we may extract a convergent subsequence from $(x_{n_k})_{k=1}^{\infty}$. Although the limit of this sequence is not in H_ε, it must belong to H, which is clearly absurd. The curves Γ_n are therefore contained in H from a certain n on.

We now observe that the distance between the two disjoint compact sets H_1 and H_2 is certainly positive and that, therefore, for ε sufficiently small, even H_ε is composed of two parts with positive distance between them; the same therefore is true for the curves Γ_n in H.

The hypothesis that $T^{-1}(y)$ can be decomposed into the disjoint union of two closed sets X_1 and X_2 has the consequence that, for n sufficiently large, the curves Γ_n are broken into two parts with positive distance between the parts. Since the image of Γ_n under T_n is the segment S_n, this contradicts the hypothesis that the T_n^{-1} are continuous. This completes the proof of the theorem.

We now turn to the case of the initial value problem for a system. We shall use the previous theorem to prove the following.

Theorem on the Peano Phenomenon. Let $f_1,\ldots,f_n$ be bounded and continuous in the cylinder $[a,b] \times \mathbf{R}^n$. The set S of solutions of

$$y_i' = f_i(x,y_1,\ldots,y_n), \qquad y_i(a) = y_i^0$$

is a continuum in the Banach space $C([a,b],\mathbf{R}^n)$ of continuous functions from $[a,b]$ to $\mathbf{R}^n$ with the sup norm.

This theorem implies the Peano phenomenon as stated in the beginning of the section: the function that maps each $f \in C([a,b],\mathbf{R}^n)$ to the value $f(x)$ of f at x is a continuous function and therefore maps a continuum into a continuum.

Proof of the Theorem: Set $X = C([a,b],\mathbf{R}^n)$. define ϕ by $\phi(x) = y^0$, and let $\delta = b - a$. We define a transformation $F\colon X \to X$ by associating with each y in X the continuous function $F(y)$ whose value at x is given, coordinatewise, by

$$F_i(y)(x) = \int_a^x f_i(t,y(t))\,dt.$$

It is clear that $y \in S$ if and only if $y = \phi + F(y)$, that is, if and only if $y - F(y) = \phi$. Then, if $T = I - F$, where I is the identity mapping on X, we have $S = T^{-1}(\phi)$. We must therefore prove that $T^{-1}(\phi)$ is a continuum. In order to apply the preceding theorem, we must first verify that T is continuous and proper and that there is a sequence $(T_m)_{m=1}^{\infty}$ of transformations satisfying the hypotheses of the theorem. To show that T is proper, we suppose that $\lim_{m\to\infty} T(y_m) = y$ and verify that $\{y_m : m \geq 1\}$ is relatively compact. Let $(y_{m_k})_{k=1}^{\infty}$ be any subsequence; we shall prove the existence of a convergent subsequence $(y_{m_{k_i}})_{i=1}^{\infty}$. From the definition of T it follows that

$$y_{m_k} = T(y_{m_k}) + F(y_{m_k}). \tag{2.5}$$

Since the f_i are bounded, we see, by means of a technique which we have already used many times in this chapter, that $(F(y_{m_k}))_{k=1}^{\infty}$ is equicontinuous and equibounded. By Ascoli's theorem we obtain a convergent subsequence $F(y_{m_{k_i}}))_{i=1}^{\infty}$. From (2.5), we see that $(y_{m_{k_i}})_{i=1}^{\infty}$ converges, so T is proper. The continuity of T can be proved immediately by means of the theorems on passing to the limit under the integral sign. It remains only to define the T_m. To do this, we define, in a way similar to F, transformations $F_m : X \to X$ by setting

$$F_{m,i}(y)(x) = \begin{cases} 0 & \text{for } a \leq x \leq a + \dfrac{\delta}{m} \\ \displaystyle\int_a^{x-\frac{\delta}{m}} f_i(t,y(t))\,dt & \text{for } a + \dfrac{\delta}{m} \leq x \leq b. \end{cases}$$

We set $T_m = I - F_m$. By working successively in the intervals $[a, a + \frac{\delta}{m}]$, $[a + \frac{\delta}{m}, a + \frac{2\delta}{m}], \ldots, [b - \frac{\delta}{m}, b]$ as in the proof of the comparison theorem of Sec. 1.3, we see that the T_m are bijections from X to X. The condition that ϕ be an interior point of $\bigcap_{n=1}^{\infty} T_m(X) = X$ is thus verified. The continuity of $T_m^{-1} : X \to X$ can be demonstrated in a way similar to our proof of the fact that T is proper. We are therefore able to apply the preceding theorem with $A = X$, $y = \phi$, and T and T_m as defined above. The theorem is thus completely proved.

Exercise 1. Show that the equation $y' = \sqrt{1 - y^2}$ does not have a unique solution at $(x,-1)$ and $(x,1)$ and so has the brush of Peano to the right of $(x,-1)$ and to the left of $(x,1)$. Study $y' = x\sqrt{1 - y^2}$ in the same way.

Exercise 2. This exercise will serve to show that when a system exhibits the Peano phenomenon, the intersection of the solutions with the hyperplane $x = \overline{x}$ may not be simply connected. Consider the problem

$$\begin{cases} y_1' = y_1 - \dfrac{y_2\sqrt{|y_2|}}{\sqrt{y_1^2 + y_2^2}} & y_1(0) = 1 \\ y_2' = y_2 - \dfrac{y_1\sqrt{|y_2|}}{\sqrt{y_1^2 + y_2^2}} & y_2(0) = 0. \end{cases}$$

We proceed as follows: If we change to polar coordinates $y_1 = \rho \cos \theta$, $y_2 = \rho \sin \theta$, we have

$$2(y_1y_1' + y_2y_2') = (\rho^2)' = 2\rho^2,$$

so that $\rho = \rho_0 e^t$. Also,

$$\theta' = \rho^{3/2}\sqrt{|\sin \theta|} = \rho_0^{3/2} e^{3t/2}\sqrt{|\sin \theta|}$$

This last equation exhibits Peano's phenomenon at all the values $\theta_0 = k\pi$. Therefore, at the time T_0 when the maximal solution achieves the value 2π, we have a section of Peano's brush that is not simply connected. (See Figure 3.1.) Prove all the claims above.

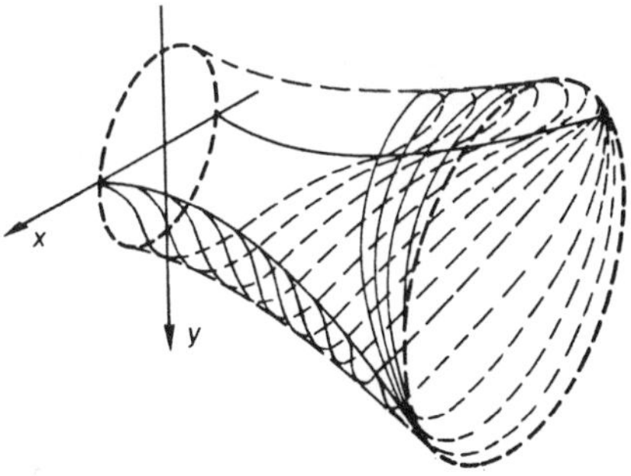

Figure 3.1

Exercise 3. Let $f_1,\ldots,f_n$ be continuous functions on the rectangle R defined by $a \le x \le b$, $a_i \le y_i \le b_i$. Prove that for every interior point y^0 of R, there is $0 < \delta \le b-a$ such that all the solutions of

$$y_i' = f_i(x,y), \quad y_i(a) = y_i^0$$

exist in $[a,a+\delta]$ and form a continuum in $C([a,a+\delta],\mathbb{R}^n)$.

2.4. Maximal Solutions, Differential Inequalities, and Global Existence

In this section, we shall use the existence of maximal solutions to compare the solutions of a given system with those of a single equation on the real line. To obtain this result, we shall have to use the following

Theorem on Differential Inequalities. Let $I \subseteq \mathbb{R}$ be an arbitrary interval, and let $\omega\colon [a,b] \times I \to \mathbb{R}$ be continuous. If $v\colon [a,b] \to I$ is a C^1-function such that

$$v' \le \omega(x,v), \quad v(a) \le u_0$$

and if u is the maximal solution of

$$u' = \omega(x,u), \quad u(a) = u_0,$$

then $v \le u$ on $[a,b]$.

Proof: We suppose that the statement is not true, for example, that $v(x_0) > u(x_0)$, and arrive at a contradiction. Let

$$\alpha = \min\{a \le t \le x_0 \mid v > u \text{ in } [t,x_0]\}.$$

By continuity we have $u(\alpha) = v(\alpha)$. We define

$$\bar{\omega}(x,y) = \begin{cases} \omega(x,y) & \text{for } y \ge v(x) \\ \omega(x,v(x)) & \text{for } y \le v(x) \end{cases}$$

for all x and y for which this makes sense. Since $\bar{\omega}$ is continuous, Peano's local existence theorem guarantees the existence of a solution y of

$$y' = \bar{\omega}(x,y), \quad y(\alpha) = v(\alpha)$$

in an interval $[\alpha,\alpha+\delta] \subseteq [\alpha,x_0]$. We want to show that $y \ge v$. If this were false, $\alpha \le x_1 < x_2 \le \alpha + \delta$ would exist such that $y(x_1) = v(x_1)$ and $y < v$ in $]x_1,x_2]$, and we would have the following absurd result.

$$0 < v(x_2) - y(x_2) = \int_{x_1}^{x_2} v'(t)dt - \int_{x_1}^{x_2} y'(t)dt$$

$$\le \int_{x_1}^{x_2} \omega(t,v(t))dt - \int_{x_1}^{x_2} \bar{\omega}(t,y(t))dt$$

$$= \int_{x_1}^{x_2} \omega(t,v(t))dt - \int_{x_1}^{x_2} \omega(t,v(t))dt \qquad \text{(since } y \le v \text{ in } [x_1,x_2])$$
$$= 0.$$

Having thus shown that $y \ge v$, we now consider the function

$$w(t) = \begin{cases} u(t) & \text{for } a \le t \le \alpha \\ y(t) & \text{for } \alpha \le t \le \alpha + \delta \end{cases}.$$

We see that w is a solution of

$$w' = \omega(x,w), \quad w(a) = u_0$$

and therefore $w \le u$ in $[a,\alpha+\delta]$, since u is the maximal solution. But then we have $u(x_2) < v(x_2) \le y(x_2) = w(x_2) \le u(x_2)$, which is absurd. This completes the proof.

Corollary. If ω is Lipschitzian and if

(i) $v' \le \omega(x,v)$ and $u' = \omega(x,u)$ in $[a,b]$

(ii) $v(a) \le u(a)$

then for any $x_0 \in]a,b]$ either $v(x_0) < u(x_0)$ or $v = u$ in $[a,x_0]$.

Proof: By the theorem on differential inequalities, the function $w = u - v$ is nonnegative in $[a,b]$. Assume that v and u are not identically equal in $[a,x_0]$ with $a < x_0 \le b$. Then there is $x_1 \in [a,x_0]$ such that $v(x_1) < u(x_1)$. We have

$$w'(x) \ge \omega(x,u(x)) - \omega(x,v(x)) \ge -Lw(x)$$

where L is the Lipschitz constant of ω. It follows that

$$\frac{d}{dx} e^{Lx}w(x) = e^{Lx}(w'(x) + Lw(x)) \ge 0$$

and so the function $e^{Lx}w(x)$ is increasing. Consequently, we have

$$w(x_0) \ge w(x_1)e^{-L(x_0-x_1)} > 0$$

which gives $u(x_0) > v(x_0)$.

We now apply the above theorem to get a global existence theorem which allows us to study a system by comparing it with a single equation on the real line. We shall provide some applications of it in the exercise.

Global Existence Theorem. Let $f_1,\ldots,f_n$ be continuous functions in the cylinder $[a,b] \times \mathbb{R}^n$, and let ω be a continuous function on $[a,b] \times \mathbb{R}$ such that

$$\sum_{i=1}^{n} f_i(x,y_1,\ldots,y_n)\cdot y_i \leq \omega(x,||y||)\,||y||$$

for every $a \leq x \leq b$, $y \in R^n$. If the initial value problem

$$u' = \omega(x,u), \quad u(a) = u_0$$

has a maximal solution u in $[a,b]$, then the solutions of

$$y_i' = f_i(x,y), \quad y_i(a) = y_i^0$$

exist in $[a,b]$ for $||y^0|| \leq u_0$ and satisfy

$$||y(x)|| \leq u(x)$$

there.

If we use the notation customary in linear algebra to indicate the scalar product,

$$(x|y) = \sum_{i=1}^{n} x_i y_i,$$

the relation between $f = (f_1,\ldots,f_n)$ and ω can be rewritten in the following way:

$$(f(x,y)|y) \leq \omega(x,||y||)||y||.$$

The proof of the global existence theorem is a direct consequence of the theorem on a priori bounds of Sec. 1.3 and of the preceding result on differential inequalities. In fact, it is easily verified that u^2 is the maximal solution of

$$w' = 2\omega(x,\sqrt{w})\sqrt{w}, \quad w(a) = u_0^2.$$

If $y_1,\ldots,y_n$ is a solution of the given initial value problem, then the function $v(t) = ||y(t)||^2$ satisfies

$$v' \leq 2\omega(x,\sqrt{v})\sqrt{v}, \quad v(a) \leq u_0^2,$$

as is easily seen upon observing that $||y(t)||^2 = y_1^2(t) + \ldots + y_n^2(t)$. Thus $v \leq u^2$ by virtue of the theorem on differential inequalities. We have thereby proved that the inequality

$$||y(x)|| \leq u(x)$$

holds whenever the $y_1,\ldots,y_n$ exist. We may now invoke the theorem on a priori bounds to complete the proof.

Exercise 1. Prove the theorem on differential inequalities by comparing v with the solutions of

$$y_n' = \omega(x,y_n) + \frac{1}{n}, \quad y_n(a) = u_0$$

Hint: Use the incremental ratio.

Exercise 2. Assume $\omega(x,\cdot)$ is increasing. Prove the theorem on differential inequalities by using the comparison theorem between the extremal solutions of Sec. 1.2 with $f_1 = \bar{\omega}$ and $f_2(x,y) = v'(x)$, where $\bar{\omega}$ is the function introduced in the proof of the theorem in the text.

Exercise 3. Prove global existence under the hypothesis

$$||f(x,y)|| \leq \omega(x,||y||),$$

where ω is as in the theorem on global existence.

Exercise 4. If f is real and decreasing in y, then every initial value problem

$$y' = f(x,y), \quad y(a) = y^0$$

has a global solution to the right of a.

Exercise 5. What conditions on $\int_a^b h(t)dt$, $\int_a^b \frac{dt}{g(t)}$ insure that $(f(x,y)|y) \leq h(x)g(||y||)||y||$ implies global existence for all the initial value problems of $y_i' = f_i(x,y)$?

3. QUESTIONS OF UNIQUENESS

In the introduction to the chapter, we saw an example of an initial value problem which has infinitely many different solutions. Actually, we may construct differential equations for which every initial value problem has infinitely many solutions; see Lavrentieff [34] and Hartman [30]. We may therefore ask under what conditions a given initial value problem will have a unique solution. Our interest in this derives from the facts that uniqueness implies continuous dependence on the data, as we shall see in Sec. 3.1, and that uniqueness is always of interest in physical applications. In this section, we shall discuss uniqueness theorems, continuous dependence, and the "size" of the set of ordinary differential equations that have a unique solution for a given initial value problem.

3.1. Continuous Dependence

In order to show the importance of uniqueness in initial value problems, we shall begin by examining one of its applications; uniqueness implies continuous dependence. The following theorem is in fact true.

Theorem (Continuous Dependence). Let $f_{m1},\dots,f_{mn}$ be functions that are continuous and bounded in the cylinder $[a,b] \times R^n$, let $y^m \in R^n$, and let $x_m \in [a,b]$. If $\lim_{m\to\infty} f_m = f_0$ uniformly on compact sets, if $\lim_{m\to\infty} y^m = y^0$ and $\lim_{m\to\infty} x_m = x_0$, then the solutions of

$$y'_{mi} = f_{mi}(x,y), \quad y_{mi}(x_m) = y_i^m$$

converge uniformly to the solution of

$$y'_{0i} = f_{0i}(x,y), \quad y_{0i}(x_0) = y_i^0$$

provided that this last initial value problem has unique solution.

This result shows that continuous dependence does not depend on any particular regularity of the second members (as, for example, satisfying a Lipschitz condition as in Chapter I) but exclusively on uniqueness.

The proof of this theorem is simply accomplished by using Ascoli's theorem to show that every subsequence $(y_{n_k})_{k=1}^{\infty}$ has in turn a subsequence that converges uniformly to y. Since we have already used this type of argument quite frequently in this chapter, we leave the details to the reader.

In the case of a single equation, $n = 1$, one can deduce the theorem from a more general result concerning certain sets of curves contained in a rectangle.

This proof makes it very clear that continuous dependence on data relies on uniqueness only. One sees that the assumption of boundedness for the f_m's is not needed. Denote the rectangle $a \leq x \leq b$, $\alpha \leq y \leq \beta$ by R, and let Γ be a family of curves γ contained in R with $\gamma \equiv (x,\gamma(x))$ a continuous function defined in an interval contained in $[a,b]$. We will assume that every curve γ has its extrema on the boundary of R, while all the remaining points of the curves are in the interior of R. We shall now prove the following theorem.

Theorem. Let L be a mapping that associates with every point (x,y) of R one and only one curve γ of the family Γ that passes through it. Let L map every point of γ to itself. If γ_0 is the curve corresponding to the point (x_0,y_0) and $\bar{\gamma}$ the one corresponding to the point $(x_0,\bar{y})$, then, for every $\varepsilon \geq 0$, it is possible to determine $\delta > 0$ such that for $|y_0 - \bar{y}| < \delta$ the domain of $\bar{\gamma}$ contains the domain of γ_0 and

$$|\gamma_0(x) - \bar{\gamma}(x)| < \varepsilon$$

in the domain of γ_0.

Note that it follows from the hypothesis of the theorem that the curves corresponding to two distinct points of R either coincide or are disjoint.

To prove the theorem, fix the point (x_0,y_0) and the curve γ_0 in Γ passing through it. Let $[c,d] \subseteq [a,b]$ be the domain of γ_0. We assume $y_0 < \beta$ and prove the conclusion of the theorem only for $x \geq x_0$. The remaining cases can be handled analogously. Fix $\varepsilon > 0$ such that $y_0 + \varepsilon \leq \beta$. For each $\delta \in]0,\varepsilon]$, let γ_δ be the curve in Γ that passes through the point $(x_0,y_0+\delta)$. Since the curves that correspond to different values of δ cannot meet, it follows that

$$\delta' < \delta'' \Rightarrow \gamma_{\delta'} < \gamma_{\delta''} \tag{3.1}$$

in the intersection of the domains of $\gamma_{\delta'}$ and $\gamma_{\delta''}$. We claim that there exists $\delta_0 \in]0,\varepsilon]$ such that, for every $\delta \in]0,\delta_0]$, γ_δ is defined in $[x_0,d]$ and

$$\gamma_0(x) \leq \gamma_\delta(x) < \gamma_0(x) + \varepsilon \qquad (x_0 \leq x \leq d). \tag{3.2}$$

If the claim is false, then sequences $\delta_n \downarrow 0$, $t_n \in [x_0,d]$ exist such that $\gamma_{\delta_n}(t_n) = \gamma_0(t_n) + \varepsilon$. Define

$$x_n = \min\{t \in [x_0,d]: \gamma_{\delta_n}(t) = \gamma_0(t) + \varepsilon\}.$$

By (3.1) the sequence $(x_n)_{n=1}^{\infty}$ is increasing. Let $x_\infty = \lim_n x_n$ and let γ_∞ be the curve in Γ that passes through the point $(x_\infty,\gamma_0(x_\infty) + \frac{\varepsilon}{2})$ which is in R. For n sufficiently large, x_n is in the domain of γ_∞. By

$$\lim_n \gamma_\infty(x_n) = \gamma_\infty(x_\infty) < \gamma_0(x_\infty) + \varepsilon = \lim_n \gamma_{\delta_n}(x_n),$$

we have

$$\gamma_\infty(x_n) < \gamma_{\delta_n}(x_n)$$

for n large enough. Since different curves in Γ cannot meet, we have

$$\gamma_0 < \gamma_\infty < \gamma_{\delta_n}$$

in the common interval of definitions for n large. From this and the fact that the curves in Γ have their extremities on the boundary of R and the remaining points in the interior of R, we deduce that γ_∞ is defined at x_0 and that $\gamma_\infty(x_0) = y_0$. But then we have the contradiction that γ_0 and γ_∞ are both passing through (x_0,y_0). This states (3.2).

The reverse inequality

$$\gamma_0(x) - \varepsilon < \gamma_\delta(x) \le \gamma_0(x) \qquad (x_0 \le x \le d)$$

can be proved in a similar way if $y_0 > \alpha$ (we have nothing to prove if $y_0 = \alpha$ or β). The theorem is thus completely proved.

We now drop the hypothesis of uniqueness; then it is possible that there are infinitely many solutions originating from one initial point P. It is now natural to inquire whether the set of such integrals, or more precisely, the inferior or superior integral originating from P, varies continuously with respect to P.

The answer is no. We observe that this is no contradiction to the previously proved theorem since the family of curves corresponding to the superior (inferior) integrals originating from the points of R does not satisfy the hypotheses of that theorem. To see this, let γ_0 be the superior (inferior) integral originating from $P_0 = (x_0,y_0)$ and let $P_1 = (x_1,y_1)$ be a point of γ_0 distinct from P_0. If $x_1 > x_0$, the superior (inferior) integral originating from P_1 coincides with γ_0 for $x \ge x_1$, but not, in general, for $x < x_1$. If, on the other hand, $x_1 < x_0$, the superior (inferior) integral originating from P_1 coincides with γ_0 for $x \le x_1$ but not, in general, for $x > x_1$. Then the family of curves whose elements are the superior (inferior) integrals originating from points in R does not satisfy the hypothesis that the same curve γ_0 corresponds to each point of a curve γ_0 of the family.

In order to illustrate what happens in this case by a simple example, we study the behavior of the integrals of the equation

$$y' = \sqrt{|y|}, \tag{3.3}$$

which we have considered before. If $y_0 \neq 0$, a unique integral of Eq. (3.3) originates from each point $P_0 = (x_0,y_0)$ of the (x,y) plane since the function on the right-hand side is differentiable there, but if $y_0 = 0$, there are infinitely many solutions, as we saw in the introduction to the chapter. If $y_0 > 0$, the unique integral originating from P_0 is the function $y(x,P_0)$ defined by

$$y(x,P_0) = \begin{cases} 0 & \text{if } x \le x_0 - 2\sqrt{y_0} \\ \frac{1}{4}(x-x_0 + 2\sqrt{y_0})^2 & \text{if } x \ge x_0 - 2\sqrt{y_0}, \end{cases}$$

while, if $y_0 < 0$, the unique integral is

$$y(x,P_0) = \begin{cases} -\frac{1}{4}(x-x_0 - 2\sqrt{-y_0})^2 & \text{if } x \le x_0 + 2\sqrt{-y_0} \\ 0 & \text{if } x \ge x_0 + 2\sqrt{-y_0}. \end{cases}$$

When $y_0 = 0$, the superior integral originating from P_0 is the function $G(x,P_0)$ defined by

$$G(x,P_0) = \begin{cases} 0 & \text{if } x \le x_0 \\ \frac{1}{4}(x-x_0)^2 & \text{if } x \ge x_0, \end{cases}$$

while the inferior integral originating from P_0 is the function $g(x,P_0)$ defined by

$$g(x,P_0) = \begin{cases} -\frac{1}{4}(x-x_0)^2 & \text{if } x \le x_0 \\ 0 & \text{if } x \ge x_0. \end{cases}$$

If we now set $Q_0 = (x_0,0)$, a point on the x-axis, and $P_0 = (x_0,y_0)$, a point not on that axis, then we immediately see that

$$\lim_{y_0 \downarrow 0} y(x,P_0) = G(x,Q_0); \quad \lim_{y_0 \uparrow 0} y(x,P_0) = g(x,Q_0).$$

This means that if the initial point $P_0 = (x_0,y_0)$, not on the x-axis, tends to a point Q_0 on the x-axis, the unique solution of (3.3) originating from P_0 tends to the maximal solution originating from Q_0 if $y_0 > 0$ and to the minimal solution if $y_0 < 0$. Therefore, neither the maximal nor the minimal solutions vary continuously with the initial data.

This does not depend on the particular equation that we chose, but is an instance of a general result which we now set out to establish.

Let us now consider the equation

$$y' = f(x,y), \tag{3.4}$$

where f is defined at all points of the rectangle R given by $a \le x \le b$, $\alpha \le y \le \beta$. (Note that since we have suppressed the hypothesis that f is continuous, the existence theorems that we proved above are no longer valid.)

If $y = \gamma(x)$ is the equation of a continuous curve γ contained in R, and $[c,d]$ is the domain of $\gamma(x)$, we shall henceforth call the region

$$c \le x \le d; \quad \gamma(x) \le y \le \beta, \qquad (c \le x \le d; \quad \alpha \le y \le \gamma(x)).$$

the *superior (inferior) region of the curve* γ.

If a curve whose endpoints are on the boundary of R but whose other points are in the interior of R has the property that it is the graph of a function differentiable on its domain of definition and satisfying (3.4) there, we shall call that curve an *integral curve* of Eq. (3.4); if an integral curve originating from P_0 belongs to the *superior (inferior)* region of every other possible integral curve originating from P_0, we shall say that it is the superior (inferior) integral with initial point

P_0. The superior (inferior) integral is necessarily unique when it exists.

We now have the following theorem; it reduces to the preceding one when we assume uniqueness.

Theorem. Any set of conditions sufficient to insure the existence of the inferior and superior integrals of the differential equation (3.4) is also sufficient to establish the continuous dependence of the superior (inferior) integral on the initial point varying in its superior (inferior) region.

We shall proceed in a manner similar to the one we followed above and deduce this theorem from a more general theorem on the properties of a family of continuous curves that satisfy suitable conditions. Precisely, we shall denote by Γ a family of continuous curves with Cartesian equations, each with its endpoints on the boundary of R and all other points in the interior of R. Let there be a mapping C that assigns to each point P_0 of R, the so-called *initial point*, not on a horizontal side of R, a unique curve of Γ through that point; we denote by Γ_1 the set of curves of Γ corresponding to the points of R under C. We shall assume the following properties for the curves in Γ_1:

a) Each curve belongs to only one of the two regions into which any of the other curves partitions R.

b) No curve may have in common with any other curve belonging to its inferior region points to the left of its own initial point or to the right of the initial point of the other. (We include the initial point among those points of a curve that are to the right (left) of its own initial point.)

Let $P_0 = (x_0,y_0)$ be a fixed point in R, let $y = \gamma(x,x_0,y_0)$ be the equation of the curve in Γ_1 with initial point P_0, and let $[c,d]$ be the interval of definition of the function $\gamma(\cdot,x_0,y_0)$. We denote by $y = \gamma(x,x_1,y_1)$ the equation of the curve in Γ_1 whose initial point is a generic point $P_1 = (x_1,y_1)$ of R. Then we prove the following results.

(i) Given $\varepsilon > 0$, we can determine a neighborhood J of P_0 contained in the right superior region of γ with respect to P_0 (i.e., the domain common to the superior region of γ_0 and the half plane $x \geq x_0$) and a neighborhood $\overline{J}$ of P_0 belonging to the left inferior region of γ_0 with respect to P_0 in such a way that if $P_1 = (x_1,y_1)$ and $\overline{P}_1 = (\overline{x}_1,\overline{y}_1)$ belong respectively to J and $\overline{J}$, the corresponding functions $\gamma(\cdot,x_1,y_1)$ and $\gamma(\cdot,\bar{x}_1,\bar{y}_1)$ are defined in $[x_0,d-\varepsilon]$ and $[c+\varepsilon,x_0]$ respectively and satisfy there

$$0 \leq \gamma(x;x_1,y_1) - \gamma(x;x_0,y_0) < \varepsilon \tag{3.5}$$

$$0 \leq \gamma(x;x_0,y_0) - \gamma(x;\bar{x}_1,\bar{y}_1) < \varepsilon. \tag{3.6}$$

We can establish this result with arguments similar to those used in the proof of the preceding theorem; it is enough to observe that - as far as curves in Γ_1 are concerned that have initial point $(x_0,y_0+\delta)$ with $\delta > 0$ ($\delta < 0$) on the right (left) of $P_0 = (x_0,y_0)$ - hypotheses a), b) play the role of the uniqueness hypothesis in the argument used to establish that theorem.

If we retain hypothesis a) but replace b) by:

b') No curve may have points in common with any other curve belonging to its inferior region which are to the right of its own initial point and to the left of the initial point of the other;

then we can show the following result in the same way:

(ii) Given $\delta > 0$, we may determine a neighborhood J of P_0 belonging to the left superior region of γ_0 with respect to P_0 and a neighborhood $\bar{J}$ of P_0 belonging to the right inferior region of γ_0 with respect to P_0 in such a way that if $P_1 \equiv (x_1,y_1)$ and $\bar{P}_1 \equiv (\bar{x}_1,\bar{y}_1)$ belong to J and $\bar{J}$ respectively, the corresponding functions $\gamma(\cdot;x_1,y_1)$ and $\gamma(\cdot;\bar{x}_1,\bar{y}_1)$ are defined in $[c+\varepsilon,x_0]$ and $[x_0,d-\varepsilon]$ respectively and satisfy (3.5) and (3.6) there.

We shall now prove that results (i) and (ii) imply continuous dependence of the superior and inferior integrals on the initial point.

For this purpose, we suppose that the equation (3.4) has superior and inferior integrals at every point of R and that the corresponding integral curves satisfy the condition assumed for the family Γ.

Let Γ_1 be the family of the integral curves of equation (3.4) that coincide to the right of their initial point with the superior integral and to the left with the inferior integral. Let Γ_2 be the family of the integral curves that coincide with the inferior integral to the right of their initial point and with the superior integral to the left. The curves in Γ_1 obviously satisfy hypotheses a), b) while those in Γ_2 satisfy a), b').

If we denote by $G(x;x_0,y_0)$ the superior integral originating from $P_0 = (x_0,y_0)$ and by $[c,d]$ its domain of definition, it is possible, by virtue of (i) and (ii), to determine $\delta > 0$ such that the curves with initial point $P_\delta = (x_0,y_0+\delta)$ belonging to the families Γ_1 and Γ_2 have equations $y = \gamma_1(x;x_0,y_0+\delta)$ and $y = \gamma_2(x;x_0,y_0+\delta)$ respectively, with γ_1 defined in $[x_0,d-\varepsilon]$ and satisfying

$$0 \leq \gamma_1(x;x_0,y_0+\delta) - G(x;x_0,y_0) < \varepsilon$$

there and γ_2 defined in $[c+\varepsilon,x_0]$ and satisfying

$$0 \leq \gamma_2(x;x_0,y_0+\delta) - G(x;x_0,y_0) < \varepsilon$$

there.

It follows that the superior integral $G(x;x_0,y_0+\delta)$ originating from P_δ is defined in the interval $c + \varepsilon \leq x \leq d - \varepsilon$ and satisfies

$$0 \leq G(x;x_0,y_0+\delta) - G(x;x_0,y_0) < \delta.$$

there.

We now consider the inferior integral $g(x;x_0,y_0+\delta)$ originating from P_δ. Let $[x_0-h,x_0+h]$ be a neighborhood of x_0 in which we have

$$g(x;x_0,y_0+\delta) > G(x,x_0,y_0).$$

We shall denote by J a neighborhood of P_0 contained in the domain

$$x_0 - h \leq x \leq x_0 + h,$$

$$G(x;x_0,y_0) \leq y \leq g(x;x_0,y_0+\delta).$$

It can be immediately verified that the superior integral originating from an arbitrary point of J is defined in $[c+\varepsilon,d-\varepsilon]$ and lies between $G(x;x_0,y_0)$ and $G(x;x_0,y_0+\delta)$ there.

This proves that the superior integral is a continuous function of the initial point in its superior region; in a similar way we prove that the inferior integral is a continuous function of the initial point in its inferior region.

3.2. Uniqueness Theorems

In this section, we shall establish the uniqueness of the integrals of the equation $y' = f(x,y)$ that satisfy an initial condition $y(x_0) = y_0$ under a hypothesis more general than the Lipschitz condition considered in Chapter I.

We begin by proving a criterion that is valid for single equations on the real line.

<u>Theorem of Cafiero</u>. Let $f(x,y)$ be defined and continuous in the strip $S: a \le x \le b, |y| < \infty$. Assume that for every point $P = (\xi,\eta)$ of S, one can determine a function $F_P(x,u)$, continuous for $x > \xi, u > 0$, such that the following conditions are satisfied.

a) In a right neighborhood of P, $\xi \le x \le \xi + \delta$, $\eta - k \le y \le \eta + k$, we have

$$f(x,y_1) - f(x,y_2) \le F_P(x,y_1-y_2) \quad \text{for } y_1 > y_2. \tag{3.8}$$

b) Given $\varepsilon > 0$, we can determine an $h > 0$ such that for every ξ_0 in the interior of the interval $[\xi,\xi+h]$, the superior integral of

$$u' = F_P(x,u), \quad u(\xi_0) = h \tag{3.9}$$

is less than or equal to ε to the right of ξ_0.

Then the solutions of the equation $y' = f(x,y)$ are, to the right of the initial point, uniquely determined by the initial value.

We prove by contradiction. Suppose that two integrals $y = g_1(x)$ and $y = g_2(x)$ of the equation $y' = f(x,y)$ originate from the point $P_0 = (x_0,y_0)$ of S and that they differ at $\bar{x} > x_0$. Let $g_1(\bar{x}) > g_2(\bar{x})$. Denote by ξ the maximum of those values of x in the interval $x_0 \le x < \bar{x}$ for which $g_1(x) = g_2(x)$, and put $g_1(\xi) = g_2(\xi) = \eta$.

It is possible to determine a right neighborhood

$$\xi \le x \le \xi + \delta, \quad \eta - k \le y \le \eta + k$$

of $P = (\xi,\eta)$ such that (3.8) is valid there and

$$\eta - k < g_2(x) < g_1(x) < \eta + k \quad \text{for } \xi < x \le \xi + \delta.$$

If we put $w(x) = g_1(x) - g_2(x)$, it follows from (3.8) that

$$w'(x) = f(x,g_1(x)) - f(x,g_2(x)) \le F_P(x,w(x)). \tag{3.10}$$

Given $\varepsilon < w(\xi + \delta)$, let h be the number that corresponds to ε in hypothesis b) above, and let ξ_0 be an interior point of the interval $[\xi,\xi+h]$ chosen sufficiently close to ξ so that $0 < w(\xi_0) \le h$. We denote by u the superior integral of the equation $u' = F_P(x,u)$ that originates from $(\xi_0,w(\xi_0))$. Since (3.10) holds, it follows from the theorem on differential inequalities that u must be greater than or equal to w to the right of ξ_0, so

$$u(\xi + \delta) \ge w(\xi + \delta) > \varepsilon,$$

which contradicts hypothesis b).

A particular case of the theorem just proved is given by the following

Corollary. Let f be defined and continuous in the strip S, and suppose that it satisfies

$$f(x,y_1) - f(x,y_2) \leq \phi(x)\omega(y_1 - y_2) \quad \text{for} \quad y_1 > y_2$$

there, where $\phi(x)$ is continuous in $a < x \leq b$, $\omega(u)$ is continuous and nonnegative for $u > 0$, and

$$\lim_{\xi \downarrow 0^+} \int_{\xi}^{u_0} \frac{du}{\omega(u)} = +\infty, \qquad (u_0 > 0); \tag{3.11}$$

$$\int_{x_1}^{x_2} \phi(x)dx \leq M = \text{a constant} \qquad (a < x_1 < x_2 \leq b). \tag{3.12}$$

Then the integrals of the equation $y' = f(x,y)$ are, on the right of the initial point, uniquely determined by the initial value.

It will suffice to prove that if to every point P of S there corresponds the function

$$F_P(x,u) = \phi(x)\omega(u),$$

then the hypothesis b) of the preceding theorem is satisfied. To show this, let $\bar{u}(x)$ be the superior integral of the equation

$$u' = \phi(x)\omega(u), \quad u(\xi_0) = h.$$

It follows that

$$\frac{\bar{u}'(t)}{\omega[\bar{u}'(t)]} = \phi(t), \quad \bar{u}(\xi_0) = h,$$

so

$$\int_{\xi_0}^{x} \frac{\bar{u}'(t)}{\omega[\bar{u}(t)]} dt = \int_{\xi_0}^{x} \phi(t)dt.$$

If we substitute $\bar{u}(t) = u$ in the integral on the left, we obtain

$$\int_{h}^{\bar{u}(x)} \frac{du}{\omega(u)} = \int_{\xi_0}^{x} \phi(t)dt,$$

whence, in view of the hypothesis (3.12), we have

$$\int_h^{\bar{u}(x)} \frac{du}{\omega(u)} < M.$$

On the other hand, if we fix $\varepsilon > 0$, hypothesis (3.11) allows us to determine $h > 0$ such that

$$\int_h^{\varepsilon} \frac{du}{\omega(u)} > M.$$

For such a value of h, we shall have $\bar{u}(x) < \varepsilon$, which is what we wanted to prove.

Remark I. It is possible to obtain another version of the theorem of Cafiero by substituting the following hypothesis for b).

b') Given $\varepsilon > 0$, one can find an $h > 0$ so that for each δ in the interval $0 < \delta \leq h$ the superior integral of the equation

$$u(x) = \delta h + \int_{\xi+\delta}^{x} F_p[t,u(t)]dt$$

is, to the right of $\xi + \delta$, less than or equal to ε.

The proof of this result is similar, except for slight modifications, to that of the theorem; if we repeat exactly the steps that lead to inequality (3.10), we observe that, since the two functions $y = g_1(x)$ and $y = g_2(x)$ have equal derivatives at $x = \xi$, the function $w(x) = g_1(x) - g_2(x)$ is infinitesimal of order greater than one as $x \downarrow \xi$, so that we have

$$\lim_{\lambda \downarrow 0} \frac{w(\xi + \lambda)}{\lambda} = 0. \tag{3.13}$$

Given $\varepsilon < w(\xi + \delta)$, let h be the number that corresponds to ε in the hypothesis b'); by virtue of (3.13), it is possible to choose a point $\xi + \lambda$ belonging to both $[\xi,\xi+\delta]$ and $[\xi,\xi+h]$ and close enough to ξ so that

$$w(\xi + \lambda) < h\lambda. \tag{3.14}$$

Now let u be the superior integral of the equation

$$u'(x) = F_p[x,u(x)]$$

originating from the initial point $(\xi+\lambda,h\lambda)$. From (3.10) and (3.14) we see that the theorem on differential inequalities implies that the function u must always be greater than or equal to w to the right of $\xi + \lambda$, so

$$u(\xi + \delta) \geq w(\xi + \delta) \geq \varepsilon,$$

which contradicts the hypothesis.

Remark II. It follows from Cafiero's result that we get a uniqueness theorem if we assume that $f(x,y)$ is decreasing in y; it is enough to take $F_p(x,y) = 0$.

Remark III. If we pick $\omega(u) = u$ and $\phi(x) = A$, a constant, we obtain from the corollary the uniqueness theorem of Chapter I, where $f(x,y)$ satisfies a Lipschitz condition with respect to y.

The preceding theorem can be generalized to systems. Since there is no total ordering in R^n as there is in R, we must determine what condition to associate with inequality (3.8) in the theorem. If we multiply (3.8) by $y_1 - y_2$ for $y_1 \geq y_2$, we get

$$(f(x,y_1) - f(x,y_2))(y_1-y_2) \leq F_p(x,y_1-y_2)(y_1-y_2).$$

If $y_1 \leq y_2$ we have

$$(f(x,y_2) - f(x,y_1))(y_2-y_1) \leq F_p(x,y_2-y_1)(y_2-y_1).$$

From this we see that

$$(f(x,y_1) - f(x,y_2))(y_1-y_2) \leq F_p(x,|y_1-y_2|)|y_1 - y_2|$$

whatever y_1 and y_2 are. This suggests that we can generalize Cafiero's uniqueness theorem by means of the scalar product. For simplicity of notation, we shall consider only local uniqueness; it is clear that if we have local uniqueness at every point, then we have global uniqueness for the solutions of every initial value problem.

Uniqueness Theorem. If the f_i are continuous in a neighborhood of $(x_0,y_1^0,\ldots,y_n^0)$ and

$$(f(x,y_1) - f(x,y_2)|y_1-y_2) \leq \omega(t,||y_1-y_2||)||y_1 - y_2||$$

with $\omega\colon [x_0,b] \times [0,c] \to R$ continuous and such that the unique solution of

$$u' = \omega(x,u), \quad u(x_0) = 0$$

is $u \equiv 0$, then the initial value problem

$$y_i' = f_i(x,y), \quad y_i(x_0) = y_i^0$$

has at most one solution in $[x_0,x_0+\delta]$.

The proof is similar to that of the global existence theorem of Section 2.4. Observe first that the maximal solution of

$$w' = 2\omega(x,\sqrt{|w|})\sqrt{|w|}\ , \quad w(x_0) = 0$$

is $w \equiv 0$. If the given problem had two solutions $y_i, \bar{y}_i$, then the function

$$v(t) = ||y(t) - \bar{y}(t)||^2$$

would satisfy

$$v' \leq 2\omega(x,\sqrt{v})\sqrt{v}, \qquad v(x_0) = 0$$

by virtue of $||z(t)||^2 = z_1^2(t) + \ldots + z_n^2(t)$. By the theorem on differential inequalities, $v \leq 0$. The theorem is proved.

Exercise 1. Prove that there is uniqueness for $y_i' = f_i(x,y)$ if

$$||f(x,y_1) - f(x,y_2)|| \leq \omega(t,||y_1 - y_2||)$$

and ω is as in the uniqueness theorem.

Exercise 2. Prove the uniqueness theorem with ω having the same properties as the F_P considered above.

3.3. How Many Differential Equations Have the Uniqueness Property?

We have seen that a given initial value problem either has a unique solution or has infinitely many different solutions. A natural question that arises is to determine "how many" equations have unique solutions and "how many" infinitely many solutions. More precisely, if P_0 is a fixed initial point, we wish to determine how "large" the set F_U is of all continuous functions f of two variables such that the corresponding differential equation $y' = f(x,y)$ has a unique integral originating from P_0, and how "large" the set F_P is of the f for which there are infinitely many integrals of $y' = f(x,y)$ originating from P_0 (in which case one has the Peano phenomenon at P_0). The set F_U is the complement of F_P and vice versa, so a determination of one implies a determination of the other.

The following theorem answers this question by showing that there are "many more" equations with uniqueness than without; the set F_P is "much smaller" than F_U.

Theorem. Let S be the strip $[a,b] \times \mathbb{R}$ and X the Banach space of real-valued, continuous, bounded functions on S with the sup norm. Let

$y^0 \in \mathbf{R}$ and let F_P be the set of all the $f \in X$ such that the initial value problem

$$y' = f(x,y), \quad y(a) = y^0$$

has infinitely many different solutions (i.e., exhibits the Peano phenomenon). Then with every measure μ on X we can associate, in a canonical way, a measure $\bar{\mu}$ of X such that $\bar{\mu}(F_P) = 0$.

The measure $\bar{\mu}$ associated with μ is defined at the beginning of the proof.

We recall that a measure on a set X is defined by specifying first a family $\mathcal{M}$ of subsets of X with the properties

(i) $\emptyset \in \mathcal{M}$,

(ii) $\mathcal{M}$ is closed under complementation, that is, $A \in \mathcal{M} \Rightarrow CA \in \mathcal{M}$,

(iii) $\mathcal{M}$ is closed under countable unions, that is, if $(A_n)_{n=1}^{\infty}$ is a sequence in $\mathcal{M}$, then $\bigcup_{n=1}^{\infty} A_n \in \mathcal{M}$,

and then defining a function $\mu: \mathcal{M} \to [0,\infty]$ such that

(iv) $\mu(\emptyset) = 0$,

(v) $\mu(\bigcup_{n=1}^{\infty} A_n) = \sum_{n=1}^{\infty} \mu(A_n)$ for every sequence $(A_n)_{n=1}^{\infty}$ of pairwise disjoint elements of $\mathcal{M}$,

(vi) $\mu(\{x\}) = 0$ for every $x \in X$,

(vii) there is a sequence $(M_n)_{n=1}^{\infty}$ in $\mathcal{M}$ such that $\mu(M_n) < \infty$ for every n and $X = \bigcup_{n=1}^{\infty} M_n$.

μ is called a *measure* on X and the elements of $\mathcal{M}$ are called *measurable sets* with respect to μ. Properties (vi) and (vii) are not always required in the definition of measure, but we shall consider only those measures that have all the properties in the list. When X is a topological space, it is always required that the open sets (and therefore the closed ones) be measurable. Thus, for all measures on the space X of continuous and bounded functions $f: S \to \mathbf{R}$, the open and closed sets are measurable. For a discussion of measures on Banach spaces, see Parthasarathy [41].

Proof of the Theorem: We fix a measure μ on X. Let X_0 be the set of all $f \in X$ that have value 0 at $P_0 = (a,y^0)$. Every f in X can be uniquely represented by

$$f = \bar{f} + \lambda$$

with $\overline{f} \in X_0$ and $\lambda \in \mathbf{R}$, λ being given by $\lambda = f(P_0)$. X can thus be identified with the product space $X_0 \times \mathbf{R}$. If we denote by μ_0 the restriction of μ to the set X_0 and by ν the Lebesgue measure on $\mathbf{R}$, we can define a new measure $\overline{\mu}$ on X by $\mu_0 \times \nu$, where the product is taken in the sense of the following theorem of Fubini, whose proof can be found in standard books on measure theory: for every measurable set E of X, the sets

$$E_f = \{\lambda \in \mathbf{R} | (f,\lambda) \in E\} \quad \text{and} \quad E_\lambda = \{f \in X_0 | (f,\lambda) \in E\}$$

are measurable with respect to ν and μ_0 respectively, and we have

$$\overline{\mu}(E) = \int_{X_0} \nu(E_f)\,d\mu_0(f) = \int_{\mathbf{R}} \mu_0(E_\lambda)\,d\nu(\lambda).$$

We begin by proving that F_P is measurable with respect to $\overline{\mu}$. For every integer n, let F_n be the set of all $f \in X$ which exhibit the Peano phenomenon at $P_0 = (a,y^0)$ with a difference greater than or equal to $1/n$, i.e.,

$$||M_f - m_f|| = \sup_{a \leq x \leq b} |M_f(x) - m_f(x)| \geq \frac{1}{n},$$

where M_f is the maximal and m_f the minimal solution of the initial value problem associated with f,

$$y' = f(x,y), \quad y(a) = y^0.$$

(Recall that we have global existence when the f are bounded.) Evidently

$$F_P = \bigcup_{n=1}^{\infty} F_n,$$

and therefore F_P is measurable if every F_n is. To prove that every F_n is measurable, it is enough to show that every F_n is closed. We therefore consider a sequence $(f_k)_{k=1}^{\infty}$ in F_n that converges uniformly to f_0 and prove that $f_0 \in F_n$; this is equivalent to showing that F_n is closed (cf. Sec. 4.1 of Chapter I). Since f_0 is bounded and $\lim_{k\to\infty} f_k = f_0$ uniformly, all the f_k are equibounded. We can therefore verify, by means of a procedure used many times already in this chapter, that the sequences $(M_{f_k})_{k=1}^{\infty}$ and $(m_{f_k})_{k=1}^{\infty}$ are equicontinuous and equibounded. By Ascoli's theorem, therefore, both possess convergent subsequences. The respective limits ϕ and ψ are solutions to the initial value problem

$$y' = f_0(x,y), \quad y(a) = y^0,$$

so we have

$$m_{f_0} \leq \psi \leq \phi \leq M_{f_0}.$$

But $||M_{f_k} - m_{f_k}|| \geq \frac{1}{n}$ implies that $||\phi - \psi|| \geq \frac{1}{n}$, since the norm of a Banach space is a continuous function. We thus deduce that $||M_{f_0} - m_{f_0}|| \geq \frac{1}{n}$, hence $f_0 \in F_n$. F_n is therefore closed, and F_P is measurable. In order to prove that $\bar{\mu}(F_P) = 0$, we consider the representation $f = \bar{f} + \lambda$ and rewrite the given initial value problems in the following way:

$$y' = \bar{f}(x,y) + \lambda, \quad y(a) = y^0.$$

We shall use the following theorem of Nakano [38]:

If Γ is a family of first order ordinary differential equations in normal form whose second member is continuous and bounded on S, and if for every pair of equations

$$y' = f(x,y), \quad y' = g(x,y)$$

we have $f(x,y) \neq g(x,y)$ for every (x,y) in S, then all but at most countably many of the equations in Γ admit one and only one integral through a fixed point $P \in S$.

We now fix $\bar{f} \in X_0$. If $\lambda' \neq \lambda''$, then the two functions $\bar{f} + \lambda'$ and $\bar{f} + \lambda''$ differ at every point of S. The family Γ of all equations

$$y' = \bar{f} + \lambda$$

therefore satisfies the hypotheses of Nakano's theorem, so there are at most countably many λ such that $(\bar{f},\lambda) = \bar{f} + \lambda \in F_P$. This implies that the set

$$F_{P,\bar{f}} = \{\lambda | (\bar{f},\lambda) \in F_P\}$$

is countable, so its measure is 0 in view of (v) and (vi). We have therefore proved that

$$\nu(F_{P,\bar{f}}) = 0, \qquad (\bar{f} \in X_0).$$

But then

$$\int_{X_0} \nu(F_{P,\bar{f}}) d\mu(\bar{f}) = 0$$

and hence $\overline{\mu}(F_P) = 0$ by Fubini's theorem. This completes the proof of the theorem.

Remark. One can find in Cafiero [16] similar results both for the case of equations depending on parameters and for the case in which $P_0 = (a, y^0)$ varies in a strip S. For a related question, see Cafiero [17], where the equation is considered fixed and P_0 is allowed to vary. The same question can also be examined from a topological point of view, as can be seen from the bibliographical notes.

4. ELEMENTS OF G-CONVERGENCE

In this section, we discuss more deeply the material treated in §5 of Chapter II. This is justified by the expansion of the theory of G-convergence in recent years, particularly due to the work of DeGiorgi and his students (cf. the bibliographical notes). The theory has not yet been much studied for nonlinear ordinary differential equations; nevertheless, it appears useful to us to present it in a simplified version that reveals its fundamental elements.

4.1. Introduction

The usual procedure of approximation, of which we have seen numerous examples in the first and the present chapters, may be described as follows.

The second members of the equations converge in a certain topology (for example, uniformly); correspondingly, the solutions converge in a stronger topology (for example, uniformly in the first derivatives). These are central to the hypotheses on the second members, and the behavior of the solutions follows as a consequence. In G-convergence, however, one begins with the observation that in many applications of differential equations, what really is of interest is the convergence of the solutions, and it appears useful to identify all those classes of equations that may even appear different, whose solutions have a qualitative behavior similar to that of a limit equation.

4.2. G-Convergence for Equations Satisfying the Lipschitz Condition

Under the Lipschitz assumption, it is possible to continue with the definition given at the end of Chapter II. This definition is applicable in all the cases in which the equations admit unique solutions to the initial value problems. The problem becomes complicated as soon as one wants to prove a compactness theorem, since in such a case it is necessary to have conditions for uniqueness sufficiently strong to insure that even the limit equation has a unique solution. In order not to complicate the problem further, we shall make the hypothesis that all the second members that we consider satisfy the following conditions:

$f(x,y)$ is continuous in $y \in \mathbb{R}^n$ and measurable in $x \in [a,b]$, and there exists a constant M such that for every x and y we have

$$||f(x,y)|| \leq M(||y|| + 1). \qquad (4.1)$$

Under such hypotheses, we can show that every initial value problem associated with the equation

$$A_f[y] = y' - f(x,y) = 0$$

admits at least one solution defined in $[a,b]$. (Prove it as an exercise.)

Definition 4.1. Let

$$A_n[y_n] \equiv A_{f_n}[y_n] = 0$$

be a sequence of differential equations, and let $A[y] = A_f[y] = 0$ be a differential equation. Suppose (4.1) is satisfied uniformly, and that for every initial datum there is a unique solution to the Cauchy problem. We say that A_n G-converges to A if for every pair $(x_0,y_0) \in [a,b] \times \mathbb{R}^m$ the sequence $(y_n^{(0)})_n$ consisting of the solutions $y_n^{(0)}$ of the Cauchy problems

$$\begin{cases} A_n[y_n^{(0)}] = 0 \\ y_n^{(0)}(x_0) = y_0 \end{cases}$$

converges uniformly in $[a,b]$ to the solution $y^{(0)}$ of the Cauchy problem

$$\begin{cases} A[y^{(0)}] = 0 \\ y^{(0)}(x_0) = y_0. \end{cases}$$

Note that this definition is based on the initial value problem, and that it is not clear that G-convergence for initial value problems must imply G-convergence for other types of problems like those considered in Chapters IV and V.

We begin with a simple example. Let f be a function continuous on $[a,b] \times \mathbb{R}^m$ satisfying the Lipschitz condition

$$||f(x,y_1) - f(x,y_2)|| \leq L||y_1 - y_2||. \tag{4.2}$$

Let f_n satisfy a Lipschitz condition in y and converge uniformly in $[a,b] \times R^m$ to f so that for every $\varepsilon > 0$, there is a $\nu_\varepsilon > 0$ such that for each $n > \nu_\varepsilon$, we have

$$||f_n(x,y) - f(x,y)|| < \varepsilon. \tag{4.3}$$

We then have, for a given initial point (x_0,y_0),

$$y_n(x) = y_0 + \int_{x_0}^{x} f_n(\xi,y_n)d\xi$$

$$y(x) = y_0 + \int_{x_0}^{x} f(\xi,y)d\xi$$

whence, upon subtracting side from side and using (4.2) and (4.3),

$$\begin{aligned} ||y_n(x) - y(x)|| &\leq \int_{x_0}^{x} ||f_n(\xi,y_n) - f(\xi,y)||d\xi \\ &\leq \int_{x_0}^{x} [||f_n(\xi,y_n) - f(\xi,y_n)|| + ||f(\xi,y_n) - f(\xi,y)||]d\xi \\ &\leq \int_{x_0}^{x} [\varepsilon + L||y_n(\xi) - y(\xi)||]d\xi. \end{aligned}$$

From Gronwall's lemma, it then follows that

$$||y_n(x) - y(x)|| \leq \frac{\varepsilon}{L}[\exp(L|x-x_0|) - 1]$$

so we have that the $y_n^{(0)}$ converge uniformly to $y^{(0)}$.

This result shows that the uniform convergence of the second member is stronger than G-convergence; actually, this fact was already noted in Chapter I, when we discussed continuous dependence on data and parameters.

4.3. Homogenization

We shall now present some cases of homogeneization in which there is only weak convergence of the second members or even no convergence.

Theorem 4.1. Let $f(y)$ be a continuous function satisfying a Lipschitz condition. Suppose that f is periodic with period M. The equations

$$y_k' = f(ky_k) \tag{4.4}$$

G-converge to the equation

$$y' = p \tag{4.5}$$

with

$$\begin{cases} p = \left\{\frac{1}{M}\int_0^M f^{-1}(y)\,dy\right\}^{-1} & \text{if } f(y) \neq 0 \text{ for every } y \\ p = 0 & \text{if there exists } y_0 \text{ such that } f(y_0) = 0. \end{cases} \tag{4.6}$$

It is not difficult to establish the theorem directly; instead, we present the following more general result.

Theorem 4.2. Let $f(x,y)$ be a C^1-function, periodic in x with period L and periodic in y with period M. Then the equations

$$y_k' = f_k(x,y_k) = f(kx,ky_k) \tag{4.7}$$

G-converge to the equation

$$y' = p, \tag{4.8}$$

where p is a number satisfying the condition

$$\min f(x,y) \le p \le \max f(x,y). \tag{4.9}$$

Proof: Let $\bar{y}$ be the solution of the problem

$$\begin{cases} \bar{y}' = f(x,\bar{y}) \\ \bar{y}(0) = 0. \end{cases} \tag{4.10}$$

For each n, set

$$\lambda_n = [\bar{y}(2^n \cdot L)] \cdot 2^{-n} L^{-1}, \tag{4.11}$$

i.e., is the mean slope of $\bar{y}$ after 2^n periods of the variable x.

We shall first of all show that λ_n is a Cauchy sequence. To do this, we need an upper bound on the solutions having initial datum different from 0. Let y_n be the solution of the problem

$$\begin{cases} y_\alpha' = f(x,y_\alpha) \\ y_\alpha(0) = \alpha; \end{cases}$$

because of the periodicity, we have, for every integer h,

$$y_{\alpha+hM}(x) - y_\alpha(x) = hM,$$

whence it follows, in particular, that

$$\begin{aligned} y_{hM}(x) &= hM + \bar{y}(x), \\ y_{hM}(2^n L) &= hM + \lambda_n \cdot 2^n L. \end{aligned} \tag{4.12}$$

If α is fixed, let $h = [\alpha/M]$; the uniqueness of the solution now implies that

$$y_{hM}(2^n L) \leq y_\alpha(2^n L) < y_{(h+1)M}(2^n L)$$

so that if we use (4.12), we get

$$\lambda_n 2^n L + hM \leq y_\alpha(2^n L) < \lambda_n 2^n L + (h + 1)M. \tag{4.13}$$

Since it is also true that

$$hM \leq \alpha = y_\alpha(0) < (h + 1)M, \tag{4.14}$$

we produce, upon subtraction of (4.14) from (4.13) and division by the length of the interval, $2^n L$, the following upper bound on the mean value of the derivatives of y_α:

$$\lambda_n - \frac{M}{L} \cdot \frac{1}{2^n} \leq \frac{y_\alpha(2^n L) - y_\alpha(0)}{2^n L} < \lambda_n + \frac{M}{L}\frac{1}{2^n}. \tag{4.15}$$

Since (4.15) is true for every α, we get, since f is periodic in x,

$$\lambda_n - \frac{M}{L}\frac{1}{2^n} \leq \frac{y_\alpha(i \cdot 2^n L) - y_\alpha((i-1)2^n \cdot L)}{2^n L} < \lambda_n + \frac{M}{L}\frac{1}{2^n}. \tag{4.16}$$

If we now sum (4.16) for $i = 1,2,\ldots,s$ and divide by s, we get

$$\lambda_n - \frac{M}{L}\frac{1}{2^n} \leq \frac{y_\alpha(s \cdot 2^n L) - y_\alpha(0)}{s \cdot 2^n L} < \lambda_n + \frac{M}{L}\frac{1}{2^n}. \tag{4.17}$$

In particular, for $\alpha = 0$, $s = 2^q$, we have, for every $q \geq 1$,

$$\lambda_n - \frac{M}{L}\frac{1}{2^n} \leq \lambda_{n+q} \leq \lambda_n + \frac{M}{L}\frac{1}{2^n},$$

that is, the λ_n form a Cauchy sequence. Let p be its limit. We now show that the equations (4.7) G-converge to (4.8) with this p.

Let $[-K,K]$ be a fixed interval. We shall first of all show that the functions $y_{k,\alpha}$ that are the solutions of the problem

$$\begin{aligned} y'_{k,\alpha} &= f_k(x, y_{k,\alpha}) \\ y_{k,\alpha}(0) &= \alpha \end{aligned}$$

converge uniformly to $y = px + \alpha$. Let ε be fixed, let n be such that $|\lambda_n - p| < \frac{\varepsilon}{K}$, and $\frac{MK}{L} \cdot \frac{1}{2^n} < \varepsilon$. By homotethy, (4.17) becomes

$$\lambda_n - \frac{M}{L} \cdot \frac{1}{2^n} \leq \frac{y_{k,\alpha}(\frac{s}{k} 2^n L) - \alpha}{\frac{s}{k} 2^n L} < \lambda_n + \frac{M}{L} \frac{1}{2^n},$$

and we have

$$|y_{k,\alpha}(\frac{s}{k} \cdot 2^n L) - \lambda_n \frac{s}{k} 2^n L - \alpha| \leq M\frac{|s|}{k}.$$

On the other hand, $|x| \leq K$, so the bounds are only needed for $|s| < \frac{kK}{2^n L}$; we have

$$|y_{k,\alpha}(\frac{s}{k} 2^n L) - \lambda_n \frac{s}{k} 2^n L - \alpha| \leq \frac{MK}{L} \cdot \frac{1}{2^n} < \varepsilon. \tag{4.18}$$

Since f is bounded, say $|f| \leq P$, and for x satisfying the inequality

$$\frac{s}{k} 2^n L \leq x < \frac{s+1}{k} 2^n L, \tag{4.19}$$

we have

$$|y_{k,\alpha}(x) - y_{k,\alpha}(\frac{s}{k} 2^n L)| \leq P \cdot \frac{1}{k} 2^n L. \tag{4.20}$$

Finally, again for those x, $|x| < K$, satisfying (4.19), we have

$$\begin{aligned} |\lambda_n \frac{s}{k} \cdot 2^n L + \alpha - (px+\alpha)| &\leq |\lambda_n \frac{s}{k} 2^n L - \lambda_n x| \\ &+ |\lambda_n - p||x| \leq |\lambda_n| \frac{1}{k} 2^n L + \frac{\varepsilon}{K} \cdot K. \end{aligned} \tag{4.21}$$

Thus, for $k > k_0$ with k_0 such that $\frac{1}{k_0} 2^n LP < \varepsilon$, $\frac{1}{k_0}|\lambda_n| \cdot 2^n L < \varepsilon$, we obtain, upon summing (4.18), (4.19), and (4.21)

$$|y_{k,\alpha}(x) - (px+\alpha)| < 4\varepsilon \quad \text{for every } x.$$

It now remains to prove uniform convergence even for those solutions with initial datum $x_0 \neq 0$. Let $y_{k,\xi,\alpha}$ be the solutions of the problems

$$\begin{cases} y'_{k,\xi,\alpha} = f_k(x, y_{k,\xi,\alpha}) \\ y_{k,\xi,\alpha}(\xi) = \alpha \end{cases}$$

with $|\xi| < K$. For every $\varepsilon > 0$, consider

$$y_1 = px + \alpha_1 \qquad \alpha_1 = \alpha - \varepsilon - p\xi$$

$$y_2 = px + \alpha_2 \qquad \alpha_2 = \alpha + \varepsilon - p\xi.$$

Because of the uniform convergence of $(y_{k,\alpha})_{k=1}^{\infty}$, we have, for $k > k_0$, $|y_{k,\alpha_1}(x) - (px+\alpha_1)| < \varepsilon$, $|y_{k,\alpha_2}(x) - (px+\alpha_2)| < \varepsilon$; in particular, $y_{k,\alpha_1}(\xi) < \alpha < y_{k,\alpha_2}(\xi)$ so that, for $k > k_0$, we also have

$$y_{k,\alpha_1}(x) < y_{k,\xi,\alpha}(x) < y_{k,\alpha_2}(x)$$

and therefore

$$px - p\xi + \alpha - 2\varepsilon < y_{k,\xi,\alpha}(x) < px - p\xi + \alpha + 2\varepsilon .$$

This shows that $(y_{k,\xi,\alpha})_{k=1}^{\infty}$ is a Cauchy sequence for the uniform convergence; it tends to $px - p\xi + \alpha$, a solution of (4.8). Finally, (4.9) clearly follows from (4.11). This completes the proof of Theorem 4.2.

We now return to Theorem 4.1. It remains to prove that p is really determined by (4.6). It will suffice to prove that the solutions of the problem

$$\begin{cases} y_k' = f_k(y_k) \\ y_k(0) = 0 \end{cases}$$

tend to px, or, since we already know that the limits are linear functions, to prove that if $\bar{y}$ is the limit of $(y_k)_{k=1}^{\infty}$, there is a point $\bar{x}$ for which

$$\frac{\bar{y}(\bar{x})}{\bar{x}} = p \tag{4.22}$$

where p is given by (4.6).

We consider the case in which there is y_0, $0 \le y_0 < M$, such that $f(y_0) = 0$. If $y_0 = 0$, we have $y_k(x) = 0$ and so $\bar{y}(x) = 0$. If $y_0 \neq 0$ we have, because of the uniqueness of the solution,

$$\frac{y_0 - L}{k} < y_k(x) < \frac{y_0}{k} ,$$

whence $\bar{y}(x) = 0$; (4.22) is therefore satisfied because in this case (4.6) yields $p = 0$.

If, now, $f(y) \neq 0$ for every y, set

$$L = \int_0^M f^{-1}(y)dy;$$

we get, because of periodicity and homotethy,

$$\frac{y_k(L)}{L} = \frac{y_1(kL)}{kL} = \frac{kM}{kL} = \left\{\frac{1}{M}\int_0^M f^{-1}(y)\,dy\right\}^{-1}$$

whence, upon passing to the limit with respect to k,

$$\frac{\overline{y}(\overline{x})}{\overline{x}} = p = \left\{\frac{1}{M}\int_0^M f^{-1}(y)\,dy\right\}^{-1},$$

which concludes the proof of Theorem 4.1.

Theorem 4.1 and the examples at the end of Chapter II can be generalized by the following theorem.

Theorem 4.3. Let $f(x,y)$ be a C^1-function, periodic with period L in the variable x. If we put $f_k(x,y) = f(kx,y)$, the sequence of equations

$$y_k' = f_k(x,y_k)$$

G-converges to the autonomous equation

$$y' = g(y)$$

with

$$g(y) = \frac{1}{L}\int_0^L f(x,y)\,dx.$$

Theorem 4.4. Let $f(x,y)$ be a C^1-function periodic with period M in the variable y. Put $f_k(x,y) = f(x,ky)$. Then the sequence of equations $y_k' = f_k(x,y_k)$ G-converges to the equation

$$y' = g(x)$$

where

$$g(x) = \begin{cases} \left[\dfrac{1}{M}\displaystyle\int_0^M \frac{dy}{f(x,y)}\right]^{-1} & \text{if } \displaystyle\int_0^M \frac{dy}{|f(x,y)|} < +\infty \\[2ex] 0 & \text{if } \displaystyle\int_0^M \frac{dy}{|f(x,y)|} = +\infty. \end{cases}$$

These theorems can be proved by approximating the functions f with functions that are piecewise constant with respect to the variable for which there is no periodicity, and then applying respectively the techniques of Chapter II and of Example 1 to the approximating functions. For a complete proof, see [65].

We now consider the natural question of determining p in Theorem 4.2. The proof provides a process for calculating p, but it does not seem easy to produce an explicit formula for p except in particular cases. We examine one such case.

Suppose that $f(x,y)$ is periodic with period L in x and with period M in y. Let $y_{\alpha\beta}(x)$ be the solution of the problem

$$\begin{cases} y' = f(x,y) \\ y(\alpha) = \beta. \end{cases}$$

If there exist two integers q and r for which

$$y(\alpha + qL) = \beta + rM,$$

then $p = rM/qL$. We shall give the proof for the case when $\alpha = \beta = 0$ and leave the proof of the general case to the reader. When $\alpha = \beta = 0$, we have, by periodicity and homotethy,

$$\frac{y_k(qL)}{qL} = \frac{y_1(kqL)}{kqL} = \frac{krM}{kqL}$$

whence, upon passing to the limit with respect to k, we have the desired result.

We now present an interesting generalization of this case. We suppose that $f(x,y)$ is periodic as above and that there are two integers q and r and two numbers α and β with $0 \leq \alpha < \beta < M$ such that the following condition is satisfied. If y_2 is the solution of the problem

$$y_2' = f(x,y_2); \quad y_2(0) = \alpha$$

and y_1 the solution of the problem

$$y_1' = f(x,y_1), \quad y_1(0) = -M + \beta,$$

let

$$(r-1)M + \beta \leq y_1(qL) < y_2(qL) \leq rM + \alpha.$$

Then

$$p = \frac{rM}{qL}.$$

The proof is practically the same as that given in the preceding example and is left as an exercise.

In order to give some numerical applications of these criteria, we consider the equation

$$y' = f(x,y),$$

where

$$f(x,y) = \begin{cases} \alpha > 0 & \text{if } [x] + [y] \text{ is even} \\ \beta > 0 & \text{if } [x] + [y] \text{ is odd.} \end{cases}$$

The coefficients are discontinuous, but the theory is still valid; we leave it to the reader to explain why. Prove, for a simple exercise, that if $\alpha \le 1$, $\beta \ge 1$, then $p = 1$. Observe that, surprisingly, (4.9) can be true even with equality, for example, if $y' = \cos(x-y)$; in this case $y = y_k = x$ is a solution of the equation

$$y_k' = \cos(kx - ky)$$

and so also of the limit equation $y' = 1$, where $1 = \max f$.

4.4. G-Compactness

We shall now present, under somewhat simplified hypotheses, a compactness theorem for G-convergence. Let $\mathscr{L}_M$ be the class of functions $f(x,y)$ that are measurable in x, satisfy the relation

$$||f(x,y)|| \le M(||y|| + 1) \tag{4.23}$$

and satisfy a Lipschitz condition in y according to the relation

$$||f(x,y_1) - f(x,y_2)|| \le M||y_1 - y_2||. \tag{4.24}$$

(Observe that the Lipschitz constant used here is "global." In such a case, we say that the functions are "equilipschitzian.") Let $\mathscr{E}_M$ be the class of ordinary differential equations associated with the functions of $\mathscr{L}_M$. We have the following compactness theorem.

Theorem 4.5. Let $(A_n)_{n=1}^{\infty}$ be a sequence of differential equations belonging to the class $\mathscr{E}_M$. For every interval $[a,b]$, it is possible to extract a subsequence $(A_{n_k})_k$ that G-converges to an equation $A \in \mathscr{E}_M$.

Proof: We have already observed that hypotheses (4.23) and (4.24) ensure that the solutions of the Cauchy problems for every equation A_n are unique and defined in all of $[a,b]$, since they satisfy

$$||y(x)|| \le (||y_0|| + \tfrac{1}{M})\exp(M|b-a|). \tag{4.25}$$

Moreover, it also follows that for all the initial data y_0, $||y_0|| < K$, we have, for almost all x,

$$||y'(x)|| \le (MK + 1)\exp(M|b-a|) + M, \tag{4.26}$$

and it follows that they are equicontinuous. By the theorem of Ascoli-Arzela, it is possible to extract a uniformly convergent subsequence from every sequence of functions (y_k) that are solutions of $A_k[y_k] = 0$ originating from the same initial value. We note that the limit y of such a sequence still satisfies (4.25) and (4.26).

We cannot, however, repeat this argument for all possible initial values, since there are uncountably many of them. We therefore fix a countable set $(x_i)_{i=1}^{\infty}$ dense in R and a countable set $(y_j)_{j=1}^{\infty}$ dense in R^m. If we use the diagonal procedure, we can extract a subsequence from A_n (which, for simplicity, we shall also call A_n) such that the solutions of the initial value problems $y_n(x_i) = y_j$ converge uniformly to functions $y_{ij}(x)$ also satisfying (4.25) and (4.26).

To get uniform convergence of the solutions associated with the other initial values, we note that under our hypotheses, we can invoke the theorems on continuous dependence on the data. For every fixed K, $\varepsilon > 0$, there is a $\delta > 0$ such that if $x', x'' \in [a,b]$, $||y_0'|| < K$, $||y_0''|| < K$, $|x' - x''| < \delta$, and $||y' - y''|| < \delta$, then we have

$$||y_{1,k}(x) - y_{2,k}(x)|| < \varepsilon, \quad (x \in [a,b],\ k \geq 1) \tag{4.27}$$

where $y_{1,k}$ is the solution relative to the problem $y' = f_k(x,y)$ with initial datum $y(x') = y_0'$, and $y_{2,k}$ the solution relative to that problem with initial datum $y(x'') = y_0''$. By uniform convergence, these relations hold in the limit. If x_0, y_0 do not belong to the dense sets of the x_i and y_j, consider two sequences $x_{i_k} \to x_0$, $y_{i_k} \to y_0$; if y_k is the solution of the problem $y' = f_k(x,y_k)$ with initial datum $y_k(x_{i_k}) = y_{j_k}$, the sequence $(y_k)_{k=1}^{\infty}$ is a Cauchy sequence because of the uniform convergence and converges to a function $\bar{y}$ with $\bar{y}(x_0) = y_0$.

In this way, we construct limit functions passing through every initial datum. We must now show that they actually are solutions of a differential equation. We need the following consistency property in order to do this.

If (x_1,y_1) and (x_2,y_2) belong to the graph of a solution, the solution associated with the point (x_1,y_1) must pass through (x_2,y_2) and vice versa. This is a result of the fact that the property is true for every k and that (4.27) holds. Thus, the family of functions that has been obtained is referred back to a family of curves with m parameters (the initial data for $x_0 = a$, $y_0 \in R^m$, for example); we can then associate with these a differential system of order m. Since all the curves are equilipschitzian in x, they generate a field of directions almost everywhere, and their derivatives, where they are defined, still satisfy (4.23). To obtain (4.24), note that (4.27) can be rewritten in the following way,

$$y_{2,k}(x) - y_{1,k}(x) - (y_0''-y_0') = \int_{x_0}^{x} [f_k(t,y_{2,k}) - f_k(t,y_{1,k})]dt,$$

whence, by (4.24),

$$||y_{2,k}(x) - y_{1,k}(x) - (y_0''-y_0')|| \le \int_{x_0}^{x} M||y_{2,k}(t) - y_{1,k}(t)||dt$$

$$\le M\int_{x_0}^{x} \Big\{||y_{2,k}(t) - y_{1,k}(t) - y_0'' + y_0'|| + ||y_0'' - y_0'||\Big\}dt.$$

If we apply Gronwall's lemma, we get

$$||y_{2,k}(x) - y_{1,k}(x) - (y_0'' - y_0')|| \le ||y_0'' - y_0'||\left(e^{M|x-x_0|} - 1\right).$$

If we observe that these inequalities have a limit as $k \to \infty$, and if we keep in mind that in the limit equation, we have, for almost all x,

$$f(x_0,y_0') = \lim_{x\to x_0} \frac{1}{x-x_0}(y_1(x)-y_0'), \quad f(x_0,y_0'') = \lim_{x\to x_0} \frac{1}{x-x_0}(y_2(x)-y_0''),$$

then we get

$$||f(x_0,y_0'') - f(x_0,y_0')|| = \lim_{x\to x_0} \frac{1}{|x-x_0|}||y_2(x)-y_1(x)-(y_0''-y_0')||$$

$$\le \lim_{x\to x_0} \frac{1}{|x-x_0|}||y_0''-y_0'||\left(e^{M|x-x_0|} - 1\right)$$

$$= M||y_0'' - y_0'||.$$

4.5. G-Convergence and the Peano Phenomenon

Definition 4.1 completely loses significance as soon as one considers equations that present phenomena of nonuniqueness of solutions. In order to study G-convergence in this case, it is necessary to give a weaker definition than that in 4.1; the convergence will then be weaker but still equivalent to the one defined by assuming uniqueness of the solutions of the Cauchy problems. We shall not enter here into the details of the more general definition. We shall merely make some introductory remarks and refer the reader to [64] and [66] for the precise definition and the compactness theorems.

The following example clarifies the reasons why the definition given in Sec. 4.2 cannot be used when the Peano phenomenon is exhibited. For every positive integer m, set

$$f_m(y) = \begin{cases} 0 & \text{for} \quad y \le 0 \\ my & \text{for} \quad 0 < y \le 1/m^2 \\ \sqrt{y} & \text{for} \quad y > 1/m^2. \end{cases}$$

It is clear that $(f_m)_{m=1}^{\infty}$ converges uniformly to

$$f(y) = \begin{cases} 0 & \text{for} \quad y \le 0 \\ \sqrt{y} & \text{for} \quad y > 0. \end{cases} \tag{4.28}$$

We consider the solutions $(y_m)_{m=1}^{\infty}$ of the problems

$$\begin{cases} y_m' = f_m(y_m) \\ y_m(0) = 0. \end{cases}$$

Since the f_m satisfy a Lipschitz condition, it follows that $y_m(x) = 0$ so $(y_m)_m$ converges uniformly to $\overline{y}(x) \equiv 0$, which is the minimal solution of the problem

$$\begin{cases} y' = f(y) \\ y(0) = 0. \end{cases}$$

On the other hand, let us now consider the sequence of functions

$$g_m(y) = \begin{cases} 0 & \text{if} \quad y \le -1/m^2 \\ \frac{m}{2}y + \frac{1}{2m} & \text{if} \quad -1/m^2 < y \le 1/m^2 \\ \sqrt{y} & \text{if} \quad y > 1/m^2 \end{cases}$$

which also converges uniformly to f given by (4.28). The solutions of the problems

$$\begin{cases} y_m' = g_m(y_m) \\ y(0) = 0 \end{cases}$$

are given by

$$y_m(x) = \begin{cases} \frac{1}{m^2}[\exp(\frac{m}{2}x) - 1] & \text{if} \quad 0 \le x < \frac{2}{m}\,\ell g\, 2 \\ \left(\frac{x}{2} - \frac{1}{m}\,\ell g\, 2 + \frac{1}{m}\right)^2 & \text{if} \quad x \ge \frac{2}{m}\,\ell g\, 2 \end{cases}$$

and converge uniformly to $\overline{y}(x) = \frac{1}{4}x^2$, the maximal solution of the problem

$$\begin{cases} y' = f(y) \\ y(0) = 0. \end{cases}$$

The fact that not even the uniform convergence of the second members is enough to ensure convergence of the solutions independently from the previously chosen approximations makes it quite clear that Definition 4.1 is unsuitable.

In substance, when the Peano phenomenon is exhibited, the convergence of single solutions no longer makes sense; one must consider the convergence of sets of solutions or, alternatively, use weaker topologies for the data and allow perturbations. See [66].

5. BIBLIOGRAPHICAL NOTES

The approximation of the solutions of a given initial value problem is a most important problem not only because it is not always possible to give a formula for the solution but also because we now have quite refined methods of numerical calculus that permits us to approximate solutions as accurately as we want. The method given in Sec. 2.1 for the approximation of all the solutions is due to Cafiero [15]. For a variant, see Baiada [6]. We shall return to procedures of numerical calculus in Chapter V.

The treatment of the Peano phenomenon given here goes back to Stampacchia [48]. See also Aronszajn [3], and, for an exposition of various related questions, Vidossich [52]. For an extension of the theorem on the connectedness of the set of fixed points, see Petryshyn [43]. In the case of first order equations on the real line, no hypothesis of boundedness of the second member is required to get connectedness for the set of solutions, as has been demonstrated by Vidossich [53].

In recent years, differential inequalities have been used with increasing frequency to study questions of uniqueness, global existence, and asymptotic behavior of solutions. Among those works that first employed differential inequalities to study these matters, we may cite Cafiero [18], [19], and Wazewski [60]. For a complete treatment, see Lakshmikantham-Leda [33], Szarski [50], and Walter [58]. In the case of equations of order greater than 1, a satisfactory theory of comparison seems yet to be lacking. The various matters relating to the Cauchy problem have thus far been treated with *ad hoc* methods; see, for example, Bebernes - Ingram [9] for the existence of maximal solutions, Baker [4] and Cartwright and Swinnerton-Dyer [23] for the existence of bounded solutions, Baxley [8] for uniqueness and global existence, and Bernfield and Yorke [12] for the asymptotic behavior of the solutions. For studying the asymptotic behavior of the solutions, Wazewski's topological method has proved very useful, but we have not discussed it here.

The uniqueness theorem we presented for equations on the real line is due to Cafiero [19], while the uniqueness and global existence theorems in which the scalar product is used are to be found in Vidossich [54].

All other known criteria (Sansone [45], Sansone - Conti [46], Lakshmikantham and Leela [33], Walter [58]) can be reduced to these. For a different sort of uniqueness theorem, see Bownds [13] and Wend [61]. The theorems of Cafiero [16], [17] on "uniqueness almost everywhere" may be compared with a similar topological result due to Orlicz [40], according to which the set of equations without uniqueness is of the first category. This topological result can be approached through the study of fixed points; cf. Vidossich [55].

For continuous dependence in the topology of uniform convergence, see Cafiero [20], [21], while for a study of the various topologies that insure continuous dependence, see Artstein [4], [5] and the references given there.

For a different treatment of the questions related to the initial value problem, see Strauss - Yorke [49].

There are criteria that characterize global existence as well as uniqueness; cf. Bernfeld [10] and Hartman [31] respectively.

It has recently been found necessary to study differential equations under hypotheses different from those we have chosen. One area of study deals with the possibility that the initial point belongs to the boundary of the domain of the functions of the second member of the given system. This situation occurs in functional analysis; see Crandall [24], where one starts from a case of this type to solve a problem about fixed points of nonexpansive transformations. Another case in which this situation occurs is that of a mathematical model for the two body problem in classical electrodynamics; cf. Travis [51]. Existence and uniqueness theorems for equations whose domain is a closed set are examined in Crandall [25], Hartman [32], and Yorke [62]; Bernfeld, Driver, and Lakshmikantham [11] have studied the case when the domain is open, but the initial point is on the boundary; see the references they give. This last type of initial problem is related, by means of a change of variable, to the terminal value problem, that is, the problem with datum $\lim_{x\uparrow\infty} y(x)$; cf. Vidossich [53], [56].

One gets a further generalization by allowing discontinuity in x or y. Discontinuity in x leads to the hypotheses of Carathéodory [22], while discontinuity in y causes us to consider $f(x,y)$ as a set instead of a point in R^n. These questions are of fundamental importance in the theory of control. See Persson [42], Filippov [26], [27], and Antosiewicz - Cellina [1], [2].

Another generalization is obtained by considering delays, that is, equations of the type $y'(x) = f(x,y(x-\tau(x)))$. These equations occur in physical and biological problems where the present state of the system depends on its past, as, for example, when a disease manifests itself after an incubation period. For an introduction to the theory of equations with delay, see Hale [29], Lakshmikantham - Leela [33, vol. 2], and Halanay - Yorke [28]. Ordinary differential equations have also been in Banach spaces, cf. Martin [36], Roseau [44], and the references they give. Ordinary differential equations have been studied also in Banach spaces, cf. Martin [36], Roseau [44] and the references they give.

For a panorama of the concrete applications of the theory of ordinary differential equations to the experimental sciences, see Braun [14].

The properties of the initial value problem of ordinary equations have been used to solve questions in other areas of mathematics, for example, partial differential equations (cf. Lions [35] and the references given there, and Nečas [39]), integral equations (cf. Scott [47]), and functional analysis (Vidossich [57]).

Many of the results of this chapter have been extended to Volterra equations; cf. Miller [37].

For a full bibliography on G-convergence, see DeGiorgi [63]. The literature almost exclusively treats problems in partial differential equations or in the calculus of variations. We have presented here the fundamentals of a similar theory for ordinary differential equations. For the case in which the Peano phenomenon appears, see Piccinini [64]; for further problems of homogenization, see Piccinini [65].

A more complete foundation of the theory and an exposition of the more recent results can be found in [66] and [67].

[1] H. A. Antosiewicz and A. Cellina, Continuous extensions: their construction and their application in the theory of ordinary differential equations, in H. A. Antosiewicz (ed.): *International Conference on Differential Equations,* Academic Press, New York, 1975, pp. 537-556.

[2] H. A. Antosiewicz and A. Cellina, Continuous selections and differential relations, *J. Diff. Eq., 19*(1975), 386-398.

[3] N. Aronszajin, Le correspondant topologique de l'unicité dans la théorie des équations différentielles, *Ann. Math., 43*(1942), 730-738.

[4] Z. Artstein, Continuous dependence on parameters: on the best possible results, *J. Diff. Eq., 19*(1975), 214-225.

[5] Z. Artstein, Topological dynamics of ordinary differential equations and Kurzweil equations, *J. Diff. Eq., 23*(1977), 224-243.

[6] E. Baiada, Le approssimazioni nella risoluzione delle equazioni differenziali ordinarie - I. Teorema di esistenza, *Rend. Accad. Naz. Lincei 2*(1947), 261-268.

[7] J. W. Baker, On the continuation and boundedness of solutions of a nonlinear differential equation, *J. Math. Anal. Appl., 55*(1976), 644-652.

[8] J. V. Baxley, Global existence and uniqueness for second-order ordinary differential equations, *J. Diff. Eq.*, *23*(1977), 315-334.

[9] J. W. Bebernes and S. K. Ingram, Existence and non-existence of maximal solutions for $y'' = f(x,y,y')$, *Anal. Polon. Math.*, *25*(1971), 125-138.

[10] S. R. Bernfeld, Liapunov functions and global existence without uniqueness, *Proc. Amer. Math. Soc.*, *25*(1970), 571-577.

[11] S. R. Bernfeld, R. D. Driver and V. Lakashmikantham, Uniqueness for ordinary differential equations, *Math. System Theory*, *9*(1975-76), 359-367.

[12] S. R. Bernfeld and J. A. Yorke, The behavior of oscillatory solutions of $x''(t) + p(t,g(x(t))) = 0$, *SIAM J. Math. Anal.*, *3*(1972), 654-667.

[13] J. M. Bownds, A uniqueness theorem for $y' = f(x,y)$ using a certain factorization of f, *J. Diff. Eq.*, *7*(1970), 227-231.

[14] M. Braun, *Differential Equations and Their Applications*, Springer-Verlag, Berlin, 1975.

[15] F. Cafiero, Sull' approssimazione mediante poligonali degli integrali del sistema differenziale $y' = F(x,y)$, $y(x_0) = y_0$, *Giorn. Mat. Battaglini*, *77*(1947-48), 28-35.

[16] F. Cafiero, Sulla classe delle equazioni differenziali ordinarie del primo ordine, i cui punti de Peano costituiscono un insieme di misure lebesguiana nulla, *Rend. Accad. Sci. Fis. Mat. Napoli*, *17*(1950), 127-137.

[17] F. Cafiero, Sul Fenomeno di Peano nelle equazioni differenziali ordinarie del primo ordine, *Rend. Accad. Sci. Fis. Mat. Napoli*, *17*(1950), 51-61 and 123-126.

[18] F. Cafiero, Su un problema ai limiti relativo all'equazione $y' = f(x,y)$, *Giorn. Mat. Battaglini*, *77*(1947), 145-163.

[19] F. Cafiero, Sui teoremi d'unicità relativi dad un'equazione differenziale ordinaria del primo ordine, *Giorn. Mat. Battaglini*, *78*(1948), 10-41 and 193-215.

[20] F. Cafiero, Un'osservazione sulla continuità, rispetto ai valori iniziali, degli integrali dell'equazione $y' = f(x,y)$, *Rend. Accad. Naz. Lincei*, *3*(1947), 479-482.

[21] F. Cafiero, Su di un teorema di Montel relativo alla continuità, rispetto al punto iniziale, dell'integrale superiore ed inferiore di una equazione differenziale, *Sem. Mat. Univ. Padova*, *17*(1948), 186-200.

[22] C. Caratheodory, *Variationsrechnung und Partielle Differentialgleichungen Ester Ordnung*, Teubner, Leipzig, 1935.

[23] M. L. Cartwright and H. P. F. Swinnerton-Dyer, Boundedness theorems for some second order differential equations, I, *Ann. Polon. Math.*, *29*(1974), 233-258.

[24] D. M. G. Crandall, Differential equations on convex sets, *J. Math. Soc. Japan, 22*(1970), 443-455.

[25] D. M. G. Crandall, A generalization of Peano's theorem and flow invariance, *Proc. Amer. Math. Soc., 36*(1972), 151-155.

[26] A. F. Filippov, Differential equations with discontinuous right hand side, *Mat. Sbornik, 51*(1960), 99-128.

[27] A. F. Filippov, On the existence of solutions of multivalued differential equations, *Mat. Zametki, 10*(1971), 307-313.

[28] A. Halanay and J. A. Yorke, Some new results and problems in the theory of differential-delay equations, *SIAM Rev., 13*(1971), 55-80.

[29] J. K. Hale, *Functional Differential Equations,* Springer-Verlag, New York, 1971.

[30] P. Hartman, A differential equation with non-unique solutions, *Amer. Math. Monthly, 70*(1963), 255-259.

[31] P. Hartman, On uniqueness and differentiability of solutions of ordinary differential equations, *Proceedings Symp. Nonlinear Problems,* Madison, 1963, pp. 219-232.

[32] P. Hartman, On invariant sets and a theorem of Wazewski, *Proc. Amer. Math. Soc., 32*(1972), 511-520.

[33] V. Lakshmikantham and S. Leela, *Differential and Integral Inequalities, Vols. I and II,* Academic Press, New York, 1969.

[34] M. Lavrantieff, Sur une équations différentielle du premier ordre, *Math. Z., 23*(1925), 197-209.

[35] J. L. Lions, *Quelques méthods de résolution des problémes aux limites non linéaries,* Dunod, Paris, 1969.

[36] R. H. Martin, Jr., *Nonlinear Operators and Differential Equations in Banach Spaces,* Wiley, New York, 1976.

[37] R. H. Miller, *Nonlinear Volterra Integral Equations,* Benjamin, Menlo Park, 1971.

[38] H. Nakano, Ueber die Verteilung der Peanoschen Punkte einer Differentialgleichung y' = f(x,y), *Proc. Phys. Math. Soc. Japan, 14*(1932), 41-43.

[39] J. Nečas, Sur une méthode genèrale pour la solution des problèmes aux limites non linéaires, *Ann. Sc. Norm. Sup. Pisa, 20*(1966), 655-674.

[40] W. Orlicz, Zur theorie der Differentialgleichung y' = f(x,y), *Bull. Acad. Polon. Sci.,* (1932), 221-228.

[41] K. R. Parthasarathy, *Probability Measures on Metric Spaces,* Academic Press, New York, 1967.

[42] J. Persson, A generalization of Carathéodory existence theorem for ordinary differential equations, *J. Math. Anal. Appl., 49*(1975), 496-503.

[43] W. V. Petryshyn, Structure of the fixed points sets of k-set-contractions, *Arch. Rat. Mech. Anal.*, *40*(1971), 312-328.

[44] M. Roseau, *Equations Différentielles*, Masson, Paris, 1976.

[45] G. Sansone, *Equazioni Differenziali Nel Campo Reale, Vols. I and II*, Zanichelli, Bologna, 1956.

[46] G. Sansone and R. Conti, *Equazioni Differenziali Nonlineari*, Cremonese, Roma, 1956.

[47] M. R. Scott, *Invariant Imbeddings and its Applications to Ordinary Differential Equations, An Introduction*, Addison-Wesley, Reading, Massachusetts, 1973.

[48] G. Stampacchia, Le trasformazioni che presentano il fenomeno di Peano, *Rend. Accad. Naz. Lincei*, *7*(1949), 80-84.

[49] A. Strass and J. A. Yorke, On the fundamental theory of ordinary differential equations, *SIAM Rev.*, *11*(1969), 236-246.

[50] J. Szarski, *Differential Inequalities*, PWN, Warsaw, 1965.

[51] S. P. Travis, A one-dimensional two-body problem of classical electrodynamics, *SIAM J. Appl. Math.*, *28*(1975), 611-632.

[52] G. Vidossich, Applications of topology to analysis: On the topological properties of the set of fixed points of nonlinear operators, *Confer. Sem. Mat. Univ. Bari*, *126*(1971), 1-62.

[53] G. Vidossich, Two remarks on global solutions of ordinary differential equations in the real line, *Proc. Amer. Math. Soc.*, *55*(1976), 111-115.

[54] G. Vidossich, Existence, comparison and asymptotic behavior of solutions of ordinary differential equations in finite and infinite dimensional Banach spaces, Notas de Matemática n. 24(1972), Universidade de Brasilia.

[55] G. Vidossich, Existence, uniqueness and approximations of fixed points of nonlinear operators as a generic property, *Bull. Soc. Mat. Brasil*, *5*(1974), 17-29.

[56] G. Vidossich, Solution of Hallam problem on terminal value comparison principle for ordinary differential equations, *Trans. Amer. Math. Soc.*, *220*(1976), 115-132.

[57] G. Vidossich, How to get zeros of nonlinear operators using the theory of ordinary differential equations, *Atas Semana Análise Funcional*, Sao Paulo, 1974.

[58] W. Walter, *Differential and Integral Inequalities*, Springer-Verlag, Berlin, 1970.

[59] T. Wazewski, Sur une principe topologique de l'examen de l'allure asymptotique des integrales des équations différentielles ordinaires, *Ann. Polon. Math.*, *20*(1947), 279-313.

[60] T. Wazewski, Systèmes de équations e des inégalités différentielles ordinéaires aux deuxième membres monotone et leurs applications, *Ann. Polon. Math., 23*(1950), 112-196.

[61] D. V. V. Wend, Existence and uniqueness of solutions of ordinary differential equations, *Proc. Amer. Math. Soc., 23*(1969), 27-33.

[62] J. A. Yorke, Invariance for ordinary differential equations, *Math. System Theory, 1*(1967), 353-372.

[63] E. DeGiorgi, Γ-convergenza e G-convergenza, *Boll. Un. Mat. It., 14A*(1977), 213-220.

[64] L. C. Piccinini, G-convergenza for ordinary differential equations with Peano Phaenomenon, *Rendiconti Sem. Matem. Padova,* 58(1977), 65-8

[65] L. C. Piccinini, Homogeneization ofor ordinary differential equations, *Rend. Circ. Mat. Palermo,* 27(1978), 95-112.

[66] L. C. Piccinini, Linearity and nonlinearity in the theory of G-convergence, in R. Conti (ed.): *Recent Advances in Differential Equations,* Academic Press, New York, 1981.

[67] E. Schechter, Necessary and sufficient conditions for convergence of temporally irregular evolution, *Nonlinear Anal., TMA,* to appear.

Chapter IV
Boundary Value Problems

In the previous chapters, we studied various kinds of questions concerning the initial value problem. We now propose to investigate other types of problems, in which the desired solution depends either on the values that it assumes at various points in its domain or on geometrical conditions (e.g., intersecting two given curves or being tangent to two lines), or on periodic conditions. Before studying such problems, we will discuss certain notions and results about continuous mappings on Euclidean spaces.

1. CONTINUOUS MAPPINGS ON EUCLIDEAN SPACES

Studying boundary value problems, we are interested in determining when a continuous function f defined on a subset of $\mathbf{R}^n$ has *zeros*, that is, points mapped to 0, or when it maps points to themselves, that is, points x such that $x = f(x)$. These points are called *fixed points* of the transformation f. Actually, we have already worked with these notions in Sec. 4 of Chapter I as well as in Chapter III.

In the case of real functions of a real variable, these questions are very simple because they depend on the fact that a continuous function has a zero if it assumes both a positive and a negative value at two points in an interval. This fact allows us to see at once that a continuous function

$$f: [a,b] \to [a,b]$$

must have at least one fixed point, for the continuous function $g(x) = f(x) - x$ satisfies the conditions

$$g(a) \geq 0, \quad g(b) \leq 0,$$

and therefore has a zero at some point, which will be a fixed point of f.

It is not so easy for functions of several variables. In order to be able to determine whether there are zeros or fixed points, we introduce the notion of topological degree. This is not the only way to establish the existence of fixed points, but it seems to be the most general one when we consider the other applications that it has and, most of all, the fact that it lies at the foundation of nonlinear functional analysis.

1.1. The Topological Degree

The topological degree of a mapping $f: A \subseteq \mathbb{R}^n \to \mathbb{R}^n$ is an integer that gives an estimate, invariant under certain parameters, of the number of points $f^{-1}(p)$ consists of for $p \in \mathbb{R}^n$. It does not determine the cardinality of $f^{-1}(p)$ exactly, but gives a qualitative measure of it.

Let $A \subseteq \mathbb{R}^n$ be a bounded open set, and let $f: \bar{A} \to \mathbb{R}^n$ and $p \notin f(\partial A)$. *The topological degree of* f *with respect to* A *and* p is an integer denoted by

$$d(f,A,p)$$

with the following four characteristic properties:

(G_1) If I is the identity transformation, then

$$d(I,A,p) = \begin{cases} 1 & \text{if } p \in A \\ 0 & \text{if } p \notin A. \end{cases}$$

(G_2) If $d(f,A,p) \neq 0$, then $f^{-1}(p) \neq \emptyset$, that is, the equation $p = f(x)$ has at least one solution.

(G_3) ADDITIVITY: If $A_1,\ldots,A_n$ is a disjoint sequence of open subsets of A such that $\bar{A} = \bar{A}_1 \cup \ldots \cup \bar{A}_n$, then

$$d(f,A,p) = \sum_{i=1}^{n} d(f,A_i,p)$$

provided that every term is defined.

(G_4) INVARIANCE UNDER HOMOTOPY: If $(f_t)_{t\in[0,1]}$ is a homotopy from $\bar{A}$ to $\mathbb{R}^n$ and $p(t)$ is a continuous mapping from $[0,1]$ to $\mathbb{R}^n$ such that $p(t) \notin f_t(\partial A)$ for each t, $0 \leq t \leq 1$, then

$$d(f_t,A,p(t)) = \text{a constant} \qquad (0 \leq t \leq 1).$$

The condition

$$p \notin f(\partial A)$$

is the only restriction we make in defining the topological degree for continuous functions in $\mathbb{R}^n$. The last phrase of (G_3) means that $p \notin f(\partial A_i)$ and $p \notin f(\partial A)$. Finally, when in (G_4), we say that $(f_t)_{t\in[0,1]}$ is a homotopy from $\overline{A}$ to $\mathbb{R}^n$ we mean that $(t,x) \to f_t(x)$ is a continuous mapping from $[0,1] \times \overline{A}$ to $\mathbb{R}^n$.

There are several different ways of defining the function $d(f,A,p)$; among them we mention

(i) using combinatorial topology,
(ii) using algebraic topology,
(iii) using differential calculus in $\mathbb{R}^n$, and
(iv) using integral calculus in $\mathbb{R}^n$.

We shall limit ourselves to describing method (iii). All the methods lead to the same result, namely, the function $d(f,A,p)$ is essentially unique, that is, there is a unique topological degree with properties $(G_1),\dots,(G_4)$, as was recently shown by Amann and Weiss [3].

We first of all suppose that f is a C^1-mapping; let Z be the set of critical points of f:

$$Z = \{x \in A \mid Jf(x) = 0\}$$

where $Jf(x)$ denotes the Jacobian of f at the point x. If the point p does not belong to $f(Z)$, then, for every $x \in f^{-1}(p)$ we have $Jf(x) \neq 0$, and so f is locally invertible at x by the implicit function theorem. Thus, all the points of $f^{-1}(p)$ are isolated points. Since $f^{-1}(p)$ is closed by virtue of the continuity of f, $f^{-1}(p)$ is compact. Thus $f^{-1}(p)$ is finite, since its points, being isolated, form an open covering for it from which we can extract a finite subcovering with the same union. It therefore makes sense to write

$$d(f,A,p) = \sum_{x\in f^{-1}(p)} \operatorname{sgn} Jf(x),$$

where $\operatorname{sgn} Jf(x)$ is $+1$ if $Jf(x) > 0$ and -1 if $Jf(x) < 0$. Note that $Jf(x) \neq 0$ because $p \notin f(Z)$, and that $d(f,A,p) = 0$ if $f^{-1}(p)$ is empty.

Using certain theorems from differential calculus, we can show that the function $d(\cdot,\cdot,\cdot)$ has the four characteristic properties (G_1), (G_2), (G_3), (G_4) so long as f is of class C^1 and $p \notin f(Z)$.

We can now extend the degree to points $p \in f(Z)$, that is, to the case in which $f^{-1}(p)$ has critical points. The procedure is as follows. By virtue of Sard's theorem, $f(Z)$ has measure 0 and thus cannot have interior points. It follows that if $p \in f(Z)$, then p is an accumulation point of $R^n \setminus f(Z)$. Since ∂A is compact, $f(\partial A)$ is closed, and so

$$\mathrm{dist}(p,f(\partial A)) > 0,$$

since $p \notin f(\partial A)$ by hypothesis. We can further prove that there is a ball B with center p and radius less than $\mathrm{dist}(p,f(\partial A))$ such that for every $q \in B \setminus f(Z)$, $d(f,A,q)$ is defined and constant with respect to q. Thus $\lim_{q \to p, q \in B \setminus f(Z)} d(f,A,q)$ exists and we may define

$$d(f,A,p) = \lim_{q \to p, q \in B \setminus f(Z)} d(f,A,q).$$

Properties $(G_1),\ldots,(G_4)$ can now be shown for all functions f of class C^1. Moreover, we can show that for any continuous function $f\colon \overline{A} \to R^n$ and $p \notin f(\partial A)$, $d(g,A,p)$ is constant as g varies among all C^1-mappings satisfying the condition

$$||f(x) - g(x)|| < \mathrm{dist}(p,f(\partial A)), \qquad (x \in \partial A).$$

This last property allows us to extend the degree to continuous functions $f\colon \overline{A} \to R^n$; to do this, note that f can be uniformly approximated by C^1-functions g, and the above mentioned property allows us to affirm that

$$\lim_{g \to f, g \text{ of class } C^1} d(g,A,p)$$

exists, so we can define

$$d(f,A,p) = \lim_{g \to f, g \text{ of class } C^1} d(g,A,p).$$

Properties $(G_1),\ldots,(G_4)$ can then be proved.

For greater detail on the construction of the degree using differential calculus, see the work of Nagumo [78] or the books of Schwartz [100], Berger and Berger [8], and Milnor [73]. For the construction of the degree using integral calculus, see the paper by Heinz [41], while for the construction using combinatorial topology, see the work of Leray and Schauder [65], which first made use of the degree in functional analysis. For the construction based on algebraic topology, see Berger and Berger [8], Spanier [101], and Thompson [107].

Property (G_2) is what makes the topological degree a useful instrument in mathematical analysis; it enables us to recognize when the equation

$$f(x) = p$$

has at least one solution x. In applications, the principal problem is that of actually calculating the degree. As we see from the definition, to calculate $d(f,A,p)$ it is enough to approximate f conveniently with a C^1-function g and consider $d(g,A,p)$. To calculate $d(g,A,p)$, it is sufficient to know $d(g,A,q)$ for all q such that $g^{-1}(q)$ does not contain critical points of g. Finally, we can use the following result.

Theorem. Let $A \subseteq \mathbb{R}^n$ be an open bounded set, $f\colon \mathbb{R}^n \to \mathbb{R}^n$ a C^1-function, and $p \notin f(\partial A)$ be such that $f^{-1}(p)$ does not contain critical points of f in A. then there exists an $\varepsilon_0 > 0$ such that for every $\varepsilon \in]0,\varepsilon_0]$ and for every continuous $\phi_\varepsilon\colon \mathbb{R}^n \to \mathbb{R}^n$ with

(i) $\displaystyle\int_{\mathbb{R}^n} \phi_\varepsilon(x)dx = 1$

and

(ii) $\operatorname{supp}(\phi_\varepsilon) \subseteq \{x \in \mathbb{R}^n \mid ||p-x|| \leq \varepsilon\}$,

we have

$$d(f,A,p) = \int_{\bar{A}} \phi_\varepsilon(f(x))Jf(x)dx.$$

For a different integral formula due to Kronecker and for its use in numerical calculation of the degree, see the work of O'Neil and Thomas [84], which also contains references to other procedures for the calculation of the degree.

Since the integral cannot always be calculated, it is in general preferable to compare the given function somehow (e.g., by a convenient homotopy) with another whose degree is known. To do this, one often uses the following:

Theorem of Borsuk. If $A \subseteq \mathbb{R}^n$ is a bounded open set that is symmetrical with respect to the origin (that is, $A = -A$) and contains the origin, and if $f\colon \bar{A} \to \mathbb{R}^n$ is continuous and antipodal

$$f(x) = -f(-x) \qquad (x \in \partial A)$$

and if $0 \notin f(\partial A)$, then $d(f,A,0)$ is an odd number (and thus nonzero).

The degree is used to study the topological properties of $\mathbb{R}^n$ and of continuous functions. See, for example, the book of Dugunji [27], where the degree is defined in a simple way for functions whose domain is the sphere.

We finally observe that if we want to replace R^n with an infinite-dimensional Banach space, then the topological degree cannot be defined for all continuous functions without losing at least one of the characteristic properties $(G_1),\dots,(G_4)$. As an example, we consider the closed unit ball B of ℓ^2 and the continuous mapping $f: B \to \ell^2$ defined by

$$f(x) = (\sqrt{1 - ||x||}, x_1, x_2, \dots)$$

where the x_i are the coordinates of x, $x = (x_1, x_2, \dots)$. One can easily check that f does not have fixed points. We now define a homotopy by

$$f_t(x) = x - tf(x)$$

and let $p(t) = 0$ for every t, $0 \le t \le 1$. If the degree could be defined in ℓ^2 for all continuous mappings from B into ℓ^2 in such a way that properties $(G_1),\dots,(G_4)$ hold, then (G_4) and (G_2) would imply that

$$d(f_t, B, 0) = 1 \qquad (0 \le t \le 1).$$

But this means (as the proof of the theorem of Brouwer to be presented in the next section shows) that f has at least one fixed point, which contradicts what we just proved. The problem of determining what classes of functions in infinite dimensional Banach spaces permit development of a theory of topological degree is still the subject of research. See, for example, Browder [11], Lloyd [123], and the references in their bibliographies.

The property of invariance under homotopy has been taken by Granas [35] as the point of departure for the study of the zeros of certain functions in Banach space.

Exercise 1. Let A be an $n \times n$ invertible matrix (that is, A has nonzero determinant), let $b \in R^n$, and let $f: R^n \to R^n$ be the mapping defined by

$$f(x) = A\cdot x + b.$$

Prove that for every bounded open set $V \subseteq R^n$ such that $-A^{-1}\cdot b \notin \partial V$, $d(f,V,0)$ is defined and

$$d(f,V,0) = \begin{cases} \operatorname{sgn} \det A & \text{for } -A^{-1}\cdot b \in V \\ 0 & \text{for } -A^{-1}\cdot b \notin V. \end{cases}$$

Exercise 2. Using invariance under homotopy, prove that:

(a) the topological degree is locally constant,

(b) the topological degree is continuous in f with respect to the topology of uniform convergence, and

(c) if $||f(x) - g(x)|| < ||g(x)||$ for all $x \in \partial A$, then $d(f,A,0) = d(g,A,0)$.

Exercise 3. Prove that if $A_0 \subseteq A$ is an open set containing $f^{-1}(p)$, then $d(f,A,p) = d(f,A_0,p)$.

Exercise 4. Prove that the degree depends solely on the values on the boundary in the sense that if f and g are continuous mappings from $\bar{A}$ to R^n such that $g|_{\partial A} = f|_{\partial A}$, then $d(f,A,p) = d(g,A,p)$ provided that the degrees are defined.

Exercise 5. Prove that $d(f,A,p) = d(f-p,A,0)$ provided that the degrees are defined, where $f-p$ is the function defined by $x \to f(x) - p$.

1.2. The Theorems of Brouwer and Miranda

We shall now derive from the theory of the topological degree a theorem of Brouwer on fixed points and one of Miranda on zeros.

Theorem of Brouwer. Let B be the closed ball in R^n with center at the origin and radius $\varepsilon > 0$, and let $f\colon B \to R^n$ be a continuous function such that

$$f(\partial B) \subseteq B.$$

Then f has at least one fixed point.

It is clear that the theorem remains true if $f(B) \subseteq B$ and B is any set homeomorphic to a closed ball in R^n. In particular, it is valid if $f(B) \subseteq B$ and B is a convex compact set in R^n. To see this, we use a well-known fact: if B is convex and compact, then there is an s, $1 \leq s \leq n$, such that B is homeomorphic to a convex and compact subset B' of R^s having interior points, and so in turn homeomorphic to the unit ball in R^s.

However, the theorem of Brouwer is not valid for an arbitrary domain B of R^s. For example, the rotation of a torus B in R^n around its axis is a continuous transformation that sends B into B, but it has no fixed point.

Proof of Brouwer's Theorem: We suppose that f does not have fixed points in ∂B, since otherwise there is nothing to show. We then consider the homotopy from B to R^n defined by

$$f_t(x) = tf(x)$$

and show that

$$0 \notin (I - f_t)(\partial B),$$

where I is the identity transformation. When $t = 1$, this follows from the hypothesis just made, while for $t < 1$, it follows from the fact that $0 = x - tf(x)$ implies that $||x|| = t||f(x)||$, so, for $x \in \partial B$, we would have

$$\varepsilon = t||f(x)||$$

which is impossible since $||f(x)|| \leq \varepsilon$ (by the hypothesis of the theorem) and $t < 1$. Because of property (G_4) on the invariance of the topological degree under homotopy, we have

$$d(I - f_0, B, 0) = d(I - f_1, B, 0),$$

that is

$$d(I, B, 0) = d(I - f, B, 0).$$

But by property (G_1), $d(I,B,0) = 1$. Therefore

$$d(I - f, B, 0) = 1.$$

Then property (G_2) implies that the equation

$$x - f(x) = 0$$

has at least one solution; this is a fixed point of f, and the theorem is proven.

We show that Brouwer's theorem is equivalent to the

Theorem of Miranda. Let

$$f_i(x_1, x_2, \ldots, x_n) \qquad (i = 1,2,\ldots,n)$$

be n functions of n variables continuous in the hypercube

$$R: |x_i| \leq L \qquad (i = 1,2,\ldots,n)$$

and satisfying the conditions

$$\left.\begin{array}{l} f_i(x_1,\ldots,x_{i-1},-L,x_{i+1},\ldots,x_n) \geq 0 \\ f_i(x_1,\ldots,x_{i-1},L,x_{i+1},\ldots,x_n) \leq 0 \end{array}\right. \qquad (i = 1,2,\ldots,n). \tag{1.1}$$

on the faces of R. Then there exists at least one point of R at which the n functions f_i all have value 0.

If $n = 1$, this theorem reduces to the elementary theorem on the zeros of a continuous function of one variable whose domain is an interval.

To prove that this result is a consequence of Brouwer's theorem, we first of all suppose that the inequalities (1.1) hold in the strict sense (i.e., without equality), and we consider the mapping

$$x_i' = x_i + \varepsilon_i f_i(x_1, x_2, \ldots, x_n) \qquad (i = 1,2,\ldots,n),$$

where $\varepsilon_1, \varepsilon_2, \ldots, \varepsilon_n$ are positive constants. It will suffice to show that for a proper choice of the constants ε_i, each point P of R is mapped by the given mapping to a point P' still in R; then by Brouwer's theorem, at least one fixed point $\overline{P} = (\overline{x}_1, \overline{x}_2, \ldots, \overline{x}_n)$ exists in R, so that

$$\overline{x}_i = \overline{x}_i + \varepsilon_i f_i(\overline{x}_1, \overline{x}_2, \ldots, \overline{x}_n)$$

and therefore

$$f_i(\overline{x}_1, \overline{x}_2, \ldots, \overline{x}_n) = 0$$

for $i = 1,2,\ldots,n$. To establish this, we denote by M_i and m_i the maximum (positive) and minimum (negative) of f_i on R, and by δ_i', δ_i'' respectively, the distance (positive) of the set $f_i < 0$ from the hyperplane $x_i = -L$ and the distance (positive) of the set $f_i > 0$ from the hyperplane $x_i = L$, and we choose ε_i less than the smaller of the two numbers $-\delta_i'/m_i$ and δ_i''/M_i. We then can easily verify that for every point $P = (x_1, x_2, \ldots, x_n)$ of R we have

$$|x_i'| = |x_i + \varepsilon_i f_i(x_1, x_2, \ldots, x_n)| \le L,$$

for, if $f_i(P) = 0$, the statement if obvious, while if $f_i(P) > 0$, we have

$$-L \le x_i < x_i' \le x_i + \varepsilon_i M_i < x_i + \delta_i'' < L;$$

and if $f_i(P) < 0$ we have

$$L \ge x_i > x_i' \ge x_i + \varepsilon_i m_i > x_i - \delta_i' > -L.$$

The hypothesis that the functions f_i satisfy (1.1) in the strict sense can at last be removed by observing that if the f_i satisfy those inequalities merely in the weak sense, the functions

$$f_i'(x_1, x_2, \ldots, x_n) = f_i(x_1, x_2, \ldots, x_n) - \varepsilon x_i$$

with $\varepsilon > 0$ satisfy them in the strict sense. From what we have seen, there corresponds to each $\varepsilon > 0$ at least one point $P_\varepsilon = (x_1^{(\varepsilon)}, x_2^{(\varepsilon)}, \ldots, x_n^{(\varepsilon)})$ such that

$$f_i'(x_1^{(\varepsilon)}, x_2^{(\varepsilon)}, \ldots, x_n^{(\varepsilon)}) = 0$$

that is,

$$f_i(x_1^{(\varepsilon)}, x_2^{(\varepsilon)}, \ldots, x_n^{(\varepsilon)}) = \varepsilon x_i^{(\varepsilon)},$$

for $i = 1,2,\ldots,n$. Choosing any sequence $\varepsilon_k \downarrow 0$, let $P_k = (x_1^{(k)}, x_2^{(k)}, \ldots, x_n^{(k)})$ be a point P_ε corresponding to $\varepsilon = \varepsilon_k$; it is then possible to extract from the sequence P_k a subsequence P_{k_r} that converges to a point $\overline{P} = (\overline{x}_1, \overline{x}_2, \ldots, \overline{x}_n)$ of R. If we pass to the limit as $r \to \infty$ in the equality

$$\varepsilon_{k_r} x_1^{(k_r)} = f_i(x_1^{(k_r)}, x_2^{(k_r)}, \ldots, x_n^{(k_r)})$$

we get

$$0 = f_1(\overline{x}_1, \overline{x}_2, \ldots, \overline{x}_n) \qquad (i = 1,2,\ldots,n)$$

and the proof is complete.

We have thus proved Miranda's theorem, deriving it from Brouwer's theorem. The equivalence of the two can now be immediately demonstrated by proving that Brouwer's result follows from that of Miranda. To do this, let

$$T\colon y_i = F_i(x_1, x_2, \ldots, x_n) \qquad (i = 1,2,\ldots,n)$$

be a continuous transformation that maps the hypercube R into a subset of itself. The functions

$$f_i(x_1, x_2, \ldots, x_n) = F_i(x_1, x_2, \ldots, x_n) - x_i$$

obviously satisfy the inequalities (1.1). By Miranda's theorem, there is at least one point $\overline{P}$ in R where all the f_i have value 0; this point is mapped to itself by T.

Remark. Miranda's theorem is valid even if the domain of the f_i is a set homeomorphic to a rectangle where one preserves an appropriate correspondence between the boundaries.

Exercise 1. Prove Altman's fixed point theorem: If A is a neighborhood of the origin in $\mathbb{R}^n$ and $f\colon \overline{A} \to \mathbb{R}^n$ is a continuous function such that

$$||x - f(x)||^2 \geq ||f(x)||^2 - ||x||^2 \qquad (x \in \partial A)$$

then f has a fixed point. The significance of this theorem rests in the fact that it does not assume the hypothesis in Brouwer's theorem that $f(\partial A) \subseteq A$. Hint: Use the invariance of the topological degree under homotopy as we did in the proof of Brouwer's theorem.

Exercise 2. Prove the fixed point theorem of Granas: If $f: R^n \to R^n$ is such that

$$\limsup_{||x|| \to \infty} \frac{||f(x)||}{||x||} < 1,$$

then f has a fixed point. Hint: Apply Brouwer's theorem in a ball with center at the origin and radius sufficiently large.

Exercise 3. Prove Miranda's theorem using the topological degree. Hint: First show that it is enough to consider (1.1) in the strict sense, and then consider the homotopy $(f_t)_t$ such that the function f has for its i-th coordinate

$$f_{t,i}(x_1,\ldots,x_n) = -(1 - t)x_i + tf_i(x).$$

Exercise 4. Does Miranda's theorem remain true if the inequalities in (1.1) are changed around?

2. GEOMETRIC BOUNDARY VALUE PROBLEMS

We now take up boundary value problems, both for equations and differential systems. They are quite different from the initial value problems studied in Chapter I. Of these problems, the first to be considered were *Picard's problem* and *Nicoletti's problem*, which we shall examine in Sec. 2.1; we shall then connect them with many other special problems and look at them all from a more general point of view.

2.1. The Boundary Value Problems of Picard and Nicoletti

The problem of Picard consists in determining a solution of the differential equation

$$y^{(n)} = f(x,y,y',\ldots,y^{(n-1)}) \tag{2.1}$$

satisfying the conditions

$$y(x_1) = y_1^0; \quad y(x_2) = y_2^0 ;\ldots; y(x_n) = y_n^0,$$

where $x_1,x_2,\dots,x_n$ are given distinct points in an interval $[a,b]$ and $y_1^0,y_2^0,\dots,y_n^0$ are assigned values.

Let S be the strip in R^{n+1} bounded by the hyperplanes $x = a$, $x = b$ $(a < b)$. We shall show that the given problem admits at least one solution if the function on the right in (2.1) is *continuous and bounded.*

We shall first of all assume that f satisfies a Lipschitz condition locally with respect to the variables $y,y',\dots,y^{(n-1)}$.

For every given n-tuple of constants $\lambda_1,\lambda_2,\dots,\lambda_n$, there exists one and only one integral $y(x,\lambda_1,\lambda_2,\dots,\lambda_n)$ of Eq. (2.1) satisfying the initial conditions

$$\begin{cases} y(a) = \lambda_1 \\ y'(a) = \lambda_2 \\ \dots\dots\dots\dots \\ y^{(n-1)}(a) = \lambda_n \end{cases},$$

as was proved in Chapter I. It follows from what was said in Chapter I that $y(x,\lambda_1,\lambda_2,\dots,\lambda_n)$ is defined in all of $[a,b]$ and is a continuous function in the initial values $\lambda_1,\lambda_2,\dots,\lambda_n$. If we put

$$\phi_i(\lambda_1,\lambda_2,\dots,\lambda_n) = y(x_i,\lambda_1,\lambda_2,\dots,\lambda_n) - y_i^0 \qquad (i = 1,2,\dots,n),$$

what we have to show is equivalent to proving the solvability of the system

$$\phi_i(\lambda_1,\lambda_2,\dots,\lambda_n) = 0 \qquad (i = 1,2,\dots,n)$$

where the functions on the left are continuous in all of R^n. To accomplish this, we observe that if M is the sup (by hypothesis finite) of $|f|$ on S, we can write, by Taylor's formula,

$$y(x,\lambda_1,\lambda_2,\dots,\lambda_n) = \lambda_1 + \lambda_2(x-x_0) + \dots + \lambda_n \frac{(x-x_0)^{n-1}}{(n-1)!} + R_n$$

and

$$\phi_i(\lambda_1,\lambda_2,\dots,\lambda_n) = \lambda_1 + \lambda_2(x_i-x_0) + \dots + \lambda_n \frac{(x_i-x_0)^{n-1}}{(n-1)!} + R_n - y_i^0,$$

with

$$|R_n| \le M \frac{(b-a)^n}{n!}.$$

We now consider the $2n$ hyperplanes π_{ij} in the space R^n whose equations are

$$\pi_{ij}: \lambda_1 + \lambda_2(x_i-x_0) + \dots + \lambda_n \frac{(x_i-x_0)^{n-1}}{(n-1)!} + (-1)^j\rho = 0$$

$$(i = 1,2,\dots,n;\ j = 1,2),$$

where the constant ρ is chosen so that the n inequalities

$$\rho > |y_i^0| + M\frac{(b-a)^n}{n!} \qquad (i = 1,2,\dots,n) \tag{2.2}$$

are satisfied.

These $2n$ hyperplanes come in n pairs (π_{i1},π_{i2}), each formed by two parallel hyperplanes, while hyperplanes from different pairs (i.e., corresponding to different values of the first index) are not parallel. Moreover, the hyperplanes $\pi_{11},\pi_{21},\dots,\pi_{n1}$ are independent in the sense that their locus of intersection is a single point of S_n; it is sufficient to observe that the determinant

$$\begin{vmatrix} 1 & x_1-x_0 & \cdots\cdots & \dfrac{(x_1-x_0)^{n-1}}{(n-1)!} \\ 1 & x_2-x_0 & \cdots\cdots & \dfrac{(x_2-x_0)^{n-1}}{(n-1)!} \\ & \cdots\cdots & \cdots\cdots & \cdots\cdots \\ 1 & x_n-x_0 & \cdots\cdots & \dfrac{(x_n-x_0)^{n-1}}{(n-1)!} \end{vmatrix}$$

is certainly not zero, since it coincides, except for a non-zero factor, with the Vandermonde determinant relative to the numbers $x_1,x_2,\dots,x_n$ which are different by hypotheses.

It follows from this that the $2n$ hyperplanes enclose a domain (which we could call a parallelopiped) in $\mathbb{R}^n$ homeomorphic to a hypercube.

It is immediately evident that at all the points of the hyperplane π_{i1} we have $\phi_i > 0$, while $\phi_i < 0$ at all points of π_{i2}, $i = 1,2,\dots,n$; the wanted result $\phi(\lambda) = 0$ therefore follows from Miranda's theorem by virtue of the observation made about possible extensions to sets that are homeomorphic to rectangles.

We now consider the general case in which f is merely continuous and bounded. Let B be the closed ball with center at the origin and radius ρ given by (2.2). Let $r: \mathbb{R}^n \to B$ be the orthogonal projection. It is known that r satisfies a Lipschitz condition with constant 1. If we approximate f uniformly on $[a,b] \times B$ by C^1-functions and compose r with these, we obtain a sequence $(f_k)_{k=1}^{\infty}$ of bounded functions satisfying a Lipschitz condition with respect to the variables $y, y', \ldots, y^{(n-1)}$ such that $(f_k)_k$ converges uniformly to f on $[a,b] \times B$, and the maximum of M_k of $|f_k|$ converges to the maximum of f on $[a,b] \times B$. Therefore, k_0 exists such that

$$|y_i^0| + M_k \frac{(b-a)^n}{n!} < \rho \qquad (k \geq k_0;\ i = 1,\ldots,n).$$

The result previously established for the Lipschitz case then guarantees that for every $k \geq k_0$, there exists a solution y_k of the Picard problem

$$y_k^{(n)} = f_k(x, y_k, y_k', \ldots, y_k^{(n-1)}), \qquad y_k(x_i) = y_i^0 \qquad (i = 1,\ldots,n)$$

such that

$$|y_k^{(i)}(a)| \leq \rho \qquad (k \geq k_0;\ i = 0,\ldots,n-1).$$

Thus, the integral representation of the solutions of the Cauchy problem and the equiboundedness of the f_k for $k \geq k_0$ imply that the sequence $(y_k^{(i)})_{k=1}^{\infty}$ is uniformly bounded and equicontinuous for each $i = 0,\ldots,n-1$. The theorem of Ascoli-Arzelà implies the existence of a subsequence $(y_{k_j})_{j=1}^{\infty}$ of $(y_k)_{k=1}^{\infty}$ such that $(y_{k_j}^{(i)})_{j=1}^{\infty}$ converges uniformly in $[a,b]$ for each $i = 0,1,\ldots,n-1$. It is easy to see that the limit is a solution of the given problem of Picard.

We now proceed to study Nicoletti's problem; it consists of determining a solution of the system

$$y_i' = f_i(x, y_1, y_2, \ldots, y_n) \qquad (i = 1,\ldots,n) \tag{2.3}$$

satisfying the conditions

$$y_1(x_1) = y_1^0;\ \ y_2(x_2) = y_2^0;\ldots;\ y_n(x_n) = y_n^0$$

where $x_1, x_2, \ldots, x_n$ are distinct points of an interval $[a,b]$ and $y_1^0, y_2^0, \ldots, y_n^0$ are assigned values.

We first of all suppose that the functions f_i on the right-hand side of (2.3) are *continuous* in the strip $S \subseteq \mathbb{R}^{n+1}$ bounded by the hyperplanes $x = a$, $x = b$, and that they are *bounded* and *satisfy a Lipschitz condition* in the variables $y_1, y_2, \ldots, y_n$; this last hypothesis will be removed later.

As in the case of Picard's problem, we consider the solution

$$y_i = y_i(x, \lambda_1, \lambda_2, \ldots, \lambda_n) \qquad (i = 1,2,\ldots,n)$$

of the systems (2.3) satisfying the initial conditions $y_i(a) = \lambda_i$; the functions y_i are defined in all of $[a,b]$ and depend continuously on the initial values $\lambda_1,\ldots,\lambda_n$ by which they are uniquely determined. If we set

$$\psi_i(\lambda_1,\lambda_2,\ldots,\lambda_n) = y_i^0 - y_i(x_i,\lambda_1,\lambda_2,\ldots,\lambda_n),$$

what we want to prove becomes equivalent to demonstrating the solvability of the system

$$\psi_i(\lambda_1,\lambda_2,\ldots,\lambda_n) = 0 \qquad (i = 1,2,\ldots,n)$$

where the functions on the left are continuous in the whole space.

To do this, we observe that the functions $y_i(x,\lambda_1,\lambda_2,\ldots,\lambda_n)$ satisfy the system of integral equations

$$y_i(x) = \lambda_i + \int_a^x f_i(t,y_1(t),\ldots,y_n(t))dt \qquad (i = 1,2,\ldots,n)$$

so that if we put

$$|f_i(x,y_1,y_2,\ldots,y_n)| \le M,$$

we get

$$\lambda_i - M(b-a) \le y_i(x,\lambda_1,\lambda_2,\ldots,\lambda_n) \le \lambda_i + M(b-a).$$

and therefore also

$$y_i^0 - \lambda_i - M(b-a) \le \psi_i(\lambda_1,\lambda_2,\ldots,\lambda_n) \le y_i^0 - \lambda_i + M(b-a).$$

If we denote by R the hypercube in the space R^n defined by the inequalities

$$|\lambda_i| \le L,$$

with

$$L > |y_i^0| + M(b-a)$$

then we immediately see that

$$\psi_i(\lambda_1,\lambda_2,\ldots,\lambda_n) > 0 \quad \text{on the face} \quad \lambda_i = -L$$

and

$$\psi_i(\lambda_1,\lambda_2,\ldots,\lambda_n) < 0 \quad \text{on the face} \quad \lambda_i = L.$$

The theorem of Miranda establishes the desired result $\psi(\lambda) = 0$.

We now drop the hypothesis that the functions f_i satisfy a Lipschitz condition, and suppose merely that they are *continuous* and *bounded* in S. We can proceed as we did in the preceding case for the problem of Picard, or fall back on an approximating procedure due to Tonelli which we previously discussed in Sec. 2.3 of Chapter III. We consider, for every natural number m, the functions

$$y_i^{(m)}(x,\lambda_1,\lambda_2,\ldots,\lambda_n)$$

defined by

$$y_i^{(m)}(x,\lambda_1,\lambda_2,\ldots,\lambda_n) = \begin{cases} \lambda_i & \text{for } a \le x \le a + \frac{1}{m} \\ \lambda_i + \int_a^{x-\frac{1}{m}} f_i(t, y_1^{(m)}(t,\lambda_1,\ldots,\lambda_n),\ldots, \\ \quad y_n^{(m)}(t,\lambda_1,\ldots,\lambda_n))dt & \text{for } a + \frac{1}{m} \le x \le b. \end{cases}$$

These functions are uniquely determined by the initial values $\lambda_1,\lambda_2,\ldots,\lambda_n$ and depend continuously on the variables $\lambda_1,\ldots,\lambda_n$; this is evident in the interval $a \le x \le a + \frac{1}{m}$ and so follows for each of the consecutive intervals of diameter $1/m$, as one can easily prove by induction.

We have, moreover,

$$\lambda_i - M(b-a) \le y_i^{(m)}(x,\lambda_1,\lambda_2,\ldots,\lambda_n) \le \lambda_i + M(b-a). \tag{2.4}$$

Reasoning as above when we assumed that the f_i satisfied a Lipschitz condition, we conclude that in the same hypercube R considered there there exists, for every value of the index m, at least one n-tuple $\lambda_1^{(m)},\lambda_2^{(m)},\ldots,\lambda_n^{(m)}$ such that

$$y_i(x_i,\lambda_1^{(m)},\lambda_2^{(m)},\ldots,\lambda_n^{(m)}) = y_i^0 \qquad (i = 1,2,\ldots,n). \tag{2.5}$$

We finally observe that for every pair x', x'' of points in $[a,b]$, we have

$$|y_i^{(m)}(x',\lambda_1,\lambda_2,\ldots,\lambda_n) - y_i^{(m)}(x'',\lambda_1,\lambda_2,\ldots,\lambda_n)| \le M|x'' - x'|$$

as we easily see from the definition of the functions $y_i^{(m)}(x,\lambda_1,\lambda_2,\ldots,\lambda_n)$ the functions are therefore equicontinuous.

It now follows from (2.4) and from the fact that the points $(\lambda_1^{(m)}, \ldots, \lambda_n^{(m)})$ all belong to the same hypercube R, that these functions are also uniformly bounded; by the theorem of Ascoli-Arzelà they form a compact set with respect to uniform convergence. It follows that we may extract from the set of n-tuples

$$\{y_i^{(m)}(\cdot,\lambda_1^{(m)},\ldots,\lambda_n^{(m)})\} \quad (i = 1,2,\ldots,n)$$

a sequence

$$\left\{y^{(m_k)}(\cdot,\lambda_1^{(m_k)},\ldots,\lambda_n^{(m_k)})\right\}_k$$

that converges uniformly in $[a,b]$ to a limit y which, by the argument in Sec. 3.1 of Chapter III, is an integral of the system (2.3). This integral, because of (2.5), satisfies the conditions $y_i(x_i) = y_i^0$, and is therefore a solution of our Nicoletti problem.

Remark I. We must point out that the Picard problem and the Nicoletti problem are essentially different, even in the case of higher order equations. The problem of Nicoletti, stated for Eq. (2.1), consists in determining an integral y of the same equation satisfying the conditions

$$y(x_1) = y_1^0, \quad y'(x_2) = y_2^0, \ldots, y^{(n-1)}(x_n) = y_n^0,$$

which are different from those of the Picard problem.

Remark II. The Picard problem for a system (2.3) consists in determining an integral y satisfying the conditions $y_1(x_i) = y_i^0$. In general, the Picard problem posed relative to the system (2.3) rather than to the Equation (2.1) does not have a solution. It suffices to consider the case when the function f_1 depends only on the variables x, y_1. The first of the equations (2.3) then becomes

$$y_1' = f(x,y_1),$$

and it is quite clear that there is in general no integral of this first order equation that satisfies the n conditions $y_1(x_1) = y_1^0$, $y_1(x_2) = y_2^0; \ldots; y_1(x_n) = y_n^0$.

Remark III. Both the Cauchy problem and the boundary value problems of Picard and Nicoletti can be thought of as particular cases of the problem of determining a solution $y_1(x), y_2(x), \ldots, y_n(x)$ of the system (2.3) satisfying the conditions

$$y_{i_1}(x_1) = c_{11}, \quad y_{i_2}(x_1) = c_{12}; \ldots ; y_{i_{\nu_1}}(x_1) = c_{1\nu_1}$$
$$y_{j_1}(x_2) = c_{21}, \quad y_{j_2}(x_2) = c_{22}; \ldots ; y_{j_{\nu_2}}(x_2) = c_{2\nu_1}$$
$$\cdots\cdots\cdots\cdots\cdots\cdots\cdots\cdots$$
$$y_{k_1}(x_m) = c_{m1}, \quad y_{k_2}(x_m) = c_{m2}; \ldots ; y_{k_{\nu_m}}(x_m) = c_{m\nu_m} .$$

where $x_1, x_2, \ldots, x_m$ are distinct points $(1 \le m \le n)$ and c_{ts} are assigned values $(\nu_1 + \nu_2 + \ldots + \nu_m = n)$. The Cauchy problem is obtained when $m = 1$ (and therefore $\nu_1 = n$), the Picard problem for $m = n$ (and therefore $\nu_1 = \nu_2 = \ldots = \nu_m = 1$) and $i_1 = j_1 = \ldots\ldots = k_1 = 1$, and the Nicoletti problem for $m = n$ and $i_1 = 1$, $j_1 = 2, \ldots, k_1 = n$.

2.2. A Geometrical Formulation of the Boundary Value Problems

We point out that the problems of Picard and Nicoletti which we studied in the preceding section can be reformulated from a more geometrical point of view. If we think of Eq. (2.1) as being written in the form of a system (2.3), we see that in both cases the integral curves $y_i = y_i(x)$ $(i = 1,2,\ldots,n)$ are curves in R^{n+1} that are met at one point only by each hyperplane perpendicular to the x-axis that originates from a point of $[a,b]$.

The conditions of the Picard problem

$$y_1(x_i) = y_i^0 \quad (i = 1,2,\ldots,n)$$

require that an integral touches the n linear spaces of dimension $n - 1$ whose equations are, respectively,

$$\begin{cases} x = x_1 \\ y_1 = y_1^0 \\ y_i = y_i \ (i \ge 2) \end{cases}, \quad \begin{cases} x = x_2 \\ y_1 = y_2^0 \\ y_i = y_i \ (i \ge 2) \end{cases}, \ldots, \begin{cases} x = x_n \\ y_1 = y_n^0 \\ y_i = y_i \ (i \ge 2) \end{cases} .$$

These spaces are all parallel to the same coordinate space R^{n-1} of points with coordinates $(y_2, y_3, \ldots, y_n)$.

The conditions of the Nicoletti problem require, on the other hand, that an integral of the system (2.3) touches the n linear spaces of dimension $n - 1$ whose equations are

$$\begin{cases} x = x_1 \\ y_1 = y_1^0, \\ y_i = y_i \ (i \neq 1) \end{cases}, \quad \begin{cases} x = x_2 \\ y_2 = y_2^0 \\ y_i = y_i \ (i \neq 2) \end{cases}, \ldots, \begin{cases} x = x_n \\ y_n = y_n^0 \\ y_i = y_i \ (i \neq n) \end{cases}.$$

These are parallel to the coordinate spaces R^{n-1} which consist of the points

$$(y_2, y_3, \ldots, y_n);\ (y_1, y_3, y_4, \ldots, y_n); \ldots; (y_1, y_2, \ldots, y_{n-1}).$$

The initial conditions of the Cauchy problem

$$y_i(x_0) = y_i^0 \qquad (i = 1,2,\ldots,n)$$

can also be interpreted as conditions that an integral touches the n linear spaces of dimension $n - 1$ given by

$$\begin{cases} x = x_0 \\ y_i = y_i^0 \end{cases} \quad (i = 1,2,\ldots,n)$$

which, this time, are contained in the hyperplane $x = x_0$ and have a point in common.

Each of the problems we have considered is thus equivalent to finding an integral curve of the given differential system which touches n linear spaces of dimension $n - 1$; the relative position of these spaces is what distinguishes one problem from the other.

We can, from this point of view, raise in a natural way a problem much more general than the ones we have thus far considered, namely, that of *determining when an integral of a given system exists that intersects* n *assigned* $n-1$ *dimensional varieties (perhaps nonlinear) of the space* R^{n+1}. This is called the *geometric boundary value problem.* Among the problems that can be formulated in this way, we find not only the problems of Cauchy, Picard and Nicoletti, but also very many other boundary value problems that appear in various circumstances of both theoretical and applied interest. Everything depends on a proper choice of the varieties. For example, the problem of determining an integral of the equation

$$y'' = f(x,y,y')$$

that is tangent to the two curves

$$y = \phi_1(x), \quad y = \phi_2(x)$$

in the (x,y) plane is equivalent to the problem of finding an integral of the system

$$\begin{cases} y' = z \\ z' = f(x,y,z) \end{cases}$$

that touches the two curves

$$\begin{cases} x = x & \qquad x = x \\ y = \phi_1(x) & \qquad y = \phi_2(x) \\ z = \phi_1'(x) & \qquad z = \phi_2'(x) \end{cases}$$

in R^3; it can therefore be expressed in terms of the preceding general formulation.

We shall now study the geometrical boundary value problem in a general setting; for simplicity of notation and in order to facilitate geometrical intuition, we shall limit ourselves to the case $n = 2$. We note, however, that all arguments and results that we shall present can be extended to the case $n > 2$.

Let us now consider the system

$$\begin{cases} y_1' = f_1(x,y_1,y_2) \\ y_2' = f_2(x,y_1,y_2) \end{cases} \tag{2.6}$$

and suppose that the functions on the right are continuous in the strip S in R^3 bounded by the planes $x = a$, and $x = b$ $(a < b)$. We shall furthermore suppose that there are four functions $\phi_{ij}(x)$, all integrable in $[a,b]$, and two functions $\phi_i(x)$, both nonnegative and integrable in $[a,b]$, such that

$$|f_i(x,y_1,y_2) - \sum_j^{1,2} \phi_{ij}(x)y_j| \leq \phi_i(x) \qquad (i = 1,2) \tag{2.7}$$

in S.

It follows from the theorem proved in Sec. 2.4 of Chapter III that the integrals of the system (2.6) are defined in all of $[a,b]$.

Although it is unnecessary, we furthermore assume the hypothesis that the functions f_i satisfy a Lipschitz condition in S with respect to the variables y_1 and y_2; this insures that the integral curves originating from a point $P_0 = (x_0,y_1^0,y_2^0)$ are uniquely determined by P_0 and depend on P_0 continuously. (In the absence of conditions insuring uniqueness and continuous dependence of the integral curves on the initial point, one could fall back on procedures of approximation of the type considered in Sec. 2.1 for the Nicoletti Problem.)

If we fix an arbitrary point c in $[a,b]$, we can make it correspond to a point P in S the point Q where the integral curve of the system (2.6) that passes through P meets the plane $x = c$. We shall say that Q is the *projection of* P *on the plane* $x = c$ *under the system* (2.6); this projection reduces to the ordinary projection when f_i is constant, $i = 1,2$.

The problem we want to study consists of determining an integral of the system (2.6) that touches two assigned curves V_1 and V_2 contained in S; we project the two curves V_1 and V_2 on the plane $x = c$ in the above mentioned way and denote by U_1 and U_2 the sets thereby obtained. Our problem has a solution if and only if U_1 and U_2 have points in common, since any possible solution must be an integral curve originating from a point in common to both U_1 and U_2. *The study of the problem is thus reduced to the search for sufficient conditions to insure that the sets* U_1 *and* U_2 *have points in common.*

We consider along with the system (2.6) the system

$$y_i' = \lambda f_i(x,y_1,y_2) + (1-\lambda) \sum_j^{1,2} \phi_{ij}(x)y_j \qquad (i = 1,2) \tag{2.8}$$

that depends on the parameter λ. If $\lambda = 1$, this coincides with the system (2.6), while if $\lambda = 0$, it reduces to the homogeneous linear system

$$y_i' = \sum_j^{1,2} \phi_{ij}(x)y_i \qquad (i = 1,2). \tag{2.9}$$

Along with the two curves V_1 and V_2 we consider two variable curves $V_1^{(\lambda)}$ and $V_2^{(\lambda)}$ that depend continuously on the parameter λ; if $\lambda = 1$, they reduce to the two given curves, $V_1^{(1)} = V_1$, $V_2^{(1)} = V_2$, while for $\lambda = 0$ they reduce to certain curves $V_1^{(0)}$ and $V_2^{(0)}$ in such a way that *every point* $V_i^{(\lambda)}$, $i = 1,2$, *as* λ *varies in* $[0,1]$, *remains in a plane perpendicular to the* x-*axis and undergoes a displacement less than a positive number* θ.

We first of all suppose that $V_1^{(0)}$ and $V_2^{(0)}$ are the two lines contained in the planes $x = x_i$ with equations

$$V_i^{(0)} = \begin{cases} x = x_i \\ \sum_j^{1,2} a_{ij}y_j = b_i \end{cases} \qquad (i = 1,2)$$

This means that the given curves V_1 and V_2 must also be contained in the same planes.

We shall define the linear homogeneous problem associated with the given one to be the problem that consists of finding an integral curve of the system (2.9) that touches the two lines with equations

$$\begin{cases} x = x_i \\ \sum_j^{1,2} a_{ij} y_j = 0 \end{cases} \qquad (i = 1,2). \tag{2.10}$$

We wish to prove the following

Theorem. The given problem has a solution if the linear homogeneous problem associated with it admits only the null solution.

If $U_i^{(\lambda)}$ is the projection of $V_i^{(\lambda)}$ on the plane $x = c$ under the system (2.8), let us measure the size of the displacement that a point of $U_i^{(0)}$ undergoes as λ varies. To do this, we first denote by $P_0 = (x_1, y_1^{(0)}, y_2^{(0)})$ a point of e.g. $V_1^{(0)}$ and by $P_\lambda = (x_1, y_1^{(\lambda)}, y_2^{(\lambda)})$ the point corresponding to P_0 on the curve $V_1^{(\lambda)}$, and for every value of λ we let

$$\begin{cases} y_1 = y_1(x,\lambda) \\ y_2 = y_2(x,\lambda) \end{cases}$$

be the equations of the integral curve of the system (2.8) that originates from P_λ. We then have

$$y_i(x,\lambda) = y_i^{(\lambda)} + \lambda\int_{x_1}^{x} \Big\{f_i[t,y_1(t,\lambda),y_2(t,\lambda)] - \sum_j^{1,2} \phi_{ij}(t) y_j(t,\lambda)\Big\}dt + \int_{x_1}^{x} \sum_j^{1,2} \phi_{ij}(t) y_j(t,\lambda)\, dt,$$

and, for $\lambda = 0$,

$$y_i(x,0) = y_i^{(0)} + \int_{x_1}^{x} \sum_j^{1,2} \phi_{ij}(t) y_j(t,0)\, dt.$$

If we subtract side from side, we get

$$\begin{aligned} y_i(x,\lambda) - y_i(x,0) = y_i^{(\lambda)} - y_i^{(0)} + \lambda\int_{x_1}^{x} \Big\{f_i[t,y_1(t,\lambda),y_2(t,\lambda)] \\ - \sum_j^{1,2} \phi_{ij}(t) y_j(t,\lambda)\Big\}dt + \int_{x_1}^{x} \Big\{ \sum_j^{1,2} \phi_{ij}(t)[y_j(t,\lambda) - y_j(t,0)]\Big\}dt. \end{aligned} \tag{2.11}$$

If $z(x,\lambda)$ is the greater of the two differences

$$|y_1(x,\lambda) - y_1(x,0)|, \quad |y_2(x,\lambda) - y_2(x,0)|,$$

then $z(x,\lambda)$ is obviously a continuous function of x in $[a,b]$, and it follows from (2.11) and (2.7), that

$$|y_i(x,\lambda) - y_i(x,0)| \leq |y_i^{(\lambda)} - y_i^{(0)}| + \int_a^b \phi_i(t)dt + \left|\int_{x_1}^x \left\{\sum_j^{1,2} |\phi_{ij}(t)| z(t,\lambda)\right\}dt\right|$$

and so also that

$$z(x,\lambda) \leq \sum_i^{1,2} |y_i^{(\lambda)} - y_i^{(0)}| + \int_a^b \left\{\sum_i^{1,2} \phi_i(t)\right\}dt + \left|\int_{x_1}^x \left\{\sum_{ij}^{1,2} |\phi_{ij}(t)| z(t,\lambda)\right\}dt\right|.$$

From Gronwall's lemma, we get

$$z(x,\lambda) \leq \left\{\sum_i^{1,2} |y_i^{(\lambda)} - y_i^{(0)}| + \int_a^b \sum_i^{1,2} \phi_i(t)dt\right\} e^{\int_a^b \sum_{ij}^{1,2} |\phi_{ij}(t)|dt}$$

and hence

$$|y_i(x,\lambda) - y_i(x,0)| \leq \left\{\sum_i^{1,2} |y_i^{(\lambda)} - y_i^{(0)}| + \int_a^b \sum_i^{1,2} \phi_i(t)dt\right\} \cdot e^{\int_a^b \sum_{ij}^{1,2} |\phi_{ij}(t)|dt} \qquad (i = 1,2).$$

If Q_0 and Q_λ are the projections of the points P_0 and P_λ on the plane $x = c$, then this last inequality and our hypothesis, $||P_0 - P_\lambda|| \leq \theta$, clearly provide us with an upper bound for the distance $||Q_0 - Q_\lambda||$:

$$||Q_0 - Q_\lambda|| \leq 2\left\{\theta + \int_a^b \sum_i^{1,2} \phi_i(t)dt\right\} e^{\int_a^b \sum_{ij}^{1,2} |\phi_{ij}(t)|dt}.$$

We now prove that the two lines (2.10) project under the system (2.9) onto two lines originating from the origin of the plane $x = c$. Therefore, to suppose that the homogeneous system associated with the given system admits only the null solution is equivalent to supposing that these last two lines are distinct.

To prove this, we denote by $Q = (c,y_1,y_2)$ a variable point in the plane $x = c$, and let

$$Y(x) = \begin{pmatrix} y_1(x) \\ y_2(x) \end{pmatrix}$$

be the integral of the system (2.9) originating from Q. We already saw in Chapter II that

$$Y(x) = \begin{pmatrix} Y_{11}(x) & Y_{12}(x) \\ Y_{21}(x) & Y_{22}(x) \end{pmatrix} \cdot \begin{pmatrix} y_1 \\ y_2 \end{pmatrix}$$

where

$$\begin{pmatrix} Y_{11}(x) \\ Y_{21}(x) \end{pmatrix}, \begin{pmatrix} Y_{12}(x) \\ Y_{22}(x) \end{pmatrix}$$

are the two integrals of the system (2.9) satisfying the initial conditions

$$\begin{cases} Y_{11}(c) = 1 \\ Y_{21}(c) = 0 \end{cases}, \quad \begin{cases} Y_{12}(c) = 0 \\ Y_{22}(c) = 1 \end{cases}.$$

The condition for the integral curve $y_1 = y_1(x)$, $y_2 = y_2(x)$ to touch the line

$$\begin{cases} x = x_i \\ \sum_j^{1,2} a_{ij} y_i = 0 \end{cases} \qquad (i = 1,2)$$

is

$$(a_{i1} \quad a_{i2}) \cdot \begin{pmatrix} Y_{11}(x_i) & Y_{12}(x_i) \\ Y_{21}(x_i) & Y_{22}(x_i) \end{pmatrix} \cdot \begin{pmatrix} y_1 \\ y_2 \end{pmatrix} = 0 \qquad (i = 1,2). \tag{2.13}$$

The equations in (2.13), linear and homogeneous in the variables y_1 and y_2, are equations of lines originating from the origin and represent - in the plane $x = c$ - the projections of the lines (2.10) under the system (2.9). The condition for the lines with equations (2.13) not to coincide is

$$\det \begin{pmatrix} a_{11}Y_{11}(x_1)+a_{12}Y_{21}(x_1) & a_{11}Y_{12}(x_1)+a_{12}Y_{22}(x_1) \\ a_{21}Y_{11}(x_2)+a_{22}Y_{21}(x_2) & a_{21}Y_{12}(x_2)+a_{22}Y_{22}(x_2) \end{pmatrix} \neq 0. \tag{2.14}$$

Observe that the matrix on the left in (2.14) is obtained by multiplying the two matrices

$$\begin{pmatrix} a_{11} & a_{12} \\ a_{21} & a_{22} \end{pmatrix}, \quad \begin{pmatrix} \gamma_{11}(x) & \gamma_{12}(x) \\ \gamma_{21}(x) & \gamma_{22}(x) \end{pmatrix}$$

and replacing x with x_1 in the first and x with x_2 in the second.

We consider the sets $U_i^{(0)}$ $(i = 1,2)$, the projections of the lines $V_i^{(0)}$ under the system (2.9); a calculation similar to that explained above shows that these projections are the two lines of the plane $x = c$ whose equations are

$$(a_{i1} \quad a_{i2}) \cdot \begin{pmatrix} \gamma_{11}(x_i) & \gamma_{12}(x_i) \\ \gamma_{21}(x_i) & \gamma_{22}(x_i) \end{pmatrix} \cdot \begin{pmatrix} y_1 \\ y_2 \end{pmatrix} = b_i \quad (i = 1,2) \tag{2.15}$$

Under hypothesis (2.14), *these lines are not parallel.*

We now summarize the results thus far obtained. The hypotheses about the functions f_i on the right of the system (2.6) and about the way the curves $V_i^{(\lambda)}$ are deformed have as a consequence that as λ varies in $[0,1]$, the point Q_λ - the projection of a point P_λ of $V_i^{(\lambda)}$ under the system (2.8) - varies in a circle with center Q_0 and radius equal to the quantity on the right hand side of (2.12). The hypothesis that the linear homogeneous problem associated with the given one has only the null solution implies that the two lines $U_i^{(0)}$, the projections of the lines $V_i^{(0)}$ under the system (2.9), are not parallel.

We must show that under these conditions, the curves $U_i \equiv U_i^{(1)}$ $(i = 1,2)$ have at least one point in common; this is equivalent, as we have seen, to the fact that our geometric boundary value problem has a solution.

To make the proof, we first denote by P_0 the point common to the two lines $U_1^{(0)}$ and $U_2^{(0)}$; next, given $R > 0$, we consider the two points P_1' and P_1'' on the line $U_2^{(0)}$ whose distance from the point P_0 is R and then the two points P_2' and P_2'' on the line $U_1^{(0)}$ whose distance from P_0 is also R. The lines through P_i' and P_i'' parallel to $U_i^{(0)}$ determine a parallelogram π with center P_0; we denote by ε_i' and ε_i'' the pairs of parallel sides containing respectively P_i' and P_i'', $i = 1,2$. It is clear that by choosing R sufficiently large, we can make the distance of ε_i' and ε_i'' from the line $U_i^{(0)}$ greater than the quantity on the right side of (2.12).

We now suppose that the given curves V_1 and V_2 are *continuous and simple Jordan curves;* it is then clear - on account of the continuous dependence of the integrals of the system (2.6) on the initial conditions - that the sets U_1 and U_2 are curves of the same type. Furthermore, by virtue of (2.12), the curve U_i does not meet any of the sides $\varepsilon_i', \varepsilon_i''$ $(i = 1,2)$.

We denote by E_1 an arc of the curve U_1 whose endpoints are on the sides $\varepsilon_2', \varepsilon_2''$ and whose other points are in the interior of π; we denote by E_2 an arc of the curve U_2 with similar properties. It is then clear (in fact, it follows from a theorem of Jordan) that E_i *separates* ε_i' *from* ε_i'' *with respect to* π in the sense that every arc of a regular curve with one endpoint on ε_i' and the other on ε_i'' and all other points in the interior of π must meet the set E_i, $i = 1,2$.

We indicate by $d_i(Q)$ the distance from a point Q of π to the set E_i, $i = 1,2$. We define two continuous functions

$$\eta_i = \eta_i(Q) \qquad (i = 1,2)$$

on π in the following way: set $\eta_i(Q) = 0$ at all points Q of E_i; when Q is not in E_i, we put $\eta_i(Q) = d_i(Q)$ if Q can be joined to a point of ε_i' by an arc of a curve that does not meet E_i and $\eta_i(Q) = -d_i(Q)$ otherwise.

We evidently have

$$\eta_i(Q) > 0 \text{ on } \varepsilon_i', \quad \eta_i(Q) < 0 \text{ on } \varepsilon_i'' \qquad (i = 1,2)$$

so that, by Miranda's theorem of Sec. 1.2, there is at least one point $\overline{Q}$ of π where

$$\eta_1(\overline{Q}) = \eta_2(\overline{Q}) = 0.$$

$\overline{Q}$ is thus a common point of the sets E_1 and E_2 (and therefore also of the two curves U_1 and U_2), and the theorem is proven.

<u>Remark</u>: The existence of a point common to the two curves U_1 and U_2 can also be proved by using Heine's theorem on the zeros of a continuous function of a real variable. In fact the function $\eta_1(Q)$, as Q varies along U_2, is a continuous function of the variable t which parametrizes U_2: $\eta_1 = \delta_1(t)$. Since U_2 meets side ε_1' as well as side ε_1'' of π, there exist two values t_0 and t_1 of t such that $\delta_1(t_0) > 0$ and $\delta_1(t_1) < 0$. It follows for at least one point $\overline{t}$ of $[t_0,t_1]$ we must have $\delta_1(\overline{t}) = 0$. The point Q of U_2 corresponding to the value $\overline{t}$ of the parameter has distance 0 from E_1 and so belongs both to U_1 and U_2.

Unfortunately, this argument does not extend to the case $n > 2$, and it is for this reason that we prefer the other proof.

2.3. Some Applications of the Geometric Formulation

We now give some applications of the results obtained in the preceding section. As a first example, we determine an integral curve of the system

$$\begin{cases} y_1' = f_1(x,y_1,y_2) \\ y_2' = f_2(x,y_1,y_2) \end{cases} \tag{2.16}$$

that touches the two lines V_1 and V_2 with equations

$$\begin{cases} x = x_i \\ \sum_j^{1,2} a_{ij}y_j = b_i \end{cases} \qquad (i = 1,2) \tag{2.17}$$

under the hypothesis that the functions f_i on the second side of the system (2.6) satisfy the estimates

$$|f_i(x,y_1,y_2)| \le \phi(x) \qquad (i = 1,2) \tag{2.18}$$

where $\phi(x)$ is integrable in $[a,b]$.

Since the given varieties V_1 and V_2 are now two lines perpendicular to the x-axis, we assume that $V_i^{(\lambda)} = V_i$ for every λ in the closed interval $[0,1]$; moreover, the hypothesis (2.18) is a particular case of (2.7) if we take the functions $\phi_{ij}(x)$ that appear there to be identically 0. The homogeneous linear problem associated with the given one is then to determine an integral curve of the system

$$\begin{cases} y_1' = 0 \\ y_2' = 0 \end{cases} \tag{2.19}$$

that touches the two lines with equations

$$\begin{cases} x = x_i \\ \sum_j^{1,2} a_{ij}y_j = 0. \end{cases} \tag{2.20}$$

The integral curves of the system (2.19) are then lines parallel to the x-axis, so the lines on the plane $x = c$ that one gets by projecting the lines with Eqs. (2.20) under the system (2.19) are distinct if and only if the lines with Eqs. (2.17) are not parallel.

It follows from the theorem proved in the preceding section that *the given problem has a solution if the lines* (2.17) *are not parallel.*

In particular, there are solutions if the two given lines V_1 and V_2 are such that one is parallel to the y_1-axis and the other to the y_2-axis, in which case the given problem coincides with the Nicoletti problem of Sec. 2.1.

For a final example, we consider the problem of determining an integral of the differential equation

$$y'' = f(x,y,y')$$

satisfying the boundary conditions

$$\begin{cases} a_{11}y(x_1) + a_{12}y'(x_1) = b_1 \\ a_{21}y(x_2) + a_{22}y'(x_2) = b_2 \end{cases}$$

under the hypothesis that

$$|f(x,y,y')| \leq \phi(x) \tag{2.21}$$

in the strip $S = [a,b] \times R^2$, with $\phi(x)$ integrable in $[a,b]$.

If we set $y = y_1$, $y' = y_2$, the problem becomes equivalent to the search for an integral curve of the system

$$\begin{cases} y_1' = y_2 \\ y_2' = f(x,y_1,y_2) \end{cases} \tag{2.22}$$

that touches the two lines with equations

$$\begin{cases} x = x_i \\ \sum_j^{1,2} a_{ij}y_j = b_i \end{cases} \qquad (i = 1,2) \tag{2.23}$$

System (2.22) satisfies the hypotheses (2.7) when we choose

$$\phi_{11}(x) \equiv 0 \qquad \phi_{12}(x) \equiv 1 \quad ; \quad \phi_1(x) \equiv 0$$

$$\phi_{21}(x) \equiv \phi_{22}(x) \equiv 0 \qquad ; \quad \phi_2(x) \equiv \phi(x).$$

The homogeneous problem associated with the given problem consists, therefore, in determining an integral of the system

$$\begin{cases} y_1' = y_2 \\ y_2' = 0 \end{cases} \tag{2.24}$$

that touches the two lines with equations

$$\begin{cases} x = x_i \\ \sum_j^{1,2} a_{ij} y_j = 0 . \end{cases}$$

The two integrals $\begin{pmatrix} \gamma_{11}(x) \\ \gamma_{21}(x) \end{pmatrix}, \begin{pmatrix} \gamma_{12}(x) \\ \gamma_{22}(x) \end{pmatrix}$ satisfying the initial conditions

$$\begin{cases} \gamma_{11}(c) = 1 \\ \gamma_{21}(c) = 0 \end{cases}, \quad \begin{cases} \gamma_{12}(c) = 0 \\ \gamma_{22}(c) = 1 \end{cases}$$

are now

$$\begin{cases} \gamma_{11}(x) = 1 \\ \gamma_{21}(x) = 0 \end{cases}, \quad \begin{cases} \gamma_{12}(x) = x - c \\ \gamma_{22}(x) = 1 \end{cases} .$$

Then condition (2.14) becomes

$$\det \begin{pmatrix} a_{11} & a_{11}(x_1 - c) + a_{12} \\ a_{21} & a_{21}(x_2 - c) + a_{22} \end{pmatrix} \neq 0.$$

In the case of the Picard problem, where $a_{11} = a_{21} = 1$, $a_{12} = a_{22} = 0$, this condition is certainly satisfied (the problem thus admits a solution) and reduced to

$$\det \begin{pmatrix} 1 & x_1 - c \\ 1 & x_2 - c \end{pmatrix} = x_2 - x_1 \neq 0.$$

Observe, however, that in the case of the Picard problem, it is immediate to verify that the associated homogeneous linear problem, i.e., the problem of determining an integral of the system (2.24) that satisfies the conditions $y_1(x_1) = y_1(x_2) = 0$, has only the null solution. In fact, the general integral of the system (2.24) is

$$\begin{cases} y_1 = c_1 x + c \\ y_2 = c_2 \end{cases}$$

where c_1 and c_2 are constants, such that the boundary conditions $y_1(x_1) = y_1(x_2) = 0$ with $x_1 \neq x_2$ can only be satisfied by choosing $c_1 = c_2 = 0$, that is, by $y_1(x) \equiv y_2(x) \equiv 0$.

3. STURM-LIOUVILLE PROBLEMS: EIGENVALUES AND EXISTENCE AND UNIQUENESS THEOREMS

We encountered the Sturm-Liouville problem at the end of the preceding section; it consists of determining a solution of a second order equation

$$y'' = f(x,y,y') \tag{3.1}$$

satisfying linear boundary conditions of the type

$$\alpha_1 y(a) + \alpha_2 y'(a) = 0; \quad \beta_1 y(b) + \beta_2 y'(b) = 0 \tag{3.2}$$

where $a < b$ are two given points in the interval in which x varies and α_i, β_i are fixed real numbers such that $\alpha_1^2 + \alpha_2^2 > 0$, $\beta_1^2 + \beta_2^2 > 0$. What is of interest is to determine conditions on f that will insure the existence and perhaps the uniqueness of solutions of (3.1) satisfying (3.2). We shall here concern ourselves with conditions that are obtained by comparing Eq. (3.1) with the linear equations

$$y'' + \lambda y = 0 \tag{3.3}$$

under the boundary conditions (3.2), where λ is a real number for which the problem admits at least one nontrivial solution. These λ are called eigenvalues of (3.3) relative to the Sturm-Liouville problem (3.2).

In the first part of this section, we shall study the eigenvalues of second order linear equations relative to a Sturm-Liouville problem and try to determine how many of them there are and how they depend on the coefficients of the equation and on the endpoints of the interval $[a,b]$. The second part is devoted to applications to uniqueness and existence theorems for non-linear Strum-Liouville problems (3.1), (3.2).

3.1. Eigenvalues and Eigenfunctions

Consider the Sturm-Liouville problem

$$(p(x)y')' + q(x,\lambda)y = 0 \tag{3.4}$$

$$\alpha_1 y(a) + \alpha_2 y'(a) = 0; \quad \beta_1 y(b) + \beta_2 y'(b) = 0 \tag{3.2}$$

depending on the parameter λ. To avoid repetitions, we state now that the coefficients p and $q(\cdot,\lambda)$ in Eq. (3.4) shall, in this section, always be functions defined on the interval $[a,b]$ satisfying the conditions that p is a *positive* C^1-function and $q(\cdot,\lambda)$ a continuous function. (It would be enough, though, merely to assume that $q(\cdot,\lambda)$ is integrable for every fixed λ.)

Note that when $\alpha_1 = \beta_1 = 1$ and $\alpha_2 = \beta_2 = 0$, this problem reduces to the two point boundary value problem or Picard's problem

$$y(a) = y(b) = 0,$$

while we get the Nicoletti problem for $\alpha_1 = \beta_2 = 1$ and $\alpha_2 = \beta_1 = 0$:

$$y(a) = y'(b) = 0.$$

In applications, the function q often has one of the forms $q(x,\lambda) = \lambda k(x)$ or $q(x,\lambda) = h(x) + \lambda k(x)$.

The *eigenvalues* of the Sturm-Liouville problem (3.4), (3.2) are those values of λ for which nontrivial solutions exist. If λ is an eigenvalue, the corresponding nontrivial solutions are called *eigenfunctions* relative to λ. The set of all eigenfunctions corresponding to a particular eigenvalue, together with the zero function, is a vector space which is called the *eigenspace* relative to λ. This means that every linear combination of solutions of (3.4), (3.2) is also a solution of (3.4), (3.2).

For an example, we consider the two-point boundary value problem

$$y'' + \lambda y = 0, \quad y(0) = y(b) = 0, \tag{3.5}$$

and determine its eigenvalues. The solutions of the equation $y'' + \lambda y = 0$ can be represented in the form

$$y(x) = c_1 y_1^{(\lambda)}(x) + c_2 y_2^{(\lambda)}(x) \tag{3.6}$$

where $y_1^{(\lambda)}$ and $y_2^{(\lambda)}$ are two linearly independent solutions and the constants c_1 and c_2 vary in the set of real numbers. The solutions of (3.5) are determined by those values of the constants c_1 and c_2 for which the function (3.6) satisfies the boundary conditions $y(0) = y(b) = 0$. These values of c_1 and c_2 are given by the solutions of the algebraic system

$$\begin{cases} c_1 y_1^{(\lambda)}(0) + c_2 y_2^{(\lambda)}(0) = 0 \\ c_1 y_1^{(\lambda)}(b) + c_2 y_2^{(\lambda)}(b) = 0 \end{cases} \tag{3.7}$$

in the unknowns c_1 and c_2. Therefore, in order to determine the eigenvalues of (3.5), we must find the values of λ for which the system (3.7) has nontrivial solutions. These values are the λ for which the determinant of the coefficients is 0:

$$\det\begin{pmatrix} y_1^{(\lambda)}(0) & y_2^{(\lambda)}(0) \\ y_1^{(\lambda)}(b) & y_2^{(\lambda)}(b) \end{pmatrix} = y_2^{(\lambda)}(b)y_1^{(\lambda)}(0) - y_1^{(\lambda)}(b)y_2^{(\lambda)}(0) = 0.$$

The eigenvalues of (3.5) are therefore the values of λ for which the equation $y'' + \lambda y = 0$ has two linearly independent solutions $y_1^{(\lambda)}$ and $y_2^{(\lambda)}$ such that

$$y_2^{(\lambda)}(b)y_1^{(\lambda)}(0) - y_1^{(\lambda)}(b)y_2^{(\lambda)}(0) = 0. \tag{3.8}$$

With this, the reader can easily determine the eigenvalues of (3.5) by constructing two linearly independent solutions of $y'' + \lambda y = 0$ and by requiring that (3.8) be satisfied. We then get

$$\lambda = \frac{n^2\pi^2}{b^2}, \quad n \geq 1.$$

We have thus found that the eigenvalues of (3.5) are positive and form an increasing sequence tending to ∞. The n-th term, $\lambda_n = n^2\pi^2/b^2$, tends to ∞ as the length b of the interval on which the boundary problem is considered tends to 0.

<u>Exercise 1</u>. We found the eigenvalues of (3.5) by considering only two particular linearly independent solutions of $y'' + \lambda y = 0$, namely, those which are obtained by using the roots of the characteristic equation $z^2 + \lambda = 0$. Why is this correct?

<u>Exercise 2</u>. Consider the Sturm-Liouville problem

$$y'' + \lambda y = 0, \quad \alpha_1 y(0) + \alpha_2 y'(0) = \beta_1 y(b) + \beta_2 y'(b) = 0.$$

Prove that the eigenvalues constitute an infinite sequence of positive real numbers. Prove that if $b = \pi$, $\alpha_1\beta_1 = \alpha_2\beta_2 = 0$ and $\alpha_1\beta_2 \neq \alpha_2\beta_1$, then the eigenvalues are the numbers $\lambda_n = (n + \frac{1}{2})^2$ with $n \geq 0$. Prove that if $b = \pi$, $\alpha_1\beta_2 = \alpha_2\beta_1$ and $\alpha_1\beta_1 \neq 0$ or $\alpha_2\beta_2 \neq 0$, then the eigenvalues are the numbers $\lambda = n^2$, $n \geq 0$.

<u>Exercise 3</u>. Consider the boundary value problem

$$y'' + \lambda y = 0, \quad y(0) - y(p) = 0; \quad y'(0) - y'(p) = 0$$

which corresponds to the problem of finding periodic solutions. (See Section 4 below.) Prove that the eigenvalues (which are defined, as before, to be the λ for which there exists at least one nontrivial solution) form a sequence of nonnegative real numbers. Which eigenfunctions correspond to $\lambda = 0$? Prove that if $p = 2\pi$, then the eigenvalues are $\lambda_n = n^2$, $n \geq 0$, and that $A \sin nt + B \cos nt$ are eigenfunctions corresponding to λ_n where A and B are constants. (This means that the eigenfunctions, together with $y \equiv 0$, form a two-dimensional vector space for which the functions $\sin nt$ and $\cos nt$ form an orthogonal basis.)

Exercise 4. Prove that if λ is not an eigenvalue of (3.4), (3.2), then the Sturm-Liouville problem

$$(p(x)y')' + q(x,\lambda)y = h(x)$$

$$\alpha_1 y(a) + \alpha_2 y'(a) = A$$

$$\beta_1 y(b) + \beta_2 y'(b) = B$$

has a unique solution for every continuous function h and every A and B in $\mathbb{R}$. Hint: Use the results on the geometric boundary value problems; alternatively, see the theorem of Green's function in Sec. 3.6.

Exercise 5. Prove that the eigenfunctions of (3.4), (3.2) are elements of the kernel of linear operators from $C([a,b])$ into itself and from $C^2([a,b])$ into $C([a,b])$.

3.2. Prüfer's Change of Variables

We shall study the eigenvalues of (3.4), (3.2) by using the change of variables introduced by Prüfer, because this method allows us to simplify and unify the proofs of various fundamental results; this change of variables transforms Eq. (3.4) into another from which it is easier to see how the eigenfunctions depend on the data.

We consider the equation

$$(p(x)y')' + q(x,\lambda)y = 0 \tag{3.4}$$

and transform it into the first order system

$$\begin{cases} y'(x) = \dfrac{1}{p(x)} z(x) \\ z'(x) = -q(x,\lambda)y(x) \end{cases} \tag{3.9}$$

that is obviously equivalent to (3.4). We then express the point $(y(x),z(x))$ in *polar coordinates* $\rho(x)$, $\theta(x)$, setting

$$y(x) = \rho(x) \sin \theta(x), \quad z(x) = \rho(x) \cos \theta(x). \tag{3.10}$$

We immediately take note of one essential point. The change of variables that we are now making is of interest to us only insofar as it allows us to determine the nontrivial solutions of the Sturm-Liouville problem (3.4), (3.2). Now by virtue of the uniqueness in initial value problems for (3.4), a solution y of (3.4) is identically 0 if and only if we have $y(x) = y'(x) = 0$ at one point x. Therefore, it follows from (3.9) and (3.10) that the nontrivial solutions of (3.4) are exactly those for which we have $\rho(x) \neq 0$ for each x. We shall therefore *assume that* $\rho(x) \neq 0$ *for each* x.

If we square both sides of (3.10) and add side by side, we get

$$\rho^2(x) = y^2(x) + z^2(x).$$

Upon differentiating the two sides of this equality, we get

$$\rho\rho' = yy' + zz'. \tag{3.11}$$

If we now differentiate the relation $\theta(x) = \arctan \frac{y(x)}{z(x)}$, we produce

$$\rho^2\theta' = zy' - yz'. \tag{3.12}$$

We now use (3.11) and (3.12) to transform the system (3.9) into the new system

$$\rho\rho' = \rho^2\left(\frac{1}{p(x)} - q(x,\lambda)\right)\sin\theta\cos\theta$$

$$\rho^2\theta' = \rho^2\left(\frac{1}{p(x)}\cos^2\theta + q(x,\lambda)\sin^2\theta\right)$$

that is, into

$$\begin{cases} \theta' = \dfrac{1}{p(x)}\cos^2\theta + q(x,\lambda)\sin^2\theta & (3.13)_1 \\[2ex] \dfrac{d}{dx}\,\ell g\,\rho = \dfrac{1}{2}\left(\dfrac{1}{p(x)} - q(x,\lambda)\right)\sin 2\theta & (3.13)_2 \end{cases}$$

where we have divided both equations by ρ^2.

The solutions of (3.13) are uniquely determined by the initial values (θ_0,ρ_0) at the point a and exist in the whole interval $[a,b]$ since the right-hand side satisfies a Lipschitz condition locally and is bounded. Since $(3.13)_1$ contains only the unknown function θ and not ρ, the solution of the Cauchy problem for (3.13) leads us to the quadrature of two first order equations; as a matter of fact, once θ is determined by $(3.13)_1$, $(3.13)_2$ allows us to compute ρ through integration:

$$\ell g \frac{\rho(x)}{\rho_0} = \frac{1}{2}\int_a^x \left(\frac{1}{p(t)} - q(t,\lambda)\right)\sin 2\theta(t)dt,$$

where ρ_0 is the initial value of ρ at the point a.

We are interested in seeing the relation between the solutions of (3.13) with initial point a and those of the Sturm-Liouville problem (3.4), (3.2). The preceding result, according to which the integration of an initial value problem for (3.13) is reduced to the integration of a first order equation, suggests that a similar result holds for the Sturm-Liouville problem, i.e., that the mere knowledge of the values of θ at a and b permits us to determine the possible nontrivial solutions of (3.4) and (3.2). To see this, we consider a nontrivial solution y of the given Sturm-Liouville problem. We have

$$y(x) = \rho(x) \sin \theta(x)$$

for a suitable pair of functions ρ and θ given by (3.10). By (3.9) and (3.10), (3.2) becomes

$$\alpha_1\rho(a)\sin \theta(a) + \alpha_2 \frac{1}{p(a)} \rho(a)\cos \theta(a) = 0;$$

$$\beta_1\rho(b)\sin \theta(b) + \beta_2 \frac{1}{p(b)} \rho(b)\cos \theta(b) = 0.$$

Since $\rho(a) \neq 0$ and $\rho(b) \neq 0$ by hypothesis, we have the following relation

$$\begin{aligned} &\alpha_1 \sin \theta(a) + \alpha_2 \frac{1}{p(a)} \cos \theta(a) = 0, \\ &\beta_1 \sin \theta(b) + \beta_2 \frac{1}{p(b)} \cos \theta(b) = 0 \end{aligned} \tag{3.14}$$

which involves only the values at a and b of the polar coordinate θ of all the possible nontrivial solutions of the Sturm-Liouville problem (3.4), (3.2). Vice versa, if θ satisfies (3.14) and $(3.13)_1$, then θ is the polar coordinate of a nontrival solution of (3.4), (3.2). We can give (3.14) an equivalent formulation. Let θ_a and θ_b be the two angles defined by

$$\theta_a = \arctan \frac{-\alpha_2}{\alpha_1 p(a)}, \qquad \theta_b = \arctan \frac{-\beta_2}{\beta_1 p(b)}$$

with the bounds $0 \leq \theta_a < \pi$, $0 < \theta_b \leq \pi$. Note that the hypothesis we made requiring that α_1 and α_2 (and β_1 and β_2) not be zero simultaneously insures that θ_a and θ_b are well defined. The equations in (3.14) then become equal, respectively, to

$$\sin(\theta(a) - \theta_a) = 0, \quad \sin(\theta(b) - \theta_b) = 0;$$

that is, to

$$\theta(a) = \theta_a + n\pi, \quad \theta(b) = \theta_b + m\pi \tag{3.15}$$

where n and m are integers. Finally, note that we can set $n = 0$ and $m \geq 0$ in (3.15): if θ is a solution of $(3.13)_1$, then $\theta + n\pi$ is a solution, too; replacing θ by $\theta + n\pi$ transforms y into $-y$.

The following theorem summarizes our discussions.

Theorem on Prüfer's Change of Variables. A necessary and sufficient condition for y to be a nontrivial solution of the Sturm-Liouville problem

$$(p(x)y')' + q(x,\lambda)y = 0, \quad \alpha_1 y(a) + \alpha_2 y'(a) = 0; \quad \beta_1 y(b) + \beta_2 y'(b) = 0$$

is that its polar coordinate θ satisfy $(3.13)_1$ and

$$\theta(a) = \theta_a, \quad \theta(b) = \theta_b + m\pi$$

where m is a nonnegative integer, and θ_a and θ_b are the angles defined by

$$\theta_a = \arctan \frac{-\alpha_2}{\alpha_1 p(a)}, \quad \theta_b = \arctan \frac{-\beta_2}{\beta_1 p(b)}$$

under the bounds $0 \leq \theta_a < \pi$, $0 < \theta_b \leq \pi$.

Note that the angles θ_a and θ_b depend exclusively on the coefficient p of Eq. (3.4) and on the coefficients α_i, β_i of the boundary conditions (3.2). *The symbols* θ_a *and* θ_b *will, in the future, always indicate the above-defined angles.*

We shall need the following two theorems, which show how θ varies as a function of the parameters of the problem.

Comparison Theorem for θ. Consider the two equations

$$\theta_1' = \frac{1}{p(x)} \cos^2\theta_1 + q_1(x,\lambda)\sin^2\theta_1$$

and

$$\theta_2' = \frac{1}{p(x)} \cos^2\theta_2 + q_2(x,\lambda)\sin^2\theta_2.$$

If $q_1 < q_2$ and if θ_1 and θ_2 are solutions of the two equations with the same initial points, that is, if $\theta_1(a) = \theta_2(a)$, then

$$\theta_1(x) < \theta_2(x) \qquad (a < x \leq b).$$

Proof: The functions on the second side of the given equations satisfy the relation

$$\frac{1}{p(x)}\cos^2 y + q_1(x,\lambda)\sin^2 y \le \frac{1}{p(x)}\cos^2 y + q_2(x,\lambda)\sin^2 y\,. \tag{3.16}$$

The inequality is strict if $\sin y \neq 0$. Then

$$\theta_1'(x) \le \frac{1}{p(x)}\cos^2\theta_1(x) + q_2(x,\lambda)\sin^2\theta_1(x).$$

Thus, the theorem on differential inequalities (Sec. 2.4 of Chapter III) gives us

$$\theta_1 \le \theta_2 \tag{3.17}$$

since our equations have a unique solution, the right hand side satisfying a Lipschitz condition in y locally. The constant function $u_n(x) = n\pi$ satisfies the condition $u_n' \le \frac{1}{p(x)}\cos^2 u_n + q_1(x,\lambda)\sin^2 u_n$. If at a point t_0, $u_n(t_0) = \theta_1(t_0)$, then $u_n(x) < \theta_1(x)$, $t_0 < x \le b$, by virtue of the corollary to the theorem on differential inequalities (Sec. 2.4 in Chap. III). From this and the fact that θ_1 is bounded it follows that at most finitely many points $x_1,\ldots,x_m$ exist in $]a,b]$ where $\sin\theta_1$ vanishes. If we had $\theta_1(x_0) = \theta_2(x_0)$ with $x_0 \in\ ]a,b] \setminus \{x_1,\ldots,x_m\}$, then (3.16) would imply that $\theta_2'(x_0) - \theta_1'(x_0) > 0$, and thus the function $\theta_2 - \theta_1$ would be strictly increasing in a neighborhood of x_0. But this, by virtue of (3.17), would contradict the equality $\theta_1(x_0) = \theta_2(x_0)$. Thus, we must have $\theta_1 < \theta_2$ in all points of $]a,b]$ which differ from x_i. Then, again by the corollary to the theorem on differential inequalities, we must have $\theta_1 < \theta_2$ in $]a,b]$, as required.

Theorem on the Variation of θ at the Point b. Let λ_0 be a real number or $\pm\infty$ and let $\theta(x,\lambda)$ be the solution of the initial value problem

$$\theta' = \frac{1}{p(x)}\cos^2\theta + q(x,\lambda)\sin^2\theta, \qquad \theta(a) = \theta_a. \tag{3.18}$$

If the function q is such that

$$\lim_{\lambda\to\lambda_0} q(x,\lambda) = -\infty \quad \text{uniformly in } [a,b],$$

then

$$\lim_{\lambda\to\lambda_0} \theta(b,\lambda) = 0.$$

If the function q is such that

$$\lim_{\lambda\to\lambda_0} q(x,\lambda) = +\infty \quad \text{uniformly in } [a,b],$$

then

$$\lim_{\lambda\to\lambda_0} \theta(b,\lambda) = +\infty.$$

Observe that the initial point in (3.18) is always constant, since $\theta_a = \arctan -\alpha_2/\alpha_1 p(a)$, while the right side of the equation is the part that varies with the parameter λ.

The theorem applies in particular when

$$q(x,\lambda) = r_1(x) + \lambda r_2(x)$$

with r_2 always positive or negative, since $\lim_{\lambda\to-\infty} q(x,\lambda) = \mp\infty$ and $\lim_{\lambda\to\infty} q(x,\lambda) = \pm\infty$ according to the sign of r_2.

Proof of the Theorem: We first examine the case in which $q(x,\lambda)$ tends to $-\infty$. We begin by proving that for every x and λ,

$$\theta(x,\lambda) \geq 0. \tag{3.19}$$

If $\theta(x,\lambda)$ were negative at some point, then, since $\theta_a \geq 0$, there would be a point x_0 such that $\theta(x_0,\lambda) = 0$ and $\theta(x,\lambda) < 0$ for $x > x_0$ in a right neighborhood of x_0. But this implies a contradiction; from the equation it follows that $\theta'(x_0,\lambda) = 1/p(x_0) > 0$ and thus $\theta(\cdot,\lambda)$ is strictly increasing in a neighborhood of x_0, which is incompatible with the sign of θ on the right of x_0. (3.19) is thus established. This implies that if $\lim_{\lambda\to\lambda_0} \theta(b,\lambda) \neq 0$, then there exist $\varepsilon > 0$ and a sequence $(\lambda_n)_{n=1}^{\infty}$ converging to λ_0 such that

$$\theta(b,\lambda_n) > \varepsilon \quad \text{for all} \quad n. \tag{3.20}$$

To prove the theorem it thus suffices to show that (3.20) leads to a contradiction. Given (3.20), we observe that ε can be chosen in such a way that we have $\pi - \varepsilon > \theta_a$. Let m be the maximum of the function $1/p$ on $[a,b]$, let $\delta > 0$ be such that

$$m - \delta \sin^2\varepsilon < 0,$$

and let n_0 be such that for $n \geq n_0$ and $a \leq x \leq b$, we have

$$q(x,\lambda_n) \leq -\delta.$$

We prove that for $n \geq n_0$ and $a \leq x \leq b$,

$$\varepsilon < \theta(x,\lambda_n) < \pi - \varepsilon. \tag{3.21}$$

If the inequality on the left in (3.21) were invalid for some $n \geq n_0$, then, by virtue of (3.20), there would exist an x_1 such that

$\theta(x_1,\lambda_n) = \varepsilon$ and $\theta(x,\lambda_n) > \varepsilon$ for $x > x_1$ in a right neighborhood of x_1. But this implies a contradiction; from the equation we get

$$\theta'(x_1,\lambda_n) \leq \frac{1}{p(x_1)} - \delta \sin^2\theta(x_1,\lambda_n) \leq m - \delta \sin^2\varepsilon < 0$$

and therefore $\theta(\cdot,\lambda_n)$ is locally decreasing at x_1, which is incompatible with the values of $\theta(\cdot,\lambda_n)$ to the right of x_1. The inequality on the left of (3.21) is thus proven. To prove the one on the right of (3.21), one proceeds in the same way using $\theta_a < \pi - \varepsilon$ and $\sin(\pi-\varepsilon) = \sin \varepsilon$.

It follows from (3.21) that for each $n \geq n_0$, we have $\sin \theta(x,\lambda_n) \geq \sin \varepsilon$, and so, from the equation we get

$$\begin{aligned} \theta'(x,\lambda_n) &\leq m + q(x,\lambda_n)\sin^2\theta(x,\lambda_n) \\ &\leq m + q(x,\lambda_n)\sin^2\varepsilon. \end{aligned}$$

This implies that for each $N > 0$, there exists an n such that $\theta'(x.\lambda_n) \leq m - N \sin^2\varepsilon$, and thus

$$\theta(b,\lambda_n) \leq \theta_a + (m - N \sin^2\varepsilon)(b - a). \tag{3.22}$$

There obviously exists an N such that the right hand side of (3.22) is less than ε; this contradicts (3.20), and the first part of the theorem is proven.

We now examine the case $\lim_{\lambda\to\lambda_0} q(x,\lambda) = \infty$. By writing

$$q(x,\lambda) = \frac{1}{p(x)} + \bar{q}(x,\lambda)$$

the given equation becomes

$$\begin{aligned} \theta' &= \frac{1}{p(x)} (\cos^2\theta + \sin^2\theta) + \bar{q}(x,\lambda)\sin^2\theta \\ &= \frac{1}{p(x)} + \bar{q}(x,\lambda)\sin^2\theta \end{aligned} \tag{3.23}$$

again with

$$\lim_{\lambda\to\lambda_0} \bar{q}(x,\lambda) = \infty.$$

We suppose that $\lim_{\lambda\to\lambda_0} \theta(b,\lambda) \neq \infty$ and arrive at a contradiction. There exist $\beta < \infty$ and a sequence $(\lambda_n)_{n=1}^{\infty}$ tending to λ_0 such that for each n,

$$\theta(b,\lambda_n) \leq \beta. \tag{3.24}$$

Let M be the minimum of the function $1/p$ on $[a,b]$, and let $(M_n)_{n=1}^{\infty}$ be a positive sequence tending to ∞. By passing to a subsequence of $(\lambda_n)_{n=1}^{\infty}$ if necessary, we may assume that

$$M_n \le q(x,\lambda_n) \quad \text{for each } n, \quad a \le x \le b.$$

Then, as we see from (3.23), the second side of the equation

$$\theta' = \frac{1}{p(x)} + \bar{q}(x,\lambda_n)\sin^2\theta$$

is greater than or equal to that of the equation

$$z' = M + M_n \sin^2 z.$$

In this last equation the variables can be separated, so the solution z_n of the initial value problem

$$z_n' = M + M_n \sin^2 z, \quad z_n(a) = \theta_a$$

is defined by

$$\int_{\theta_a}^{z_n(x)} \frac{1}{M + M_n \sin^2 u}\, du = \int_a^x dt = x - a. \tag{3.25}$$

By virtue of the theorem on differential inequalities, (Sec. 2.4 of Chapter III), the fact that the second side of the equation $\theta' = \frac{1}{p(x)} + \bar{q}(x,\lambda_n)\sin^2\theta$ is greater than or equal to that of the equation $z' = M + M_n \sin^2 z$ implies that for each n,

$$z_n(x) \le \theta(x,\lambda_n).$$

In particular, from (3.24) we have, for each n and x,

$$\theta_a \le z_n(x) \le \beta.$$

This leads to a contradiction, as we shall now see. Since the sine function has only finitely many zeros in the set $[\theta_a,\beta]$, we have

$$\lim_n \frac{1}{M + M_n \sin^2 u} = 0$$

at all but finitely many points of $[\theta_a,\beta]$. Thus, the convergence theorem of Lebesgue implies that

$$\lim_n \int_{\theta_a}^{\beta} \frac{du}{M + M_n \sin^2 u} = 0$$

which contradicts (3.25). The second part of the theorem is thus proven.

Exercise. When we treated the case $\lim_{\lambda\to\lambda_0} q(x,\lambda) = -\infty$ in the preceding proof, we did not consider $z' = m - N^0\sin^2 z$ to be a separable equation. Why?

3.3. Existence and Properties of the Eigenvalues

The topic of this section is to establish the existence of the eigenvalues and to determine their most important properties. We shall use Prüfer's change of variables with the notation $\theta(x,\lambda)$, θ_a, and θ_b of the preceding section.

Existence Theorem for Eigenvalues. Consider the Sturm-Liouville problem

$$(p(x)y')' + q(x,\lambda)y = 0 \tag{3.4}$$

$$\alpha_1 y(a) + \alpha_2 y'(a) = 0; \quad \beta_1 y(b) + \beta_2 y'(b) = 0. \tag{3.2}$$

If $q(x,\lambda)$ is continuous in both variables and strictly increasing in λ for $\lambda \in]\alpha,\beta[$, and if we have

$$\lim_{\lambda\downarrow\alpha} q(x,\lambda) = -\infty, \qquad \lim_{\lambda\uparrow\beta} q(x,\lambda) = +\infty$$

uniformly, then (3.4), (3.2) has eigenvalues in $]\alpha,\beta[$ and all eigenvalues can be enumerated in a sequence $(\lambda_n)_{n=1}^{\infty}$ such that

$$\lambda_n < \lambda_{n+1} \quad \text{and} \quad \lim_n \lambda_n = \beta.$$

The eigenspace corresponding to every eigenvalue has dimension 1, and every eigenfunction corresponding to λ_n takes the value 0 at exactly $n - 1$ points in $]a,b[$.

The fact that the eigenspace corresponding to λ has dimension 1 is expressed by saying that λ is a *simple eigenvalue*. When we speak of the eigenvalues of a given Sturm-Liouville problem, we always suppose that they are ordered in a monotone sequence whose n-th element is called the *n-th eigenvalue*.

Proof: The theorem on Prüfer's change of variables implies that λ is a eigenvalue of (3.4), (3.2) if and only if λ satisfies the identity

$$\theta(b,\lambda) = \theta_b + n\pi \tag{3.26}$$

where n is an integer and $\theta(\cdot,\lambda)$ is the solution of $(3.13)_1$ with initial value $\theta(a,\lambda) = \theta_a$. The function $\theta(x,\lambda)$ is continuous in λ by virtue of uniqueness in the initial value problems for the equation

$$\theta' = \frac{1}{p(x)} \cos^2\theta + q(x,\lambda)\sin^2\theta,$$

as was established in Sec. 3.1 of Chapter I. From the comparison theorem for θ, it follows that $\theta(b,\lambda)$ is a strictly increasing function in λ, while from the theorem on the variation of θ at the point b, it follows that the range of the function $\theta(b,\cdot)$ for $\lambda \in]\alpha,\beta[$ is the set $]0,\infty[$. (3.26) therefore has one and only one solution λ for each n. If we denote by λ_n the unique solution of (3.26) that corresponds to $n - 1$, we get a strictly increasing sequence that contains all the eigenvalues of (3.4), (3.2). Put $\lambda_\infty = \sup_n \lambda_n$. Clearly, $\lim_{n\to\infty} \lambda_n = \lambda_\infty$. Thus, to prove that $\lambda_\infty = \beta$, it is enough to show that the hypothesis $\lambda_\infty < \beta$ leads to a contradiction. The condition $\lambda_\infty < \beta$ implies $\lambda_\infty \in]\alpha,\beta[$, so the function $\theta(b,\cdot)$ is continuous at λ_∞, but this is impossible since for $n \neq m$, we have

$$|\theta(b,\lambda_n) - \theta(b,\lambda_m)| = |n - m|\pi > 0.$$

We must therefore necessarily have $\lambda_\infty = \beta$. It remains to show that the eigenvalues are simple. This follows from the fact that once the solution $\theta(x,\lambda_n)$ of $(3.13)_1$ satisfying $\theta(a,\lambda_n) = \theta_a$, $\theta(b,\lambda_n) = \theta_b + n\pi$ has been determined, it is enough, in order to get a solution of the system (3.13) itself, to fix the initial value $\rho(a)$; the eigenspace thus depends on a single parameter. Finally, let w be an eigenfunction for λ_n. Then, as we have seen, w has polar coordinates given by Prüfer's change of variables with $\theta(a) = \theta_a$, $\theta(b) = \theta_b + (n-1)\pi$. Thus, w has value 0 as often as $\sin\theta(t)$ does in the interval $]a,b[$, that is, as often as t is such that $\theta(t) = i\pi$ for an integer i. Thus, w assumes the value 0 at least $n - 1$ times, since $0 < \theta_b \leq \pi$. From the fact that θ satisfies the equation

$$\theta' = \frac{1}{p(x)} \cos^2\theta + q(x,\lambda)\sin^2\theta,$$

it follows that θ is strictly increasing in a neighborhood of every point of the form $t = i\pi$. Thus, θ cannot assume the value 0 more than $n - 1$ times in $]a,b[$ since, if it did, θ would have an interior maximum with value greater than $\theta_b + (n-1)\pi$; there would then be two consecutive points t_1 and t_2 at which θ would assume the value $n\pi$, and θ would not be strictly increasing at both. (Sketch the graph to see this.) This completes the proof of the theorem.

Comparison Theorem for Eigenvalues. Let $q_1(x,\lambda)$ and $q_2(x,\lambda)$ be continuous in both variables, strictly increasing in $\lambda \in]\alpha,\beta[$ and such that

$$\lim_{\lambda\downarrow\alpha} q_i(x,\lambda) = -\infty, \qquad \lim_{\lambda\uparrow\beta} q_i(x,\lambda) = +\infty \qquad (i = 1,2)$$

uniformly in x. If λ_n' and λ_n'' are the n-th eigenvalues of the Sturm-Liouville problem (3.2) for the two equations

$$(p(x)y')' + q_1(x,\lambda')y = 0$$

$$(p(x)y')' + q_2(x,\lambda'')y = 0,$$

then $q_1 < q_2$ implies that $\lambda_n' > \lambda_n''$.

In other words, if the coefficient q of (3.4) increases, then the eigenvalues corresponding to the same index n decrease.

Proof of the Comparison Theorem for Eigenvalues: Let $\theta_1(x,\lambda)$ and $\theta_2(x,\lambda)$ be the solutions of the initial value problems

$$\theta_1' = \frac{1}{p(x)} \cos^2\theta_1 + q_1(x,\lambda)\sin^2\theta_1, \quad \theta_1(a) = \theta_a$$

$$\theta_2' = \frac{1}{p(x)} \cos^2\theta_2 + q_2(x,\lambda)\sin^2\theta_2, \quad \theta_2(a) = \theta_a.$$

By virtue of what was established in the preceding theorem, we have

$$\theta_1(b,\lambda_n') = \theta_b + (n-1)\pi; \quad \theta_2(b,\lambda_n'') = \theta_b + (n-1)\pi. \tag{3.27}$$

From the comparison theorem for θ it follows that

$$\theta_1(b,\lambda) < \theta_2(b,\lambda) \qquad (\alpha < \lambda < \beta).$$

In particular, for $\lambda = \lambda_n'$, we get, using (3.27),

$$\theta_2(b,\lambda_n'') < \theta_2(b,\lambda_n').$$

But $\theta_2(b,\cdot)$ is strictly increasing by the comparison theorem for θ, so $\lambda_n'' < \lambda_n'$ and the theorem is proven.

The existence theorem provides only an upper bound for the eigenvalues. As for a lower bound, we may in general say that under the hypotheses of the existence theorem, there are at most finitely many negative eigenvalues when

$$\lim_{\lambda\uparrow\infty} q(x,\lambda) = \infty.$$

The following theorem furnishes a lower bound for the first eigenvalue and therefore allows us to determine conditions sufficiently strong to insure that the eigenvalues are positive.

Theorem on the Lower Bound for Eigenvalues. Consider the Sturm-Liouville problem

$$(p(x)y')' + (q(x)+\lambda)y = 0, \quad \alpha_1 y(a) + \alpha_2 y'(a) = 0, \quad \beta_1 y(b) + \beta_2 y'(b) =$$

If $\alpha_1\alpha_2 = 0$, $\beta_1\beta_2 = 0$ and $m = \min_{a\leq x\leq b} (-q(x))$, then $\lambda_1 \geq m$.

Proof: Let y be an eigenfunction corresponding to λ_1. If we multiply the identity

$$(p(x)y')' + (q(x) + \lambda_1)y = 0$$

by $y(x)$ and integrate by parts on $[a,b]$, we get

$$0 = \int_a^b (py')'y dx + \int_a^b qy^2 dx + \lambda_1 \int_a^b y^2 dx$$

$$= py'y\Big|_a^b - \int_a^b py'^2 dx + \int_a^b qy^2 dx + \lambda_1 \int_a^b y^2 dx.$$

One can easily verify that the hypotheses $\alpha_1\alpha_2 = 0$, $\beta_1\beta_2 = 0$ imply that $py'y\Big|_a^b = 0$. Thus, the identity obtained above is equivalent to

$$\int_a^b py'^2 dx - \int_a^b qy^2 dx = \lambda_1 \int_a^b y^2 dx.$$

Since the first side is not less than $-\int_a^b qy^2 dx$, we have

$$\lambda_1 \int_a^b y^2 dx \geq -\int_a^b q(x)y^2(x) dx \geq m \int_a^b y^2 dx.$$

Since $y \not\equiv 0$, $\int_a^b y^2 dx > 0$, so we deduce that $\lambda_1 \geq m$, and the theorem is proved.

Exercise 1. State and prove the analogues of the first two theorems of this section under the hypothesis that $q(x,\lambda)$ is strictly decreasing. Hint: Replace λ by $-\lambda$.

Exercise 2. Prove that the n-th eigenvalue λ_n depends continuously on p and q. Hint: The uniqueness of the fixed points implies that they vary continuously.

Exercise 3. Prove that when $q < 0$, all the eigenvalues of $y'' + (q(x) + \lambda)y = 0$, $y(0) = 0 = y(b)$ are positive.

3.4. Applications to Questions of Uniqueness for Problems Relative to Nonlinear Equations

In this section, we shall see how the theory of eigenvalues can be used to study questions of uniqueness relative to the solutions of Sturm-Liouville problems for nonlinear equations. We begin with a uniqueness theorem from which we shall deduce an existence theorem in the next section. Observe that the uniqueness theorem can be applied in particular to the two point problem.

Uniqueness Theorem. Given the Sturm-Liouville problem

$$y'' + f(x,y) = 0; \quad \alpha_1 y(a) + \alpha_2 y'(a) = 0, \quad \beta_1 y(b) + \beta_2 y'(b) = 0,$$

where f and $\partial f/\partial y$ are continuous functions, let λ_n be the n-th eigenvalue of the linear problem

$$y'' + \lambda y = 0; \quad \alpha_1 y(a) + \alpha_2 y'(a) = 0, \quad \beta_1 y(b) + \beta_2 y'(b) = 0.$$

Suppose that one of the following hypotheses is valid:

(a) There exists an n such that $\lambda_n < \frac{\partial f}{\partial y} < \lambda_{n+1}$;

(b) $\frac{\partial f}{\partial y} < \lambda_1$.

Then the given problem admits at most one solution.

Proof: Suppose that there exist two different solutions u and v; we shall derive a contradiction. If we integrate the relation

$$\frac{d}{d\xi} f(x,\xi(u-v) + v) = \frac{\partial}{\partial y} f(x,\xi(u-v) + v)\cdot(u-v)$$

between 0 and 1, we get

$$\begin{aligned} u''(x) - v''(x) &= -f(x,u(x)) + f(x,v(x)) \\ &= -\int_0^1 \frac{\partial}{\partial y} f(x,\xi(u(x)-v(x)) + v(x))\cdot(u(x)-v(x))d\xi. \end{aligned}$$

This implies that the function $w_0 = u - v$ is a solution of the linear problem

$$w'' + q(x)w = 0, \quad \alpha_1 w(a) + \alpha_2 w'(a) = 0, \quad \beta_1 w(b) + \beta_2 w'(b) = 0, \tag{3.28}$$

where

$$q(x) = \int_0^1 \frac{\partial}{\partial y} f(x,\xi(u(x) - v(x)) + v(x))d\xi.$$

Conditions (a) and (b) imply, respectively,

(a)* $\lambda_n < q(x) < \lambda_{n+1}$ $(a \le x \le b)$

(b)* $q(x) < \lambda_1$ $(a \le x \le b)$.

Assume (a) is true. Then (a)* is also true. We compare the i-th eigenvalues of the two problems

$$w'' + (q(x)+\lambda')w = 0, \quad \alpha_1 w(a) + \alpha_2 w'(a) = 0, \quad \beta_1 w(b) + \beta_2 w'(b) = 0 \tag{3.29}$$

$$z'' + (\lambda_n+\lambda'')z = 0, \quad \alpha_1 z(a) + \alpha_2 z'(a) = 0, \quad \beta_1 z(b) + \beta_2 z'(b) = 0 \tag{3.30}$$

whose equations have the functions $q_1(x,\lambda) = q(x) + \lambda$ and $q_2(x,\lambda) = \lambda_n + \lambda$ for coefficients respectively. From $q(x) + \lambda > \lambda_n + \lambda$, we get $\lambda_i' < \lambda_i''$ by virtue of the theorem on the comparison of eigenvalues. But since λ_i'' is the i-th eigenvalue of (3.30), it follows that $\lambda_n + \lambda_i''$ is the i-th eigenvalue of

$$y'' + \lambda y = 0, \quad \alpha_1 y(a) + \alpha_2 y'(a) = 0, \quad \beta_1 y(b) + \beta_2 y'(b) = 0. \tag{3.31}$$

In particular, for $i = n$, we have

$$\lambda_n + \lambda_n'' = \lambda_n,$$

and so $\lambda_n'' = 0$. Therefore $\lambda_n' < 0$. If we compare (3.29) with

$$z'' + (\lambda_{n+1}+\lambda''')z = 0, \quad \alpha_1 z(a) + \alpha_2 z'(a) = 0, \quad \beta_1 z(b) + \beta_2 z'(b) = 0,$$

we similarly find that $0 < \lambda_{n+1}'$. This implies that $\lambda = 0$ cannot be an eigenvalue of (3.29) since it is included between two consecutive eigenvalues. But this is absurd, since w_0 is a nontrivial solution of (3.28), and therefore $\lambda = 0$ is an eigenvalue of (3.29).

We now assume (b) and therefore (b)* also. As we have just seen, $\lambda = 0$ is an eigenvalue of (3.29). Then there exists an integer n_0 such that $\lambda_{n_0}' = 0$. If we compare the eigenvalues of (3.29) with those of

$$Y'' + (\lambda_1 + \lambda^{IV})Y = 0, \quad \alpha_1 Y(a) + \alpha_2 Y'(a) = 0, \quad \beta_1 Y(b) + \beta_2 Y'(b) = 0,$$

we get $0 = \lambda'_{n_0} > \lambda^{IV}_{n_0}$ from $q(x) + \lambda < \lambda_1 + \lambda$. But $\lambda_1 + \lambda^{IV}_{n_0}$ is an eigenvalue of (3.31), so $\lambda_1 + \lambda^{IV}_{n_0} \geq \lambda_1$. We therefore have the contradiction $0 > 0$, and the theorem is proved.

The following theorem is useful in establishing the total number of solutions of a given Sturm-Liouville problem. We observe that we need only assume the inequality $u < v$ to be true almost everywhere, as we see upon developing the theory above with $q(x,\lambda)$ integrable in x rather than continuous.

Theorem on the Number of Comparable Solutions. Suppose that u and v are solutions of the Sturm-Liouville problem

$$y'' + f(x,y) = 0, \quad \alpha_1 y(a) + \alpha_2 y'(a) = 0 = \beta_1 y(b) + \beta_2 y'(b)$$

such that $u < v$ in $]a,b[$. If f is continuous, has a continuous partial derivative $\frac{\partial}{\partial y} f(x,y)$ in the region $\{(x,y)\colon a \leq x \leq b,\ u(x) \leq y \leq v(x)\}$, and is strictly monotone in $y \in [u(x),v(x)]$ with the same type of monotonicity for each x, then there is no solution y of the given problem such that $u \leq y < v$ in $]a,b[$ and $u \not\equiv y$.

Proof: We suppose that there is a solution y_0 such that $u \leq y_0 < v$ in (a,b) and $u \not\equiv y_0$ and arrive at a contradiction. Upon integrating the relation

$$\frac{d}{d\xi} f(x,\xi(y-u) + u) = \frac{\partial}{\partial y} f(x,\xi(y-u) + u)(y-u)$$

we get

$$f(x,y) - f(x,u) = q(x,y-u)(y-u) \tag{3.32}$$

with

$$q(x,y) = \int_0^1 \frac{\partial}{\partial y} f(x,\xi y+u)\,d\xi.$$

We consider the Sturm-Liouville problem

$$y'' + q(x,y)y = 0, \quad \alpha_1 y_1(a) + \alpha_2 y'(a) = 0 = \beta_1 y(b) + \beta_2 y'(b). \tag{3.33}$$

It follows from (3.32) that the functions $w_1 = v - u$ and $w_2 = y_0 - u$ are solutions of (3.33). Furthermore, $0 \leq w_2 < w_1$ in $]a,b[$ and $w_2 \not\equiv 0$. We now compare the eigenvalues of the two linear problems

$$y'' + (q(x,w_1(x)) + \lambda')y = 0,$$
$$\alpha_1 y(a) + \alpha_2 y'(a) = 0 = \beta_1 y(b) + \beta_2 y'(b) \tag{3.34}$$

$$z'' + (q(x,w_2(x)) + \lambda'')z = 0,$$
$$\alpha_1 z(a) + \alpha_2 z'(a) = 0 = \beta_1 z(b) + \beta_2 z'(b). \tag{3.35}$$

Since $\partial f(x,\cdot)/\partial y$ is strictly monotone, $q(x,\cdot)$ is also and we thus have $q(x,w_1(x)) + \lambda < q(x,w_2(x)) + \lambda$ or the opposite inequality. It follows from the theorem on the comparison of eigenvalues that $\lambda'_n \neq \lambda''_n$ for each n. Since w_1 and w_2 are nontrivial solutions of (3.33), 0 is an eigenvalue of (3.34) and (3.35). Since w_1 and w_2 do not assume the value 0 in $]a,b[$, it follows from the theorem on the existence of eigenvalues that 0 is the first eigenvalue of (3.34) and (3.35). But then we have the contradiction $0 = \lambda'_1 \neq \lambda''_1 = 0$, which proves the theorem.

Exercise. Use the uniqueness theorem to prove that the problem

$$y'' = f(x,y), \quad y(a) = A, \quad y(b) = B$$

has at most one solution if f is of class C^1 and increasing in y.

3.5. Application to the Existence of Solutions for Problems Relative to Nonlinear Equations

In this section, we shall deduce some existence theorems from the theory of eigenvalues developed in the preceding sections. For simplicity, we shall limit ourselves to the two-point boundary value problem, or problem of Picard. In the proofs, we shall use the uniqueness which we established in the preceding section as well as the following result, which ties uniqueness in with existence.

Lemma of Lasota and Opial. Let $f\colon [a,b[\times \mathbb{R}^2 \to \mathbb{R}$ be a continuous function such that every initial value problem for $y'' = f(x,y,y')$ has a unique global solution. Let $r_1, r_2 \in \mathbb{R}$ and $x_1, x_2 \in [a,b[$ with $x_1 < x_2$. The two point boundary value problem

$$y'' = f(x,y,y'), \quad y(x_1) = r_1, \quad y(x_2) = r_2$$

has a unique solution if there exists a neighborhood U of x_2 such that the problem

$$y'' = f(x,y,y'), \quad y(x_1) = r_1, \quad y(t) = r$$

has at most one solution for every $t \in U$ and every $r \in \mathbb{R}$.

Proof: For each $u \in \mathbb{R}$, let $y(x,u)$ be the unique solution of

$$y'' = f(x,y,y'), \quad y(x_1) = r_1, \quad y'(x_1) = u.$$

By hypothesis, $y(x,u)$ exists in $[a,b[$. We define $T: \mathbb{R} \to \mathbb{R}$ by $T(u) = y(x_2,u)$. To prove the lemma, it is enough to show that $T(\mathbb{R}) = \mathbb{R}$. The uniqueness of the solutions of the Cauchy problem implies continuous dependence, as we know from Chapter III; T is therefore continuous. What is more, T is injective; if $T(u) = T(v)$, then $y(x_2,u) = y(x_2,v)$ so $y(\cdot,u)$ and $y(\cdot,v)$ are solutions of the Picard problem

$$y'' = f(x,y,y'), \quad y(x_1) = r_1, \quad y(x_2) = y(x_2,u)$$

and thus $y(\cdot,u) = y(\cdot,v)$. The fact that T is continuous and injective implies that $T(\mathbb{R})$ is an open interval (cf. Exercise 1). As a consequence, to prove that $T(\mathbb{R}) = \mathbb{R}$, it is sufficient to show that $\sup T(\mathbb{R}) = \infty$ and $\inf T(\mathbb{R}) = -\infty$. We shall show only the first relation because the other is established in a similar manner. We suppose that $\sup T(\mathbb{R}) = p_\infty < \infty$ and arrive at a contradiction; it is enough to prove that $p_\infty \in T(\mathbb{R})$, since we thereby contradict the fact that $T(\mathbb{R})$ is an open set. There exists a sequence $(p_n)_{n=1}^{\infty}$ of points of $T(\mathbb{R})$ such that $p_n < p_{n+1}$ and $\lim p_n = p_\infty$. Let $u_n = T^{-1}(p_n)$ and, for simplicity of notation, let $y_n = y(\cdot,u_n)$. We have

$$y_n(x) > y_1(x) \qquad (n \geq 2;\ x \in U) \tag{3.36}$$

since otherwise y_n and y_1 would be two different solutions of a two-point boundary value problem with endpoints x_1 and some point of U. For infinitely many n, we have $y_n'(x_2) \leq 0$ or alternatively $y_n'(x_2) \geq 0$. In the first case, by passing to a subsequence if necessary, we may assume that

$$y_n'(x_2) \leq 0 \quad \text{for each } n. \tag{3.37}$$

We fix $x_3 \in \,]x_2,b[\, \cap U$ arbitrarily; this is the only point at which we make use of the hypothesis that the interval is open on the right. We have

$$\begin{aligned}\frac{y_n(x_3) - y_n(x_2)}{x_3 - x_2} &\geq \frac{y_1(x_3) - y_n(x_2)}{x_3 - x_2} \quad (\text{by } (3.36)) \\ &\geq \frac{y_1(x_3) - p_\infty}{x_3 - x_2} \quad (\text{by } y_n(x_2) = p_n < p_\infty) \\ &\geq \min\left\{0, \frac{y_1(x_3) - p_\infty}{x_3 - x_2}\right\}.\end{aligned} \tag{3.38}$$

If we set

$$K = \min\left\{0, \frac{y_1(x_3) - p_\infty}{x_3 - x_2}\right\},$$

let $S_n = \{x \in [x_2,x_3] \mid K \leq y_n'(x) \leq 0\}$. The set S is not empty because if we had $y_n' < K$ in $[x_2,x_3]$, the mean value theorem would imply that

$$y_n(x_3) - y_n(x_2) = y_n'(\xi)(x_3 - x_2) < K(x_3 - x_2)$$

which contradicts (3.38). There therefore exists a point x_n such that $y_n'(x_n) \geq K$. If we had $y_n'(x_n) > 0$, then (3.37) and continuity would imply the existence of an $x_n^* \in [x_2,x_n]$ such that $y_n'(x_n^*) = 0$ and so $x_n^* \in S_n$. But if $y_n'(x_n) \leq 0$, then $x_n \in S_n$. Having so established that $S_n \neq \emptyset$, set

$$\sigma_n = \inf S_n.$$

By continuity, we have $y_n'(\sigma_n) \leq 0$. We now prove that

$$y_n'(x) \leq 0 \qquad (x_2 \leq x \leq \sigma_n). \tag{3.39}$$

If this were not true, then there would exist a $t_n \in [x_2,\sigma_n]$ such that $y_n'(t_n) > 0$. This implies that there is $t_n^* \in [x_2,t_n]$ such that $y_n'(t_n^*) = 0$, and we would therefore have $t_n^* \in S_n$, which contradicts the definition of σ_n. (3.39) is thus true and implies that y_n is decreasing in $[x_2,\sigma_n]$, so, in particular,

$$y_n(\sigma_n) < p_\infty \quad \text{for each } n.$$

From (3.36) it follows that

$$y_n(\sigma_n) \geq y_1(\sigma_n) \geq m$$

with

$$m = \inf_{x_1 \leq x \leq x_3} y_1(x).$$

Therefore

$$m \leq y_n(\sigma_n) < p_\infty \quad \text{for all } n.$$

From this, since $K \leq y_n'(\sigma_n) \leq 0$ and $x_2 \leq \sigma_n \leq x_3$, there exists a sequence $(n_k)_{k=1}^\infty$ tending to infinity such that

$$\lim_k y_{n_k}(\sigma_{n_k}) = u_0, \ \lim_k y_{n_k}'(\sigma_{n_k}) = u_1, \ \lim_k \sigma_{n_k} = \sigma_\infty.$$

The continuous dependence of the solutions of the initial value problem (cf. Sec. 3.1 of Chapter III) now insures that the sequence of functions $(y_{n_k})_{k=1}^{\infty}$ converges uniformly on compact subsets of $[a,b[$ to the unique solution y_∞ of the initial value problem

$$y'' = f(x,y,y'), \quad y(\sigma_\infty) = u_0, \quad y'(\sigma_\infty) = u_1.$$

In particular, $\lim_k y_{n_k}(x_2) = y_\infty(x_2)$. This implies, by virtue of $y_{n_k}(x_2) = p_{n_k}$, that $y_\infty(x_2) = p_\infty$. Since $y_\infty(x_1) = \lim_k y_{n_k}(x_1) = r_1$, y_∞ is also a solution of the Cauchy problem

$$z'' = f(x,z,z'), \quad z(x_1) = r_1, \quad z'(x_1) = y_\infty'(x_1)$$

and therefore $p_\infty \in T(\mathbf{R})$. The lemma is thus proved.

Existence Theorem. Given the problem

$$y'' + f(y) = h(x); \quad y(0) = 0, \quad y(b) = 0$$

where f is of class C^1, let λ_n be the n-th eigenvalue of the linear problem $y'' + \lambda y = 0$ with $y(0) = 0$ and $y(b) = 0$ (that is, $\lambda_n = n^2\pi^2/b^2$). Suppose that one of the following hypotheses holds:

(a) there exist n, μ_n, and μ_{n+1} with the property that

$$\lambda_n < \mu_n \leq f' \leq \mu_{n+1} < \lambda_{n+1};$$

(b) f' is bounded, and there exists μ_1 such that

$$f' \leq \mu_1 < \lambda_1.$$

Then the given problem possesses one and only one solution for each continuous function $h(x)$ defined on $[0,b]$.

Proof: We note that $h(x)$ can be continuously extended into an interval $[0,\beta[$ with $\beta > b$. Conditions (a) and (b) insure that all Cauchy problems for $y'' + f(y) = h(x)$ have a unique solution defined on $[0,\beta[$. If we observe that the n-th eigenvalue $\lambda_n(c)$ of

$$w'' + \lambda w = 0, \quad w(0) = 0; \quad w(c) = 0$$

is given by $\lambda_n(c) = n^2\pi^2/c^2$, it is clear that there exists a neighborhood U of b such that

$$\lambda_n(c) < f' < \lambda_{n+1}(c) \quad \text{or} \quad f' < \lambda_1(c) \qquad (c \in U)$$

according to whether (a) or (b) is true. Then the uniqueness theorem of the preceding section implies that every problem of Picard

$$y'' + f(y) = h(x), \quad y(0) = r_1, \quad y(c) = r$$

has at most one solution for each $r \in \mathbb{R}$ and each $c \in U$. The lemma of Lasota and Opial therefore implies existence for the given problem, and the theorem is thus proven.

From a physical point of view (phenomena of mechanics), one of the questions of greatest interest is the study of nonlinear equations which are almost linear near one or more eigenvalues. These cases are known under the name of *resonance problems*. For an example, we prove the following result, where a linear equation involving an eigenvalue is perturbed by a bounded function.

Theorem of Landesman and Lazer. Let λ be an eigenvalue of the Picard problem

$$y'' + \lambda y = 0, \quad y(0) = 0, \quad y(b) = 0$$

let $h: [a,b] \to \mathbb{R}$ be continuous, and let $f: \mathbb{R} \to \mathbb{R}$ be continuous and such that the limits $\lim_{y\to\pm\infty} f(y) = f_\pm$ exist and are finite. The problem

$$y'' + \lambda y = f(y) - h(x), \quad y(0) = 0, \quad y(b) = 0$$

has at least one solution if

$$\begin{aligned} f_-\int_{\{w>0\}} |w(x)|dx - f_+\int_{\{w<0\}} |w(x)|dx &< \int_a^b h(x)w(x)dx \\ &< f_+\int_{\{w>0\}} |w(x)|dx - f_-\int_{\{w<0\}} |w(x)|dx \end{aligned} \tag{3.40}$$

where w is an eigenfunction relative to λ.

Since the eigenspace relative to a Sturm-Liouville problem has dimension 1, whenever (3.40) is satisfied for one eigenfunction it is satisfied for them all. Note that the assumptions of the theorem imply that f is bounded.

Proof of the Landesman-Lazer Theorem: Since the eigenvalues of the Picard problem

$$y'' + \lambda y = 0, \quad y(0) = 0; \quad y(b) = 0 \tag{3.41}$$

have the form $\lambda_n = n^2\pi^2/b^2$, there is a sequence $(\varepsilon_n)_{n=1}^\infty$ of positive real numbers tending to zero such that $\lambda - \varepsilon_n$ is not an eigenvalue of (3.41). Thus the problem

$$y_n'' + (\lambda - \varepsilon_n)y_n = f(y_n) - h(x), \quad y_n(0) = 0, \quad y_n(b) = 0 \tag{3.42}$$

has a solution y_n by virtue of the existence theorem for geometric problems, for we get Eq. (3.42) by perturbing with the bounded function $f(y) - h(x)$ the linear homogeneous equation $y'' = -(\lambda - \varepsilon_n)y$ which has $y \equiv 0$ as the unique solution for the given boundary conditions (cf. Sec. 2.3 for more details). Let us suppose that the sequence $(y_n)_n$ is equibounded. Then it follows from (3.42) that $(y_n'')_n$ is also equibounded. In order to prove that even $(y_n')_n$ is equibounded, we fix $x \in [0,b]$ and let c be that point of $\{0,b\}$ that is farther from x, so that $|x-c| \geq \frac{1}{2}b$. From Taylor's formula we get

$$y_n'(x) = \frac{1}{x-c}\left[y_n(c) - y_n(x) - y_n''(\xi_x)\,\frac{(x-c)^2}{2}\right].$$

From this, from $1/|x-c| \leq 2/b$, and from the fact that $(y_n)_n$ and $(y_n'')_n$ are equibounded, we see that $(y_n')_n$ is equibounded. From the theorem of Arzelà and Ascoli, it now follows that there is a subsequence $(y_{n_k})_k$ that converges uniformly to a function z. If we apply the theorem of Arzelà and Ascoli once again, but this time to the subsequence $(y_{n_k}')_k$, we get a new subsequence, which we continue to denote by $(y_{n_k}')_k$, such that $\lim y_{n_k}' = v$ uniformly. From the convergence theorems for derivatives, we get that $z' = v$. If we pass to the limit in Eq. (3.42), we get the uniform convergence of $(y_{n_k}'')_k$ and, as a result, the existence of z'' as well as the fact that z solves the given problem

$$z'' + \lambda z = f(z) - h(x), \quad z(0) = 0 = z(b).$$

Now that we have seen that there is a solution to the given problem when $(y_n)_n$ is equibounded, it is enough, in order to prove the theorem, to show that the assumption that $(y_n)_n$ is not equibounded leads to a contradiction. By passing to a subsequence if necessary, we may assume that $\lim_n \|y_n\|_\infty = \infty$ when $(y_n)_n$ is unbounded ($\|\cdot\|_\infty$ represents the sup norm). By dividing Eq. (3.42) by $\|y_n\|_\infty$ and putting $z_n = y_n/\|y_n\|_\infty$, we get

$$z_n'' + (\lambda - \varepsilon_n)z_n = \frac{1}{\|y_n\|_\infty}(f(y_n) - h(x)). \tag{3.43}$$

Since $(z_n)_n$ is equibounded, we can repeat the argument used above when we assumed that $(y_n)_n$ is equibounded and show the existence of a uniformly convergent subsequence $(z_{n_k})_k$ whose limit z is a solution of the following problem of Picard (recall that the second member of (3.43) is bounded and that $\lim_n ||y_n||_\infty = \infty$):

$$z'' + \lambda z = 0, \quad z(0) = 0, \quad z(b) = 0. \tag{3.44}$$

If we multiply Eq. (3.43) with $n = n_k$ by $z(x)$ and then integrate by parts, we get

$$\int_a^b z''_{n_k} z dx + \lambda \int_a^b z_{n_k} z dx - \varepsilon_{n_k} \int_a^b z_{n_k} z dx$$
$$= \frac{1}{||y_{n_k}||_\infty} \left[\int_a^b f(y_{n_k}) z dx - \int_a^b hz\, dx \right].$$

But

$$\int_a^b z''_{n_k} z\, dx + \lambda \int_a^b z_{n_k} z\, dx = 0$$

as we see upon multiplying Eq. (3.44) by $z_{n_k}(x)$ and then integrating. We therefore have

$$-\varepsilon_{n_k} \int_a^b z_{n_k} z\, dx = \frac{1}{||y_{n_k}||_\infty} \left[\int_a^b f(y_{n_k}) z\, dx - \int_a^b hz\, dx \right]. \tag{3.45}$$

But

$$\lim_n \int_a^b z_{n_k} z\, dx = \int_a^b z^2\, dx > 0$$

since $||z||_\infty = 1$, and so, for k sufficiently large, (3.45) implies that

$$\int_a^b f(y_{n_k}) z\, dx - \int_a^b hz\, dx \le 0$$

and therefore

$$\limsup \int_a^b f(y_{n_k}) z\, dx \le \int_a^b hz\, dx. \tag{3.46}$$

By virtue of (3.44), z is an eigenfunction of (3.41). Therefore $\alpha \in R$ exists such that $z = \alpha w$. (Recall that the eigenspace in Sturm-Liouville problems has dimension 1.) Since $||z||_\infty = 1$, $\alpha \ne 0$. Let us suppose that $\alpha > 0$. Then, for each x such that $w(x) > 0$,

$$\lim_k z_{n_k}(x) = \lim_k \frac{y_{n_k}(x)}{||y_{n_k}||_\infty} = \alpha w(x) > 0,$$

so $\lim_k y_{n_k}(x) = \infty$. This implies that $\lim_k f(y_{n_k}(x)) = f_+$ for every x such that $w(x) > 0$. In a similar way we can show that $\lim_k f(y_{n_k}(x)) = f_-$ if x is such that $w(x) < 0$. If we use Lebesgue's theorem to pass to the limit in (3.46), putting αw for z, we get

$$f_+ \int_{\{w>0\}} |w|\,dx - f_- \int_{\{w<0\}} |w|\,dx = \lim_k \int_a^b f(y_{n_k}) w\,dx \leq \int_a^b hw\,dx,$$

which contradicts (3.40). In a similar manner we can show that if $\alpha < 0$, we arrive at the inequality

$$f_- \int_{\{w>0\}} |w|\,dx - f_+ \int_{\{w<0\}} |w|\,dx = \lim_k \int_a^b f(y_{n_k}) w\,dx \geq \int_a^b hw\,dx,$$

which also contradicts (3.40). The theorem is thus completely proven.

Exercise 1. Prove that if $f\colon \mathbb{R} \to \mathbb{R}$ is a continuous injection, then f is an open mapping, that is, f sends open sets to open sets. Hint: Recall that continuous injections are strictly monotone.

Exercise 2. Prove that the condition in the theorem of Landesman and Lazer is also necessary if we add $f_- < f < f_+$.

3.6. Further Properties of Eigenvalues and Eigenfunctions

To complete the discussion of the material in the preceding paragraphs, we now prove certain useful results in the general theory of eigenvalues and eigenfunctions.

Orthogonality Theorem. Consider the Sturm-Liouville problem

$$(p(x)y')' + [q(x) + \lambda r(x)]y = 0, \quad \alpha_1 y(a) + \alpha_2 y'(a) = 0,$$

$$\beta_1 y(b) + \beta_2 y'(b) = 0$$

where $r(x)$ is a positive continuous function. If $y_1(x)$ is a solution relative to the eigenvalue λ_1 and $y_2(x)$ a solution relative to the eigenvalue λ_2, then

$$\int_a^b r(x)y_1(x)y_2(x)\,dx = 0.$$

Proof: Multiply the equation relative to λ_1, y_1 by $y_2(x)$ and subtract from it the equation relative to λ_2, y_2 multiplied by $y_1(x)$. If we integrate the equation thus obtained, we get

$$\Big[p(x)[y_1'(x)y_2(x) - y_2'(x)y_1(x)]\Big]_a^b + (\lambda_1-\lambda_2)\int_a^b r(x)y_1(x)y_2(x)dx = 0.$$

On the other hand, since $\beta_1^2 + \beta_2^2 > 0$, $\alpha_1^2 + \alpha_2^2 > 0$, it is possible to express $y_i'(b)$ in terms of $y_i(b)$ (or vice versa) and $y_i'(a)$ in terms of $y_i(a)$ (or vice versa) and thus get

$$p(b)(y_1'(b)y_2(b) - y_2'(b)y_1(b)) = p(a)(y_1'(a)y_2(a) - y_2'(a)y_1(a)) = 0.$$

This completes the proof.

The next theorem is very important in the advanced theory of Sturm-Liouville problems.

Theorem on Green Function. If λ is not an eigenvalue of the Sturm-Liouville problem

$$(p(x)y')' + q(x,\lambda)y = 0 \tag{3.4}$$

$$\alpha_1 y(a) + \alpha_2 y'(a) = 0, \quad \beta_1 y(b) + \beta_2 y'(b) = 0 \tag{3.2}$$

then a unique function $G(t,s)$ exists on $[a,b] \times [a,b]$ such that

(i) G is a continuous function of (t,s);

(ii) $\frac{\partial}{\partial t} G(t,s)$ is a continuous function of (t,s) for $t \neq s$ and we have

$$\lim_{h\downarrow 0}\left[\frac{\partial G}{\partial t}(s+h,s) - \frac{\partial}{\partial t} G(s-h,s)\right] = \frac{1}{p(s)};$$

(iii) $G(t,s)$, considered as a function of t, satisfies the homogeneous equation (3.4) for each $t \neq s$;

(iv) $G(t,s)$, considered as a function of t, satisfies the boundary condition (3.2) for each $s \in [a,b]$;

(v) $G(t,s) = G(s,t)$ for all s,t;

(vi) for every continuous function $h\colon [a,b] \to R$, a function y is a solution of the nonhomogeneous Sturm-Liouville problem

$$(p(x)y')' + q(x,\lambda)y = h(x), \quad \alpha_1 y(a) + \alpha_2 y'(a) = 0 = \beta_1 y(b) + \beta_2 y'(b)$$

if and only if

$$y(x) = \int_a^b G(x,s)h(s)ds$$

for all $x \in [a,b]$.

The function G is called *Green function* corresponding to the given Sturm-Liouville problem. As will be clear from the proof, the uniqueness of Green function depends only on (i) and (vi).

The integral representation (vi) for the solution of a nonhomogeneous problem is very important and useful for various reasons. First of all, (vi) reduces the solvability of a nonhomogeneous problem to the calculation of an integral involving the nonhomogeneous term -- a fact very useful when G is explicitly known (cf. the proof of the theorem as well as the exercises below). Moreover, (vi) allows us to apply to Sturm-Liouville problems the theory of integral equations as well as the most sophisticated tools of functional analysis (cf. the bibliographical notes at the end of the chapter). Note that (vi) implies also the uniqueness of the solution of the nonhomogeneous problem whenever λ is not an eigenvalue (a fact that can be proved directly in a simple way, cf. the proof of the theorem).

Proof of the Theorem on Green Function: First of all, we state the existence of G. Let y_1 be a non-null solution to equation (3.4) such that

$$\alpha_1 y_1(a) + \alpha_2 y_1'(a) = 0$$

and let y_2 be a non-null solution to equation (3.4) such that

$$\beta_1 y_2(b) + \beta_2 y_2'(b) = 0$$

(the existence of the y_i is easily established by solving suitable Cauchy problems). Since λ is not an eigenvalue of (3.4), (3.2), the solutions y_1, y_2 are linearly independent (otherwise, we would have a non-null solution of (3.4), (3.2)). Let $W(y_1,y_2)(t)$ be the Wronskian of y_1, y_2 at t. We define $G: [a,b] \times [a,b] \to \mathbb{R}$ by

$$G(t,s) = \begin{cases} \dfrac{y_2(s)y_1(t)}{p(a)W(y_1,y_2)(a)} & \text{for } a \le t \le s \\[2ex] \dfrac{y_1(s)y_2(t)}{p(a)W(y_1,y_2)(a)} & \text{for } s \le t \le b. \end{cases}$$

It is easily seen by direct computation that G satisfies conditions (i),...,(v) [to state (ii), use Abel's formula]. If we define the function z by

$$z(t) = \int_a^b G(t,s)h(s)ds$$
$$= \frac{y_2(t)}{p(a)W(y_1,y_2)(a)} \int_a^t y_1(s)h(s)ds + \frac{y_1(t)}{p(a)W(y_1,y_2)(a)} \int_t^b y_2(s)h(s)ds$$

then we may easily verify by direct substitution that z is a solution of the nonhomogeneous Sturm-Liouville problem in (vi). Since the difference between two solutions of this nonhomogeneous problem is a solution of (3.4), (3.2), the assumption that λ is not an eigenvalue implies that the difference between two solutions of the nonhomogeneous problem is identically zero. Therefore z is the only solution of the nonhomogeneous problem and (vi) is completely established. It remains to show the uniqueness of a function G satisfying (i),...,(vi). Assume $H(t,s)$ to be another function satisfying (i),...,(vi). We have already seen that the nonhomogeneous problem in (vi) has a unique solution. Therefore from (vi) we have

$$\int_a^b G(t,s)h(s)ds = \int_a^b H(t,s)h(s)ds$$

or equivalently

$$\int_a^b \{G(t,s) - H(t,s)\}h(s)ds = 0$$

for every $t \in [a,b]$ and every continuous function $h\colon [a,b] \to \mathbb{R}$. For any fixed $t \in [a,b]$, we choose

$$h(s) = G(t,s) - H(t,s)$$

and substitute in the above equation. We get

$$\int_a^b \{G(t,s) - H(t,s)\}^2 ds = 0$$

for all $t \in [a,b]$. Since $G - H$ is continuous by (i), we conclude that $G = H$ and with this the theorem is completely proved.

Exercises. Calculate the Green function for the following problems:

1. $y'' = 0$, $y(0) = 0$, $y(1) = 0$.

Answer: $G(x,\xi) = \begin{cases} (1-\xi)x & x \le \xi \\ (1-x)\xi & x \ge \xi \end{cases}.$

2. $y'' = 0$, $y(0) = 0$, $y'(1) = 0$.

Answer: $G(x,\xi) = \begin{cases} x & x \le \xi \\ \xi & x \ge \xi \end{cases}.$

3. $y'' = 0, \quad y(0) = y'(0), \quad y(1) = -y'(1).$

Answer: $G(x,\xi) = \begin{cases} -\frac{1}{3}(x+1)(\xi-2) & x \le \xi \\ -\frac{1}{3}(\xi+1)(x-2) & x \ge \xi \end{cases}.$

4. $y'' + y = 0, \quad y(0) = 0, \quad y(\pi/2) = 0.$

Answer: $G(x,\xi) = \begin{cases} \cos\xi \sin x & x \le \xi \\ \cos x \sin\xi & x \ge \xi \end{cases}.$

4. PERIODIC SOLUTIONS

In this section, we shall consider another aspect of boundary value problems, that of the existence of periodic solutions. This is studied in the context of boundary value problems because of the fact that the existence of a periodic solution of period p is equivalent, under natural hypotheses, to the existence of a solution y such that $y(0) = y(p)$; cf. Exercise 1 in Sec. 4.1.

The equation

$$y' = 1$$

has the peculiarity that the second member is a function in every respect regular, but the equation has no periodic solution. This example might make one wonder whether the problem is of little mathematical interest. On the contrary, though, it is of enormous practical importance, since in all mechanical and engineering problems where there are oscillations (pendula, springs, etc.) or motions on closed trajectories (planets, electrons, etc.), the solutions of interest to the equations that describe the motions are exactly the periodic ones. This question has recently arisen in medicine and biology, as, for example, in the study of the number of inhabitants of a city that contract a contagious disease that has the property that whoever gets sick and then recovers is not immune from catching it again (a cold, for example). The applications are really what give direction to the development of the theory. We shall examine certain existence theorems here with the intention of providing a panorama both of the methods used in the proofs and of the results that can be applied most easily in concrete cases.

4.1. The Case of First Order Equations

There is a very simple necessary and sufficient condition for the existence of periodic solutions of first order equations on the real line; all the known criteria can be easily deduced from it.

<u>Theorem of Massera</u>. If $f\colon [a,\infty[\times \mathbb{R} \to \mathbb{R}$ is periodic in x with period p and if every Cauchy problem for $y' = f(x,y)$ has at most one solution, then the equation

$$y' = f(x,y)$$

has a periodic solution of period p if and only if it has a bounded solution on the interval $[a,\infty[$.

Before proving the theorem, we give two simple examples to show how the theorem can be applied. Other applications appear in the exercises.

If f is as in the hypothesis of the theorem, and if there are two numbers α and β such that $\alpha < \beta$ and $f(x,\alpha) \geq 0$, $f(x,\beta) \leq 0$ for $x \geq a$, then there is a periodic solution of $y' = f(x,y)$. To see this, observe that it follows from Corollary 2 of Sec. 1.3 of Chapter III that the solution of the initial value problem

$$y' = f(x,y), \quad y(a) = x_0$$

takes on all the values in $[\alpha,\beta]$ if $x_0 \in [\alpha,\beta]$.

For a second example, we consider a C^1-function $f\colon \mathbb{R}^2 \to \mathbb{R}$ that is periodic in x with period p with

$$-\infty < M \leq \frac{\partial}{\partial y} f \leq m < 0.$$

Then the equation $y' = f(x,y)$ has at least one bounded solution and therefore a periodic solution. To prove the existence of a bounded solution, one may proceed in the following manner. Let B be the Banach space of all real, continuous and bounded functions defined on $[0,\infty[$ with the sup norm, $||u||_\infty = \sup_{t\geq 0} |u(t)|$. Then the transformation T that maps every $u \in B$ to the unique solution of the initial value problem

$$y_u' = My_u + f(x,u(x)) - Mu(x), \quad y_u(0) = 0 \tag{4.1}$$

is a contraction of B into B. This fact can be verified by means of the well known representation of the solutions of (4.1)

$$y_u(x) = e^{Mx}\int_0^x e^{-Mt}(f(t,u(t)) - Mu(t))dt,$$

and by the mean value theorem. Thus T has a fixed point u_0 by the theorem of Banach and Cacciappoli. This fixed point satisfies the condition $u_0' = f(t,u_0(t))$, and therefore is the desired bounded solution.

<u>Proof of Massera's Theorem</u>: It is sufficient to show that if there

is a bounded solution y_0, then there is also a periodic solution with period p. For every integer n, we define a new function y_n by

$$y_n(x) = y_0(x + np) \qquad (x \geq a).$$

By virtue of the periodicity of f in x, every y_n is a solution of the given equation. There are two possible cases.

<u>Case 1</u>. There exists $x_0 \geq a$ such that $y_0(x_0) = y_1(x_0)$. Then $y_0 = y_1$ by virtue of the uniqueness of the initial value problem. But $y_0 = y_1$ means that y_0 is a periodic function with period p, so in this case the theorem is proven.

<u>Case 2</u>. $y_0(x) \neq y_1(x)$ for every $x \geq a$. We must then have $y_0 > y_1$ in $[a,\infty[$ or $y_0 < y_1$ in $[a,\infty[$ since otherwise the two functions would be equal at a point, and we would have case 1. Let us suppose that $y_0 > y_1$ in $[a,\infty[$. If we replace x by $x + np$ in $y_0(x) > y_1(x)$, we get

$$y_0(x + np) > y_1(x + np).$$

But by the definition of y_n, this inequality means that

$$y_n(x) > y_{n+1}(x).$$

The sequence $(y_n)_{n=0}^{\infty}$ is therefore decreasing. Since it is uniformly bounded by virtue of the boundedness of y_0, there is a function y_∞ such that

$$\lim_n y_n(x) = y_\infty(x)$$

pointwise. From $y_n' = f(x,y_n)$, it follows that the derivatives of the y_n are also uniformly bounded, so $(y_n)_n$ is an equicontinuous sequence. Thus, $\lim_n y_n = y_\infty$ uniformly on all compact subintervals of $[a,\infty[$. If we pass to the limit in

$$y_n(x) = y_n(a) + \int_a^x f(t,y_n(t))dt,$$

we get

$$y_\infty(x) = y_\infty(a) + \int_a^x f(t,y_\infty(t))dt,$$

and therefore y_∞ is a solution of the given equation. From

$$y_\infty(x) = \lim_n y_n(x) = \lim_n y_{n+1}(x) = \lim_n y_n(x+p) = y_\infty(x+p)$$

it follows that y_∞ is a periodic function with period p. The case

$y_0 < y_1$ is treated in a similar manner, and thus the theorem is completely proven.

Exercise 1. Prove that if $f(\cdot,y)$ is periodic with period p, then $y' = f(x,y)$ has a periodic solution with period p if and only if there is a solution y such that $y(0) = y(p)$.

Exercise 2. Prove the following result, of which the first example is a special case: If f is as in Massera's theorem, and if there are two bounded C^1-functions α and β such that $\alpha \leq \beta$, $\alpha' \leq f(x,\alpha)$, and $\beta' \geq f(x,\beta)$, then the given equation has a periodic solution. Is this true if $\alpha' \geq f(x,\alpha)$, $\beta' \leq f(x,\beta)$?

Exercise 3. Prove that if $f\colon \mathbb{R}^2 \to \mathbb{R}$ is periodic in x and decreasing in y, then $y' = f(x,y)$ has a periodic solution if and only if it has a bounded solution.

Exercise 4. Prove that if f is a periodic function with period p in x, then y is a periodic solution with period p of the equation $y' = f(x,y)$ if and only if

$$y(x) = y(0) + \frac{1}{p}\int_0^p f(t,y(t))dt + \int_0^x \{f(t,y(t)) - \frac{1}{p}\int_0^p f(s,y(s))ds\}dt.$$

4.2. The Case of Second Order Equations

In order to state a criterion for second order equations, we must first introduce the concept of subsolution and supersolution. A C^2-function α [β] is a *subsolution* [*supersolution*] of $y'' = f(x,y,y')$ if

$$\alpha'' \geq f(x,\alpha,\alpha') \qquad [\beta'' \leq f(x,\beta,\beta')].$$

Theorem of Knobloch. Let $f\colon [a,\infty[\times \mathbb{R} \to \mathbb{R}$ be a C^1-function that is periodic in x with period p. The equation

$$y'' = f(x,y)$$

has a periodic solution with period p if it has a subsolution α and a supersolution β that are periodic functions with period p satisfying the condition $\alpha \leq \beta$.

There is a case in which one can easily find subsolutions and supersolutions satisfying the hypotheses of the theorem, namely, when there are two constants c_1 and c_2, $c_1 \leq c_2$, such that the two functions $f(\cdot,c_1)$ and $f(\cdot,c_2)$ are of constant and opposite sign such that

$$0 \geq f(x,c_1), \quad 0 \leq f(x,c_2) \qquad (x \geq a).$$

Then the constant functions $\alpha \equiv c_1$, $\beta \equiv c_2$ are a subsolution and supersolution respectively. (Cf. Example 2 in Sec. 5.8 of Chapter I.)

Another case in which there are easily recognizable subsolutions and supersolutions is the following:

$$y'' = \lambda y^{2n+1} + q(x,y),$$

where $\lambda > 0$ and q is bounded and periodic in x with period p. Let $M > \sup|q(x,y)|$ and consider the two constant functions

$$\alpha \equiv -(M/\lambda)^{1/(2n+1)}, \qquad \beta \equiv (M/\lambda)^{1/(2n+1)}.$$

Clearly, $\alpha < \beta$. We furthermore have

$$\lambda \cdot \alpha^{2n+1}(x) + q(x,\alpha(x)) = -M + q(x,\alpha(x)) < 0 \equiv \alpha''$$

$$\lambda \cdot \beta^{2n+1}(x) + q(x,\beta(x)) = M + q(x,\beta(x)) > 0 \equiv \beta''.$$

Thus, α and β verify the hypotheses of Knobloch's theorem.

Proof of Knobloch's Theorem: We shall use an argument that is constructive as well as quite elementary and leads to an iterative method suitable for the numerical approximation of the solutions. Let $M > 0$ be greater than the maximum of the continuous function $\frac{\partial f}{\partial y}$ on the set

$$\{(x,y) \mid a \leq x \leq a + p, \quad \alpha(x) \leq y \leq \beta(x)\}.$$

From $\frac{\partial f}{\partial y} - M < 0$ it follows that $g_x(y) = f(x,y) - My$ is, for each fixed x, a decreasing function on the interval $[\alpha(x),\beta(x)]$. We define a sequence $(\alpha_{n+1})_{n=0}^{\infty}$ by induction in the following manner. $\alpha_0 = \alpha$ and α_{n+1} is the unique periodic solution with period p of

$$\alpha''_{n+1} - M\alpha_{n+1} = f(x,\alpha_n(x)) - M\alpha_n(x).$$

The element α_{n+1} is well-defined because if the eigenvalues of the problem

$$u'' + \lambda u = 0, \qquad u(a) = u(a + p), \quad u'(a) = u'(a + p)$$

are nonnegative, as we saw in Exercise 3 of Sec. 3.1, then $-M$ is not an eigenvalue and we apply Exercise 3 or, alternatively, the Fredholm alternative. Before proceeding further, we must show:

(*) $\qquad w'' - Mw \leq 0$ with w periodic implies $w \geq 0$.

To see this, note that w has a minimum since it is periodic. If w were negative somewhere, the minimum would be negative and there would therefore exist a point x_0 such that

$$w(x_0) < 0, \quad w'(x_0) = 0, \quad w''(x_0) \geq 0.$$

But we would in that case have

$$w''(x_0) - Mw(x_0) \geq -Mw(x_0) > 0$$

which contradicts the inequality (*). The claim is therefore true, and we use it to establish that

$$\alpha_n \leq \alpha_{n+1} \leq \beta$$

for each n; it is enough to prove that the two functions $u_n = \alpha_{n+1} - \alpha_n$, $v_n = \beta - \alpha_n$ satisfy the conditions $u_n \geq 0$, $v_n \geq 0$. For $n = 0$, we have

$$\begin{aligned} u_0'' - Mu_0 &= \alpha_1'' - \alpha'' - M\alpha_1 + M\alpha = f(x,\alpha) - M\alpha - \alpha'' + M\alpha \\ &= f(x,\alpha) - \alpha'' \leq 0, \end{aligned}$$

so $u_0 \geq 0$ by virtue of (*). It can similarly be established that $v_0 \geq 0$. We now assume that $u_n \geq 0$ and $v_n \geq 0$ and prove that $u_{n+1} \geq 0$ and $v_{n+1} \geq 0$. Since

$$\begin{aligned} u_{n+1}'' - Mu_{n+1} &= \alpha_{n+2}'' - \alpha_{n+1}'' - M\alpha_{n+2} + M\alpha_{n+1} \\ &= f(x,\alpha_{n+1}) - M\alpha_{n+1} - (f(x,\alpha_n) - M\alpha_n) \\ &\qquad \text{(because } g_x(y) \text{ is decreasing)} \\ &\leq 0, \end{aligned}$$

it suffices to apply (*). It is similarly established that $v_{n+1} \geq 0$. Since $(\alpha_n)_{n=1}^{\infty}$ is an increasing and equibounded sequence, there is a function α_∞ to which it converges pointwise. α_∞ is clearly a periodic function with period p. We now show that α_∞ is a solution of the given equation. From

$$\alpha_n'' = M\alpha_n + f(x,\alpha_{n-1}) - M\alpha_{n-1} \tag{4.2}$$

and from the uniform boundedness of the α_n, it follows that $(\alpha_n'')_{n=1}^{\infty}$ is equibounded. In order to show that even $(\alpha_n')_{n=1}^{\infty}$ is uniformly bounded, we fix $x \in [a,a+p]$ and let c be that one of the numbers a and $a+p$ that is farther from x; thus, $|x - c| \geq \frac{1}{2}p$. We then write Taylor's formula

$$\alpha_n(c) = \alpha_n(x) + \alpha_n'(x)(x-c) + \alpha_n''(\xi_x)\,\frac{(x-c)^2}{2}\,, \tag{4.3}$$

where ξ_x is a suitable point in the interval that has the smaller of x and c as left endpoint and the larger as right endpoint. From (4.3) we obtain

$$\alpha_n'(x) = \frac{1}{x - c}\left(\alpha_n(c) - \alpha_n(x) - \alpha_n''(\xi_x)\,\frac{(x-c)^2}{2}\right).$$

This, the fact that $1/|x-c| \leq 2/p$, and the equiboundedness of the α_n and α_n'' imply that $(\alpha_n')_{n=1}^{\infty}$ is an equibouned sequence. From the fact that the α_n, α_n', and α_n'' are equibounded, it follows that $(\alpha_n)_{n=1}^{\infty}$ and $(\alpha_n')_{n=1}^{\infty}$ are equicontinuous. The first consequence of this is that $\lim_n \alpha_n = \alpha_\infty$ uniformly on compact sets. We shall now show that α_∞' exists and that $\alpha_\infty' = \lim_n \alpha_n'$; we shall use the following property of limits: $z = \lim z_n$ if and only if every subsequence $(z_{n_k})_{k=1}^{\infty}$ has in turn a subsequence $(z_{n_{k_i}})_{i=1}^{\infty}$ that converges to z. Let $(\alpha_{n_k}')_{k=1}^{\infty}$ be a subsequence of $(\alpha_n')_{n=1}^{\infty}$. Since it is equicontinuous and equibounded, the sequence $(\alpha_{n_k}')_{k=1}^{\infty}$ has, by the theorem of Arzelà and Ascoli, a subsequence $(\alpha_{n_{k_i}}')_{i=1}^{\infty}$ which converges uniformly on compact sets to a function $\bar{\alpha}$. If we take the limit in

$$\alpha_{n_{k_i}}(x) = \alpha_{n_{k_i}}(a) + \int_a^x \alpha_{n_{k_i}}'(t)\,dt$$

we get $\alpha_\infty' = \bar{\alpha}$. Therefore α_∞' exists, and every subsequence of $(\alpha_n')_{n=1}^{\infty}$ has in turn a subsequence that converges to α_∞'. Therefore $\lim_n \alpha_n' = \alpha_\infty'$. We can similarly show that α_∞'' exists and that $\alpha_\infty'' = \lim_n \alpha_n''$. We can then pass to the limits on both sides of (4.2) and get

$$\alpha_\infty'' = M\alpha_\infty + f(x,\alpha_\infty) - M\alpha_\infty = f(x,\alpha_\infty).$$

Thus α_∞ is a solution of the given equation, and the proof of the theorem is complete.

<u>Exercise 1</u>. Prove that $y'' = 3y^3 + \cos y$ has periodic solutions with period 2π.

<u>Exercise 2</u>. Study sufficient conditions for $y'' = f(x,y)$ to have a periodic supersolution β and a periodic subsolution α, $\alpha \leq \beta$, when $f(x,\cdot)$ is increasing. Do they always exist?

<u>Exercise 3</u>. Show that $y'' + \lambda y = h(x)$ has a periodic solution of period p whenever h is continuous with period p and λ not an eigenvalue of $y'' + \lambda y = 0$, y periodic of period p. <u>Hint</u>: Consider the one-parameter family of mappings $T_t\colon \mathsf{R}^2 \to \mathsf{R}^2$ that, to each $z = (z_1,z_2) \in \mathsf{R}^2$, assigns the value $(y_t(p),y_t'(p))$ with y_t the unique solution of the Cauchy problem $y'' + \lambda y = th(x)$, $y(0) = z_1$, $y'(0) = z_2$. Show by contradiction the existence of an a priori bound for the solutions of the equation

$T_t(z) = 0$. Then use Borsuk's theorem on the topological degree and proceed as in the proof of the theorem on small perturbations in Sec. 4.3 below.

Exercise 4. Knobloch's theorem is valid even if f is only continuous. Prove this under the assumption $\alpha < \beta$. Hint: Approximate f by appropriate C^1-functions and use the theorem above.

Exercise 5. Give sufficient conditions for the existence of a periodic supersolution β and a periodic subsolution α, $\alpha \leq \beta$, for $y'' + \lambda y = f(x,y)$ where f is bounded and periodic in x.

4.3. The Case of Systems

In this section, we shall establish two conditions that are sufficient for a system to have periodic solutions. We shall begin with a question that frequently arises in practical situations; if a linear system with periodic solutions is perturbed slightly, are there still periodic solutions? The answer is yes, if the linear system has a unique periodic solution, as we see from the following.

Theorem on Small Perturbations. Let $a_{ij}: \mathbb{R} \to \mathbb{R}$ be continuous functions, periodic with period p, and let B be a closed ball in $\mathbb{R}^n$ with center at the origin. Let $f: [a,\infty[\times B \times [0,c] \to \mathbb{R}^n$ be continuous, satisfy a Lipschitz condition in y, and be periodic in x with period p such that

$$\lim_{\lambda\to 0} f(x,y,\lambda) = 0$$

uniformly in $[a,a+p] \times B$. If $y_i \equiv 0$ is the unique periodic solution with period p of

$$y_i' = \sum_{j=1}^{n} a_{ij}(x)y_j,$$

then there exists $\lambda_0 > 0$ such that the system

$$y_i' = \sum_{j=1}^{n} a_{ij}(x)y_j + f_i(x,y,\lambda).$$

has, for $|\lambda| \leq \lambda_0$, at least one solution $y_1,\ldots,y_n$ in which each y_i is periodic with period p.

This theorem covers practically all the cases that can occur with respect to "small" perturbations of linear systems. For an application, consider the following example. Let $f: \mathbb{R}^{n+1} \to \mathbb{R}^n$ satisfy a Lipschitz condition locally and be periodic in x. We can restrict f to the set $\mathbb{R} \times B$ where B is the closed unit ball in $\mathbb{R}^n$. Then the equation

$$y_i' = \sum_j a_{ij}(x)y_j + \lambda g_i(x,y)$$

with $g = f|_{R\times B}$, satisfies the assumption of the theorem on small perturbations. (Recall that a function that satisfies a Lipschitz condition locally satisfies a Lipschitz condition on compact sets; cf. Exercises 4 and 5 of Sec. 4.1, Chapter I.)

Proof of the Theorem on Small Perturbations: We shall use the continuity of the topological degree. From the fact that $y_i \equiv 0$ is a solution of the Cauchy problem

$$y_i' = \sum_{j=1}^{n} a_{ij}(x)y_j + f_i(x,y,0), \qquad y_i(a) = 0$$

and from the continuous dependence of the solutions on data, there follows the existence of $\varepsilon_0 > 0$ and $\lambda_1 > 0$ such that the closed ball $\overline{B(0,\varepsilon_0)}$ is contained in B and such that for every $y \in B(0,\varepsilon_0)$ and $|\lambda| \leq \lambda_1$, the Cauchy problem

$$\bar{y}_{\lambda i}' = \sum_{j=1}^{n} a_{ij}(x)\bar{y}_{\lambda j} + f_i(x,\bar{y}_\lambda,\lambda), \qquad \bar{y}_{\lambda i}(a) = y_i$$

has a unique solution $\bar{y}_\lambda$ that exists in the whole interval $[a,a+p]$. The function $V_\lambda := \overline{B(0,\varepsilon_0)} \to \mathbf{R}^n$ given by

$$V_\lambda(y) = \bar{y}_\lambda(a+p) - y = \bar{y}_\lambda(a+p) - \bar{y}_\lambda(a)$$

is then well defined. If there exists a point y where V_λ has value 0, then there is a solution of the given system that takes on the same value at a and $a + p$; we know that this implies the existence of a periodic solution. To prove the theorem, it is therefore sufficient to show that V_λ has at least one zero for λ sufficiently small. To do this, we shall show that there is a $\lambda_0 \in\ ]0,\lambda_1]$ such that

$$d(V_\lambda, B(0,\varepsilon_0), 0) \neq 0 \qquad (|\lambda| \leq \lambda_0) \tag{4.4}$$

holds for the topological degree, for this implies the conclusion by property (G_2) of the topological degree. In order to establish (4.4), we define an auxiliary function $U: \overline{B(0,\varepsilon_0)} \to \mathbf{R}^n$ as follows. For every $y \in \overline{B(0,\varepsilon_0)}$, let $\bar{\bar{y}}$ be the unique solution of the linear initial value problem

$$\bar{\bar{y}}_i' = \sum_{j=1}^{n} a_{ij}(x)\bar{\bar{y}}_j, \qquad \bar{\bar{y}}_i(a) = y_i. \tag{4.5}$$

We then define

$$U(y) = \bar{\bar{y}}(a + p) - y = \bar{\bar{y}}(a + p) - \bar{\bar{y}}(a).$$

From the uniformly continuous dependence of the solutions on all the data, that is, from Kamke's theorem (Sec. 1.9, Chapter I) applied to the system (4.5), it follows that

$$\lim_{\lambda \to 0} V_\lambda = U$$

uniformly in $\overline{B(0,\varepsilon_0)}$. The continuous dependence of the topological degree then implies the existence of $\lambda_0 \in]0,\lambda_1]$ such that

$$d(V_\lambda,B(0,\varepsilon_0),0) = d(U,B(0,\varepsilon_0),0) \qquad (|\lambda| \le \lambda_0).$$

It is therefore enough to show that

$$d(U,B(0,\varepsilon_0),0) \neq 0,$$

but this is true by virtue of Borsuk's theorem on the topological degree. To see this, note that U is linear, so $U(y) = -U(-y)$. Moreover, U cannot have zeros on the boundary of $B(0,\varepsilon_0)$ because that would be equivalent to the existence of a periodic solution with period p for the linear system

$$y_i' = \sum_j a_{ij}(x)y_j;$$

this solution could not be null since it would then pass through a point of $\partial B(0,\varepsilon_0)$, which would contradict the hypothesis of the theorem. Thus $d(U,B(0,\varepsilon_0)) \neq 0$ by virtue of Borsuk's theorem, and our theorem is proven.

We now examine the case of unperturbed systems. Following Krasnosel'skii, we shall say that a function $V\colon \mathbb{R}^n \to \mathbb{R}$ is a *guided function* of the system

$$y' = f(x,y)$$

if it is of class C^1 and there exists an $r_V > 0$ such that

$$(\operatorname{grad} V(y) \mid f(x,y)) < 0 \qquad (\text{all } x;\ \|y\| \ge r_V).$$

In this formula, $(\cdot|\cdot)$ represents the scalar product in $\mathbb{R}^n$:

$$(u|v) = \sum_{i=1}^{n} u_i v_i .$$

The existence of guided functions is related to the existence of periodic solutions, as we see from the following:

<u>Theorem on Guided Functions</u>. Let $f = (f_1,\ldots,f_n)$ be a C^1-mapping on the strip $[a,\infty[\times \mathbb{R}^n$ which is periodic in x with period p. If a continuous and periodic matrix $A(x)$ and finitely many guided functions $V_1,\ldots,V_m$ for the system $y' = f(x,y)$ exist such that

(i) $(\operatorname{grad} V_k(y)|A(x)y) < 0$ for all x, for all k, and for all $||y|| \geq r_{V_k}$;

(ii) $\lim_{||y||\to\infty} \max_k |V_k(y)| = +\infty$;

then there is at least one periodic solution with period p of the system $y' = f(x,y)$.

We consider two examples to see how to associate guided functions with a given system.

Let f be as in the theorem, and $V(y) = -||y||^2$. Then V satisfies (i) with $A(x)$ equal to the identity matrix as well as (ii). Thus, the system $y' = f(x,y)$ has at least one periodic solution if there is an $r > 0$ such that

$$(y \mid f(x,y)) > 0 \qquad (||y|| \geq r).$$

As a second example, we consider a system whose principal part is a potential

$$y_i' = \frac{\partial}{\partial y_i} F(y_1,\ldots,y_n) + f_i(x,y_1,\ldots,y_n)$$

where $F\colon \mathbb{R}^n \to \mathbb{R}$ is a C^1-function and the f_i are periodic in x with period p. We suppose that

$$m||y||^\alpha \leq ||\operatorname{grad} F(y)|| \leq M||y||^\alpha \qquad (||y|| \geq r) \tag{4.7}$$

where m, M, α, and r are positive constants. (The condition is satisfied if F is homogeneous of order α.) We also suppose that

$$\lim_{||y||\to\infty} \frac{|f_i(x,y)|}{||y||^\alpha} = 0. \tag{4.8}$$

(This means that if $||y||$ is sufficiently large, the principal part in the second member of our system is given by the $\partial F/\partial y_i$, that is, by grad F.) From the following inequalities

$$\begin{aligned}(-\operatorname{grad} F(y)|\operatorname{grad} F(y) + f(x,y)) &= -||\operatorname{grad} F(y)||^2 - (\operatorname{grad} F(y)|f(x,y)) \\ &\leq -||\operatorname{grad} F(y)||^2 + ||\operatorname{grad} F(y)||\ ||f(x,y)||\end{aligned}$$

$$\leq -m^2||y||^{2\alpha} + M||y||^{\alpha}||f(x,y)|| \qquad \text{(by (4.7))}$$

$$\leq -m^2||y||^{2\alpha} + M||y||^{\alpha}\,\frac{m^2}{2M}\,||y||^{\alpha}$$

$$\text{(for } ||y|| \text{ sufficiently large, because of (4.8))}$$

$$\leq -\tfrac{1}{2}m^2||y||^{2\alpha} < 0$$

it follows that $V(y) = -F(y)$ is a guided function for the given system. This will then have a periodic solution when V satisfies conditions (i) and (ii) in the theorem.

Proof of the Theorem on Guided Functions: If we put

$$M = \max_k \sup_{||y|| \leq r_{V_k}} |V_k(y)|,$$

then, by virtue of condition (ii), there exists $r > \max_k r_{V_k}$ such that

$$\max_k |V_k(y)| > M \qquad (||y|| \geq r).$$

We prove that for any possible periodic solution $y_1,\ldots,y_n$ of the given system, the inequality

$$||y(x)|| < r \qquad \text{(for all } x) \tag{4.9}$$

holds. To do this, we consider the function $v_k(x) = V_k(y(x))$. Let x_0 be a point where $|v_k(x)|$ has a maximum. Obviously, x_0 is a maximum or minimum point of v_k. Since x_0 is in the interior of the domain of v_k, we have

$$0 = v_k'(x_0) = (\text{grad } V_k(y(x_0))|y'(x_0))$$
$$= (\text{grad } V_k(y(x_0))|f(x_0,y(x_0))$$

and therefore $||y(x_0)|| \leq r_{V_k}$. It follows that

$$|v_k(x_0)| \leq \sup_{||y|| \leq r_{V_k}} |V_k(y)| \leq M.$$

Thus, y must assume all its values in the ball $\overline{B(0,r)}$, so (4.9) holds. We now define the auxiliary functions

$$\alpha(u) = \begin{cases} 0 & \text{if } 0 \leq u \leq r \\ u - r & \text{if } r \leq u \leq r + 1 \\ 1 & \text{if } r + 1 \leq u, \end{cases}$$

$$\overline{f}_i(x,y) = \frac{f_i(x,y)}{1 + \alpha(||y||)||f(x,y)||}.$$

The function $\overline{f} = (\overline{f}_1,\ldots,\overline{f}_n)$ is bounded and periodic in x with period p; it also satisfies a Lipschitz condition locally and coincides with f in $[a,\infty[\ \times \overline{B(0,r)}$. What is more, the V_k are guided functions for $y' = \overline{f}(x,y)$. For every $\lambda \in [0,1]$, we consider the following system, where the $a_{ij}(x)$ are the entries of the matrix $A(x)$:

$$y_i' = \lambda \sum_{j=1}^{n} a_{ij}(x)y_j + (1 - \lambda)\overline{f}_i(x,y). \tag{4.10}$$

For every $y \in \overline{B(0,r)}$, let y_λ be the unique solution of (4.10) that satisfies the initial condition $y_\lambda(a) = y$. Since the $\overline{f}$ are bounded, y_λ exists globally (cf. Exercise 10, Sec. 1.3, Chapter III). Therefore, the mapping $T_\lambda\colon \overline{B(0,r)} \to \mathbb{R}^n$ given by

$$T_\lambda(y) = y_\lambda(a + p) - y_\lambda(a) = y_\lambda(a + p) - y$$

is well defined. If there is a point y where T_λ is zero, then there is a solution of (4.10) which assumes the same value at a and $a + p$; we know that this implies the existence of a periodic solution with period p. Inequality (4.9) is valid for any possible periodic solution of (4.10), whatever λ is, since it depends only on the V_k. Therefore

$$0 \neq T_\lambda(y)$$

for every $\lambda \in [0,1]$ and every $y \in \partial B(0,r)$. It follows that

$$d(T_\lambda,B(0,r),0) = \text{a constant for all } \lambda$$

by virtue of the invariance of the topological degree under homotopy. In particular, we have

$$d(T_0,B(0,r),0) = d(T_1,B(0,r),0).$$

Since T_1 is linear, $T_1(y) = -T_1(-y)$ and therefore Borsuk's theorem insures that $d(T_1,B(0,r),0) \neq 0$. Thus $d(T_0,B(0,r),0) \neq 0$, and so T_0 has at least one zero. This is equivalent to the existence of a periodic solution $\overline{y}$ with period p for the system

$$y' = \overline{f}(x,y).$$

Since $||\overline{y}(x)|| < r$ by virtue of (4.9) (which, as we observed, remains valid even for $y' = \overline{f}(x,y)$), y is a solution of $y' = f(x,y)$ also, since f and $\overline{f}$ coincide on the strip $[a,\infty[\ \times B(0,r)$. This completes the proof of the theorem.

In all the following exercises, suppose the functions under consideration to be periodic in x with period p.

Exercise 1. Generalize the first example on guided functions by proving that $y' = f(x,y)$ has a periodic solution if

$$(f(x,y)|y) \leq 0 \qquad (||y|| = r).$$

Hint: Prove that all the solutions through a point of $\overline{B(0,r)}$ remain in $\overline{B(0,r)}$, and then use Brouwer's fixed point theorem.

Exercise 2. Prove that a linear system $y' = A(x)y + p(x)$ has a unique periodic solution if and only if the associated homogeneous system has $y \equiv 0$ as its unique periodic solution. Find an upper bound for the periodic solution as a function of p.

Exercise 3. Use the result of Exercise 2 to give a condition for $y' = A(x)y + f(x,y)$ to have a periodic solution if f satisfies a Lipschitz condition. Hint: Use contractions.

Exercise 4. Prove that if $y \equiv 0$ is the unique periodic solution of $y' = A(x)y$ and if $\lim_{||y||\to\infty} \frac{||f(x,y)||}{||y||} = 0$, then $y' = A(x)y + f(x,y)$ has at least one periodic solution.

Exercise 5. Prove that $V(y) = -(\frac{1}{3} a y_1^3 + b y_1^2 y_2 + c y_1 y_2^2 + \frac{1}{3} y_3^3)$ is a guided function for the system

$$\begin{cases} y_1' = a y_1^2 + 2b y_1 y_2 + c y_2^2 + f_1(x,y_1,y_2) \\ y_2' = b y_1^2 + 2c y_1 y_2 + d y_2^2 + f_2(x,y_1,y_2) \end{cases}$$

when

$$\begin{vmatrix} a & b \\ b & c \end{vmatrix} \cdot \begin{vmatrix} b & c \\ c & d \end{vmatrix} - \frac{1}{4}\begin{vmatrix} a & c \\ b & d \end{vmatrix} > 0$$

and

$$\lim_{||y||\to\infty} \frac{f_1(x,y)}{||y||^2} = 0, \qquad \lim_{||y||\to\infty} \frac{f_2(x,y)}{||y||^2} = 0.$$

Are the hypotheses of the theorem on guided functions satisfied?

Exercise 6. Consider the system

$$y' = \text{grad } F(y) + f(x,y).$$

written in vector notation. Prove that if F is homogeneous of degree α, then there exist m and M such that

$$m||y||^{\alpha} \leq ||\text{grad } F(y)|| \leq M||y||^{\alpha},$$

so that one can construct a guided function. Are the hypotheses of the theorem on guided functions satisfied?

Exercise 7. Prove that the theorem on guided functions remains valid if $V_1, \ldots, V_m$ satisfy (i), (ii), and

$$(\text{grad } V_k(y)|f(x,y)) \leq 0 \qquad (\text{for all } k,\ ||y|| \geq r_{V_k}).$$

Hint: Substitute $f + \frac{1}{n} A(x)y$ for f.

4.4. On the Structure of Periodic Solutions

We have seen enough criteria to insure the existence of a periodic solution. These criteria, however, say nothing at all about what these solutions are like; it would be possible to have a constant function. (Constant functions are periodic with period p for every $p > 0$.) Let us consider for an example the equation

$$y' = f(y) \tag{4.11}$$

where $f: \mathbb{R} \to \mathbb{R}$ is continuous and such that all the initial value problems for (4.11) have a unique solution (e.g., f might be of class C^1 or decreasing). If y is a periodic solution of (4.11), then y has maximum points and there therefore exists a point x_0 such that $y'(x_0) = 0$. From (4.11) we deduce that $f(y(x_0)) = 0$. This implies that the constant function z defined by $z(x) = y(x_0)$ is a solution of (4.11); $z'(x) = f(z(x))$. What is more, $z(x_0) = y(x_0)$. Thus, y and z are both solutions of the Cauchy problem

$$u' = f(u), \quad u(x_0) = y(x_0)$$

and must therefore coincide by virtue of the uniqueness for the initial value problems involving (4.11). This shows that all the periodic solutions of (4.11) are constant functions.

The argument we just made cannot be extended to the case of systems (that is, to equations in $\mathbb{R}^n$ with $n > 1$) since it depends strictly on the order structure of the real line. One can ask, however, under what conditions this same phenomena occurs in systems. This question also has physical interest and can be used in nonlinear functional analysis, as will be pointed out in the bibliographical references. For an answer, we shall prove the following two theorems.

Theorem on the Lipschitz Case. If $f: A \subseteq \mathbb{R}^n \to \mathbb{R}^n$ satisfies a Lipschitz condition with constant L:

$$||f(u) - f(v)|| \leq L||u - v|| \qquad (u,v \in A)$$

then the period p of all the nonconstant periodic solutions of

$$y' = f(y)$$

satisfies the condition $p \geq 4/L$.

In other words, any periodic solution of period $p < 4/L$ must be constant. The theorem cannot be extended to the case when f depends on x and

$$||f(x,u) - f(x,v)|| \leq ||u - v||,$$

as is shown by the example $y' = \sin x$.

The other theorem is valid in the case when the second member is no longer assumed to satisfy a Lipschitz condition.

Theorem on the Non-Lipschitz Case. Let $A \subseteq \mathbb{R}^n$, let $f: A \to \mathbb{R}^n$ be continuous, and let N be the greatest lower bound of the norms $||y'||_\infty = \sup_x ||y'(x)||$ of the derivatives of the nonconstant periodic solutions of the system

$$y' = f(y).$$

If $N > 0$ and at least one of the following conditions

(i) f is uniformly continuous and bounded;

(ii) there is a Lipschitz function $g: A \to \mathbb{R}^n$ such that $\sup_{x \in A} ||f(x) - g(x)|| < \frac{1}{2} N$;

is true, then $p_0 > 0$ exists such that every periodic nonconstant solution of $y' = f(y)$ has period $\geq p_0$.

Thus, the existence of a positive greatest lower bound for the norms of the *derivatives* of the non-constant periodic solutions of $y' = f(y)$ implies the existence of a positive greatest lower bound for their periods. A case in which $N > 0$ is when the origin does not belong to $\overline{f(A)}$, i.e., $0 \notin \overline{f(A)}$. Observe that by virtue of the theorem of Weierstrass on approximation by polynomials, condition (i) is included in (ii) when A is compact.

The proof of both theorems is based on the following:

Lemma. Let $v: \mathbb{R} \to \mathbb{R}^n$ be a continuous function, periodic with period p, with the following properties:

(i) there is an integrable function $u: [0, \frac{1}{2}p] \to \mathbb{R}$ such that

$$||v(t) - v(s)|| \le u(t-s) \qquad (s \le t;\ t-s \le \tfrac{p}{2});$$

(ii) $\int_0^p v_i(t)dt = 0 \qquad (i = 1,\dots,n).$

Then,

$$p||v||_\infty \le 2\int_0^{p/2} u(t)dt$$

where $||v||_\infty = \sup_t ||v(t)||$.

Proof: If we fix t and integrate with respect to s both sides of the equality

$$v(t) = v(s) + (v(t) - v(s))$$

over the interval $[t - \frac{1}{2}p, t + \frac{1}{2}p]$, we obtain

$$pv(t) = \int_{t-p/2}^{t+p/2} (v(t) - v(s))ds$$

from (ii), since $\int_a^{a+p} v = \int_0^p v$ by virtue of the periodicity of v. Therefore, for every t we have

$$\begin{aligned} p||v(t)|| &\le \int_{t-p/2}^{t+p/2} ||v(t) - v(s)||ds \\ &\le \int_{t-p/2}^{t} ||v(t) - v(s)||ds + \int_t^{t+p/2} ||v(t) - v(s)||ds \\ &\le \int_{t-p/2}^{t} u(t-s)ds + \int_t^{t+p/2} u(s-t)ds \qquad \text{(by (i))} \\ &\le 2\int_0^{p/2} u(s)ds. \end{aligned}$$

If we take the sup over t, we get the formula we want, and the lemma is proved.

Proof of the Theorem in the Case f Satisfies a Lipschitz Condition: We shall choose v and u appropriately and apply the lemma. For simplicity, we shall assume that the norm in $\mathbb{R}^n$ is the sup norm:

$$||x|| = \max_i |x_i|.$$

We leave it to the reader as an exercise to deduce the general case from this one. Let y be any nonconstant periodic solution with period p of $y' = f(y)$. We define

$$v(t) = (y_1'(t),\dots,y_n'(t)).$$

Then

$$\int_0^p v_i(t)dt = y_i(p) - y_i(0) = 0 \quad \text{(by the periodicity of } y_i)$$

so (ii) in the lemma is valid. We observe that for each $s \leq t$, we have

$$\begin{aligned}||v(t) - v(s)|| &= \max_i |y_i'(t) - y_i'(s)| \\ &= \max_i |f_i(y(t)) - f_i(y(s))| \\ &\leq L \max_i |y_i(t) - y_i(s)| \\ &\quad \text{(since } f \text{ satisfies a Lipschitz condition)} \\ &\leq L(t-s) \max_i |y_i'(\xi)| \\ &\quad \text{(because of the mean value theorem, if } \xi \text{ is a suitable point in }]s,t[) \\ &= L(t-s) \max_i |v_i(\xi)| \\ &= L(t-s)||v(\xi)|| \\ &\leq L(t-s)||v||_\infty,\end{aligned}$$

where $||v||_\infty = \sup||v(t)||$. Therefore (i) in the lemma is true, if we pick

$$u(t) = L||v||_\infty t.$$

The lemma implies the following inequality:

$$p||v||_\infty \leq 2L||v||_\infty \frac{p^2}{8}. \tag{4.12}$$

Since y is not a constant, there is at least one derivative y_i' that is not always 0. Thus, $||v||_\infty \neq 0$ and so $p \geq 4/L$ follows from (4.12) and the theorem is proved.

Proof of the Theorem when f does not Satisfy a Lipschitz Condition: Let y be a nonconstant periodic solution with period p of $y' = f(y)$.

Case (i): Let ω be the modulus of continuity of f, that is, let $\omega\colon \mathbb{R}^+ \to \mathbb{R}^+$ be the increasing function defined by

$$\omega(t) = \sup_{||u-v||\leq t} ||f(u) - f(v)||.$$

For every $s \leq t$, we have

$$\begin{aligned} ||y'(t) - y'(s)|| &= ||f(y(t)) - f(y(s))|| \\ &\leq \omega(||y(t) - y(s)||) \\ &\leq \omega(||y'||_\infty (t - s)) \end{aligned}$$

(by virtue of the mean value theorem and the fact that ω is increasing, since $||y'||_\infty = \sup_t ||y'(t)||$)

$$\leq \omega(M(t - s))$$

where $M = \sup_u ||f(u)|| \geq ||y'||_\infty$. We now apply the lemma with $v = y'$ (as in the proof of the preceding theorem) and $u(t) = \omega(Mt)$, which is integrable on $[0, \frac{1}{2}p]$ by virtue of the fact that it is increasing. We get

$$p \leq \frac{2}{||y'||_\infty} \int_0^{p/2} \omega(Mt)\,dt.$$

Since $N \leq ||y'||_\infty$, we have

$$p \leq \frac{2}{N} \int_0^{p/2} \omega(Mt)\,dt \leq \frac{2}{N}\frac{p}{2} \sup_{0 \leq t \leq p/2} \omega(Mt) = \frac{p}{N}\,\omega(M\,\frac{p}{2})$$

and so

$$N \leq \omega(M\,\frac{p}{2}).$$

Since f is uniformly continuous, $\lim_{t \downarrow 0} \omega(t) = 0$. Therefore the preceding inequality would be impossible if p were arbitrarily small.

<u>Case (ii)</u>. Let L be the Lipschitz constant of g, and

$$\varepsilon = \sup_{x \in A} ||f(x) - g(x)||.$$

For each $s \leq t$, we have

$$\begin{aligned} ||y'(t) - y'(s)|| &= ||f(y(t)) - f(y(s))|| \\ &\leq ||f(y(t)) - g(y(t))|| + ||g(y(t)) - g(y(s))|| \\ &\qquad + ||g(y(s)) - f(x(s))|| \\ &\leq 2\varepsilon + L||y(t) - y(s)|| \\ &\leq 2\varepsilon + L||y'||_\infty (t - s) \quad \text{(by the mean value theorem).} \end{aligned}$$

If we apply the lemma with $v = y'$ and $u(t) = 2\varepsilon + L||y'||_\infty t$, we get

$$p \le \frac{2}{||y'||_\infty} \int_0^{p/2} (2\varepsilon + L||y'||_\infty t)dt = \frac{2p\varepsilon}{||y'||_\infty} + \frac{L}{4}p^2 \le \frac{2p\varepsilon}{N} + \frac{L}{4}p^2.$$

Therefore

$$p \ge \frac{4}{L}\left(1 - \frac{2\varepsilon}{N}\right).$$

Since $\varepsilon < \frac{1}{2}N$, we can take $p_0 = \frac{4}{L}(1 - 2\varepsilon/N)$. This completes the proof of the theorem.

Exercise 1. Find a positive lower bound for the periods of the nonconstant periodic solutions of $x'' = f(x)$, $x'' = f(x,x')$, $x'' = f(x')$, where f satisfies a Lipschitz condition with constant L.

Exercise 2. Use the lemma of this section to prove that if $v\colon \mathsf{R} \to \mathsf{R}^n$ is periodic with period p, if $\int_0^p v(t)dt = 0$, $v(t) \ne 0$ for at least one t, and if

$$||v(t) - v(s)|| \le L||v||_\infty(t-s) \qquad (s \le t),$$

then $p \ge 4/L$.

Exercise 3. From Exercise 2 deduce another proof of the theorem for the case when f satisfies a Lipschitz condition. Hint: If y is a nonconstant periodic solution of $y' = f(y)$, consider $v(t) = y(t) - y(t + t_1 - t_2)$ with $y(t_1) \ne y(t_2)$.

5. FUNCTIONAL BOUNDARY VALUE PROBLEMS

In this section, we shall consider a type of boundary problem which, in the generality of its formulation, may be compared to the geometrical problems. Let I be an interval of real numbers, let $U \subseteq \mathsf{R}^n$ be open, and let $f\colon I \times U \to \mathsf{R}^n$. Consider the problem of finding a solution y of the equation

$$y' = f(x,y)$$

that satisfies the further condition

$$L(y) = r$$

where $r \in \mathsf{R}^n$ and L is a function whose domain is a subset X of the space of continuous functions $C(I,\mathsf{R}^n)$ and whose range is in R^n:

$$L\colon X \to \mathsf{R}^n.$$

f, L, and r are the data of the problem. In order to emphasize that their formulation depends on the function L, these problems will be called *functional boundary problems*. Observe that L can be linear or nonlinear, in which case we speak respectively of linear functional problems or nonlinear functional problems. By a solution of

$$y' = f(x,y), \quad L(y) = r$$

we mean a function $y \in C(I,\mathbb{R}^n)$ that solves the equation $y' = f(x,y)$ on all of I (we are thus dealing with a global solution) and furthermore satisfies the condition $L(y) = r$. We have already considered special cases of this general scheme, and many other new problems can be formulated in this way; everything depends on the form of L and r. For example, the initial value problem $y(x_0) = y_0$ is obtained simply by taking $L(y) = y(x_0)$ and $r = y_0$; Nicoletti's problem $y_i(x_i) = \delta_i$ is produced by putting $L(y) = (y_1(x_1),\ldots,y_n(x_n))$ and $r = (\delta_1,\ldots,\delta_n)$; the problem of periodic solutions $y(0) = y(p)$ is obtained by taking $L(y) = y(0) - y(p)$ and $r = 0$. As an example of a new problem, consider $I = [a,\infty[$ and the problem of finding solutions of $y' = f(x,y)$ whose limit is y_∞ as x tends to infinity; in this case, X is the space of continuous functions having a limit at ∞, $L(y) = \lim_{x\to\infty} y(x)$ and $r = y_\infty$. Another example consists of finding solutions of a given equation $y' = f(x,y)$ satisfying the condition

$$\sum_{n=1}^{\infty} 2^{-n} g_n(y(x_n)) = y_\infty$$

where $g_n\colon \mathbb{R}^n \to \mathbb{R}^n$ are bounded functions and $x_n \in I$; in this case, $r = y_\infty$ and $L(y) = \sum_n 2^{-n} g_n(y(x_n))$. A further example is obtained by considering the nonlinear problem of the type of Nicoletti:

$$y'' = f(x,y,y'), \quad y(0) = y_0, \quad y'(1) = y^2(1).$$

In this case, if we consider the system in $\mathbb{R}^2$ associated with the given equation of the second order, L and r are defined respectively by $L((y_1,y_2)) = (y_1(0),y_2(1) - y_1^2(1))$ and $r = (y_0,0)$.

In this section, we shall prove certain existence and uniqueness theorems formulated from the functional point of view by means of the Banach spaces introduced in Sec. 4.2 of Chapter I. We shall find among the various applications the general explanation of a fact that we have already observed in special cases: a linear problem often has a solution if f satisfies a Lipschitz condition with a sufficiently small Lipschitz constant.

5.1. Linear Functional Problems

In this section, we examine the case in which L is linear. We shall limit ourselves to treating one question only: the relation between the uniqueness and existence of solutions. This is a well known problem in linear functional analysis, where it is called Fredholm alternative. We have seen it before many times: in geometrical problems, where we saw that the existence of the solution is preserved under certain perturbations of linear systems with unique solution; in the proof of the existence theorems for the two-point boundary value problem, where we used the theory of eigenvalues; and in the treatment of periodic solutions, where we saw that if we slightly perturbed linear systems with unique periodic solution, we continued to have periodic solutions.

The first theorem we shall prove insures the existence of the solution when we have uniqueness for a whole family of boundary value problems. This means that when we examine a given problem, we must attempt to associate a suitable family of problems with it in order to apply the theorem, a procedure which we shall illustrate in the corollary of the theorem and in three concrete examples that follow it. The prototype of the theorem can be seen in the lemma of Lasota and Opial that we used in Section 3.

Before stating the first theorem, we must first consider three special Banach spaces. The first is the space $C(I,R^n)$ of continuous function on the compact interval I with values in R^n, endowed with the sup norm

$$||u||_\infty = \sup_{x\in I} ||u(x)||.$$

The second is the Banach space $C^1(I,R^n)$ of C^1-functions defined on the compact interval I with values in R^n, endowed with the norm

$$||u||_1 = ||u||_\infty + ||u'||_\infty.$$

Observe that the topology of $C^1(I,R^n)$ is finer than that of $C(I,R^n)$; this observation will permit us to apply the theorem. The third space of interest to us is that of all continuous linear operators between two Banach spaces; let X and Y be two normed spaces and let $\mathscr{L}(X,Y)$ be the space introduced in Chapter II, Sec. 1.2.

We now state the first theorem of this section.

<u>Theorem for Nonlinear Equations</u>. Let $f\colon [a,b] \times R^n \to R^n$ be a continuous mapping such that all Cauchy problems

$$y' = f(x,y), \quad y(a) = y_0$$

have a unique solution in $[a,b]$ as y_0 varies in R^n. Let X be a vector subspace of $C([a,b],R^n)$, endowed with a norm whose associated topology is finer than that of the sup norm; suppose X contains all the solutions of the equation $y' = f(x,y)$. Let U be a set of continuous linear operators $L: X \to R^n$ such that the boundary value problems

$$y' = f(x,y), \quad L(x) = r$$

have at most one solution for every $L \in U$ and every $r \in R^n$. If U is an open subset of $\mathscr{L}(X,R^n)$, then all the boundary problems

$$y' = f(x,y), \quad L(y) = r$$

have a solution (necessarily unique) as L varies in U and r in R^n.

Note that it is not required that X be a Banach space. The fact that the norm of X is finer than $||\cdot||_\infty$ means that the topology of X is finer than that of uniform convergence; we may thus take $X = C([a,b],R^n)$ or $X = C^1([a,b],R^n)$. As for the hypothesis made about the initial value problems, we recall that it is satisfied if f is of class C^1 and bounded, or, more generally, if there is satisfied locally an inequality of the form

$$(f(x,u) - f(x,v) \mid u - v) \le \omega_1(x,||u-v||)||u - v||$$

where ω_1 can vary from neighborhood to neighborhood, and if there is satisfied globally an inequality of the form

$$(f(x,y)|y) \le \omega_2(x,||y||)||y||,$$

with ω_1 such that the unique solution of

$$u' = \omega_1(x,u), \quad u(x_0) = 0$$

is $u \equiv 0$, and ω_2 is such that every initial value problem for $v' = \omega_2(x,v)$ has a maximal global solution. We saw these conditions in Chapter III.

Before passing on to the proof of the theorem, we give a few concrete examples. They are all deduced from the following corollary, which we shall prove after the theorem.

Corollary. Let $f: [a,b] \times R^n \to R^n$ be continuous and such that

$$(f(x,u) - f(x,v)|u - v) \le M||u - v||^2 \quad \text{(for all } x,u,v)$$

and let L_c be the linear operator from $C([a,b],R^n)$ into R^n which corresponds to the Cauchy problem:

$$L_c(y) = y(a).$$

Then for every $L \in \mathscr{L}(C([a,b],R^n),R^n)$ such that $||L_c-L||< e^{-M(b-a)}$

the boundary value problem

$$y' = f(x,y), \qquad L(y) = r$$

has a unique solution as r varies in $\mathbb{R}^n$. Moreover, if for every x, u, and v,

$$||f(x,u) - f(x,v)|| \leq M||u - v||,$$

then, for every $L \in \mathscr{L}(C^1([a,b],\mathbb{R}^n),\mathbb{R}^n)$ such that

$$||L_c - L|| < \frac{e^{-M(b-a)}}{1 + M},$$

the boundary value problem

$$y' = f(x,y), \qquad L(y) = r$$

has a unique solution as r varies in $\mathbb{R}^n$.

In order to avoid the confusion that may result from using the same notation, we warn that the norm $||L_c - L||$ in the first inequality is that of $\mathscr{L}(C([a,b],\mathbb{R}^n),\mathbb{R}^n)$ and in the second inequality that of $\mathscr{L}(C^1([a,b],\mathbb{R}^n),\mathbb{R}^n)$. This is possible because L_c belongs to both those spaces.

The result expressed by the corollary states that we may associate a set U of linear problems with the Cauchy problem in such a way that the hypotheses of the theorem above are satisfied; precisely, U is the ball in $\mathscr{L}(C([a,b],\mathbb{R}^n),\mathbb{R}^n)$ with center L_c and radius $e^{-M(b-a)}$, or the ball in $\mathscr{L}(C^1([a,b],\mathbb{R}^n),\mathbb{R}^n)$ with center L_c and radius $e^{-M(b-a)}/(1+M)$.

The corollary explains a phenomenon that is frequently encountered when f satisfies a Lipschitz condition (and extends it even to the case when f does not): if the Lipschitz constant is small enough, then the given problem has a solution.

We shall now examine applications to three different problems. For the first, we consider the Nicoletti problem

$$y_i' = f_i(x,y), \qquad y_i(x_i) = \delta_i \tag{5.1}$$

without making the hypothesis that the f_i are bounded. If f satisfies a Lipschitz condition,

$$||f(x,u) - f(x,v)|| \leq M||u - v|| \qquad \text{(for every } x,u,v)$$

and if, furthermore,

$$\sqrt{n}\cdot\sup|x_i - a| < \frac{e^{-M(b-a)}}{1 + M},$$

then (5.1) admits a unique solution. To see this, observe that the mean value theorem implies that for each $y \in C^1([a,b],R^n)$,

$$|y_i(x_i) - y_i(a)| \leq |x_i - a|\, ||y'||_\infty \leq |x_i - a|\, ||y||_1$$

which, since $||z|| = \sqrt{\Sigma|z_i|^2}$ for $z \in R^n$, implies that

$$||L_c - L_N|| = \sup_{||y||_1=1} ||L_c(y) - L_N(y)|| \leq \sqrt{n}\cdot\sup_i |x_i - a|,$$

where $L_c(y) = y(a)$ and $L_N(y) = (y_1(x_1),\dots,y_n(x_n))$. This shows that L_N belongs to the ball in $\mathscr{L}(C^1([a,b],R^n),R^n)$ with center L_c and radius $e^{-M(b-a)}/(1+M)$, and it is then enough to apply the corollary.

We now consider the problem (5.1) once again, this time making the hypothesis

$$(f(x,u) - f(x,v) \mid u - v) \leq M||u - v||^2 \qquad \text{(for all } x,u,v).$$

We now use the fact that $L_N \in \mathscr{L}(C([a,b],R^n),R^n)$. We have

$$\begin{aligned} ||L_c - L_N|| &= \sup_{||y||_\infty=1} ||L_c(y) - L_N(y)|| \\ &= \sup_{||y||_\infty=1} \sqrt{\Sigma_i|y_i(x_i) - y_i(a)|^2} \\ &\leq \sup_{||y||_\infty=1} \sqrt{\Sigma_i(||y||_\infty + ||y||_\infty)^2} \\ &\leq 2\sqrt{n}. \end{aligned}$$

We shall therefore have

$$||L_c - L_N|| < e^{-M(b-a)}$$

provided that M is negative and such that

$$2\sqrt{n} < e^{-M(b-a)}.$$

For such $M < 0$, the problem (5.1) admits a unique solution because of the corollary.

As a second example of an application of the corollary, we consider the two-point boundary value problem (or problem of Picard)

$$y'' = f(y)y', \quad y(a) = y_0, \quad y(b) = y_1. \tag{5.2}$$

We suppose that the Cauchy problems have unique solutions in $[a,b]$ and that

$$f(y) \leq M$$

for all y. We consider the first order system $z' = F(x,z)$ associated with the equation $y'' = f(y)y'$, with

$$F(x,(z_1,z_2)) = \left(\int_0^{z_1} f(t)dt,\ Mz_2\right).$$

It is easy to verify that in the scalar product in $\mathbb{R}^2$, we have

$$(F(x,u) - F(x,v) \mid u - v) \le M||u - v||^2$$

for all x, u, and v. Let L_c and L_p be the operators in $\mathscr{L}(C([a,b], \mathbb{R}^n)$ defined by

$$L_c(z) = (z_1(a), z_2(a)), \quad L_p(z) = (z_1(a), z_1(b)).$$

From the definition of norm, we get

$$\begin{aligned} ||L_c - L_p|| &= \sup_{||z||_\infty = 1} ||L_c(z) = L_p(z)|| \\ &= \sup_{||z||_\infty = 1} |z_2(a) - z_1(b)| \le 2. \end{aligned}$$

We therefore deduce from the corollary that (5.2) admits at least one solution (cf. Exercise 2) if M is negative and such that

$$2 < e^{-M(b-a)}.$$

For a last application, we consider the case of periodic solutions with period p. The corresponding linear operator is defined by

$$L_\#(y) = y(0) - y(p).$$

We have

$$\begin{aligned} ||L_c - L_\#|| &= \sup_{||y||_\infty = 1} ||L_c(y) - L_\#(y)|| \\ &= \sup_{||y||_\infty = 1} ||y(0) - y(0) + y(p)|| \le 1. \end{aligned}$$

The corollary therefore implies that the equation $y' = f(x,y)$ has a unique periodic solution of period p if $f\colon \mathbb{R} \times \mathbb{R}^n \to \mathbb{R}^n$ is continuous, periodic in x with period p, and

$$(f(x,u) - f(x,v) \mid u - v) \le M||u - v||^2$$

for all x, u, and v with $M < 0$. In particular, on the real line, it is sufficient that $\partial f(x,\cdot)/\partial y$ be bounded above by a negative constant.

Proof of the Theorem for Nonlinear Equations: Let us fix $L_0 \in U$. For each $c \in \mathbb{R}^n$, let y_c be the unique solution on $[a,b]$ of the Cauchy problem

$$y' = f(x,y), \quad y(a) = c.$$

We define $T: \mathbb{R}^n \to \mathbb{R}^n$ with

$$T(c) = L_0(y_c).$$

To prove that

$$y' = f(x,y), \quad L_0(y) = r$$

has a solution for each $r \in \mathbb{R}^n$, it suffices to show that $T(\mathbb{R}^n) = \mathbb{R}^n$. We shall do this by showing that $T(\mathbb{R}^n)$ is both an open and closed set. We begin by showing that $T(\mathbb{R}^n)$ is open. In the first place, T is continuous because it is the composition of two continuous functions, $c \rightsquigarrow y_c$ (which is continuous by virtue of the uniqueness of the Cauchy problem, as we know from Chapter III) and L_0. Moreover, T is injective. In fact, $c' \neq c''$ implies that $y_{c'} \neq y_{c''}$ because of the uniqueness for Cauchy problem. Hence if $T(c') = T(c'') = r$ for $c' \neq c''$, then the problem

$$y' = f(x,y), \quad L_0(y) = r$$

would have two distinct solutions, $y_{c'}$ and $y_{c''}$. Since that would contradict the hypothesis of uniqueness made on the members of U, we must necessarily have $T(c') \neq T(c'')$ when $c' \neq c''$, and thus T is injective. The set $T(\mathbb{R}^n)$ is then an open subset of $\mathbb{R}^n$ by virtue of one of the fundamental theorems of the topology of $\mathbb{R}^n$:

<u>Theorem</u>. Let A be an open subset of $\mathbb{R}^n$ and let $g: A \to \mathbb{R}^n$. If g is continuous and injective, then $g(A)$ is open in $\mathbb{R}^n$.

To prove that $T(\mathbb{R}^n)$ is closed, it is sufficient to prove that if $p_k \in T(\mathbb{R}^n)$ and $\lim_k p_k = p_0$, then $p_0 \in T(\mathbb{R}^n)$. Let c_k be such that $T(c_k) = p_k$. If $(c_k)_{k=1}^{\infty}$ is a Cauchy sequence, there is a $c_0 = \lim_k c_k$. Then the continuity of T implies that $T(c_0) = \lim_k T(c_k) = \lim_k p_k = p_0$ and so $p_0 \in T(\mathbb{R}^n)$. Hence we can prove that $p_0 \in T(\mathbb{R}^n)$ and thus that $T(\mathbb{R}^n)$ is closed if we can show that the hypothesis that $(c_k)_{k=1}^{\infty}$ is not a Cauchy sequence leads to a contradiction. Let us suppose that $(c_k)_{k=1}^{\infty}$ is not a Cauchy sequence. Then there is an $\varepsilon > 0$ such that for every k, there exists j_k with the property

$$||c_{k+j_k} - c_k|| \geq \varepsilon. \tag{5.3}$$

For each k, we now define a linear operator $L_k \in \mathscr{L}(X, R^n)$ with the two properties

$$L_k(y_{c_{k+j_k}} - y_{c_k}) = -L_0(y_{c_{k+j_k}} - y_{c_k}); \tag{5.4}$$

$$||L_k|| = \frac{||L_0(y_{c_{k+j_k}} - y_{c_k})||}{||y_{c_{k+j_k}} - y_{c_k}||}. \tag{5.5}$$

We define L_k in the following way. Set $z_k = y_{c_{k+j_k}} - y_{c_k}$. In the subspace $V_k = Rz_k$ of X generated by z_k, we can define a linear operator $S_k: V_k \to R^n$ by

$$S_k(\alpha z_k) = -\alpha L_0(z_k).$$

It is easy to show that S_k satisfies (5.4) and (5.5) when it replaces L_k there. We now extend S_k from V_k to all of X by applying the following theorem from linear functional analysis to each of its n coordinates:

Hahn-Banach Theorem. Let Y be a normed space, and let V be a vector subspace of Y with the induced norm. If $g: V \to R$ is a continuous linear transformation, then g has an extension $\bar{g}: Y \to R$ such that $\bar{g}$ is linear and $||\bar{g}|| = ||g||$.

Now that the existence of L_k is established, we observe that we have

$$||y_{c_{k+j_k}} - y_{c_k}|| \geq A||y_{c_{k+j_k}} - y_{c_k}||_\infty \geq A||y_{c_{k+j_k}}(a) - y_{c_k}(a)||$$
$$= A||c_{k+j_k} - c_k|| > \varepsilon A$$

for a suitable positive constant A; this depends on the fact that the topology on X is finer than that of uniform convergence. We then have, from (5.5):

$$||L_k|| \leq \frac{1}{A\varepsilon}||L_0(y_{c_{k+j_k}} - y_{c_k})|| = \frac{1}{A\varepsilon}||L_0(y_{c_{k+j_k}}) - L_0(y_{c_k})||$$
$$= \frac{1}{A\varepsilon}||p_{k+j_k} - p_k||.$$

This means that

$$\lim_k ||L_k|| = 0 \tag{5.6}$$

since $(p_k)_{k=1}^\infty$ is a Cauchy sequence. At this point, we use the hypothesis of the theorem that U is an open set; since $L_0 \in U$, there exists, by virtue of (5.6), a k_0 such that $L_0 + L_{k_0} \in U$. But this leads to a contradiction, since, for $r_0 = (L_0 + L_{k_0})(y_{c_{k_0}})$, the boundary value problem

$$y' = f(x,y), \quad (L_0 + L_{k_0})(y) = r_0$$

has two distinct solutions, $y_{c_{k_0}}$ (as follows from the definition of r_0) and $y_{c_{k_0+j_{k_0}}}$ (as follows from (5.4)); this is incompatible with the hypothesis of uniqueness made on the elements of U. The hypothesis that $(c_k)_{k=1}^{\infty}$ is not a Cauchy sequence is therefore absurd; we conclude that $T(\mathbb{R}^n)$ is a closed set. $\mathbb{R}^n$ is connected because every pair of points can be joined by a segment and hence the only subsets of $\mathbb{R}^n$ that are both open and closed are $\emptyset$ and $\mathbb{R}^n$. Since $T(\mathbb{R}^n)$ is not empty, the only possibility left is that $T(\mathbb{R}^n) = \mathbb{R}^n$, and the theorem is thus proved.

Proof of the Corollary: We first of all consider the case when f satisfies the inequality involving the scalar product and $L \in \mathcal{L}(C([a,b],\mathbb{R}^n),\mathbb{R}^n)$ is such that

$$||L_c - L|| < e^{-M(b-a)}. \tag{5.7}$$

By virtue of the theorem, it is enough to show that

$$y' = f(x,y), \quad L(y) = r \tag{5.8}$$

has at most one solution for each L that satisfies (5.7). Let us suppose that for some L satisfying (5.7), there are two distinct solutions u and v of (5.8); we shall arrive at a contradiction. Let $y = u - v$. Recall that the scalar product in $\mathbb{R}^n$ is defined by

$$(p|q) = \sum_{i=1}^{n} p_i q_i$$

and that $||y(x)||^2 = y_1^2(x) + \ldots + y_n^2(x)$. The function $z(x) = ||y(x)||^2$ is clearly differentiable and its derivative satisfies

$$z'(x) = 2(f(x,u(x)) - f(x,v(x))|y(x)) \leq 2M||y(x)||^2 = 2Mz(x).$$

Thus the theorem on differential inequalities of Sec. 2.4 of Chapter III implies that $z(x) \leq w(x)$, where w is the unique solution of the Cauchy problem

$$w' = 2Mw, \quad w(a) = y(a).$$

Since $w(x) = z(a)e^{2M(x-a)}$, it follows that

$$||y(x)|| \leq ||y(a)||e^{M(b-a)} \qquad (a \leq x \leq b). \tag{5.9}$$

Therefore, if $y(a) = 0$, then $y \equiv 0$, and we have the contradiction we want. If $y(a) \neq 0$, then we arrive at a contradiction through the following inequalities:

$$\begin{aligned} 0 &= ||L(y)|| \qquad (\text{since } L(u) = r = L(v)) \\ &= ||L_c(y) + (L - L_c)(y)|| \\ &\geq \Big| ||L_c(y)|| - ||L - L_c||\ ||y||_\infty \Big| \\ &\geq ||y(a)|| - ||L - L_c||\ ||y(a)||e^{M(b-a)} \quad (\text{by } (5.9)) \\ &> ||y(a)|| - e^{-M(b-a)}||y(a)||e^{M(b-a)} \\ &\qquad\qquad (\text{by } (5.7) \text{ and } y(a) \neq 0) \\ &= 0. \end{aligned} \tag{5.10}$$

The uniqueness of (5.8) is thus established for the case when f satisfies the inequality involving the scalar product. The proof for the case when f satisfies a Lipschitz condition is similar. One arrives once again at (5.9); hence

$$||y'(t)|| = ||f(x,u(x)) - f(x,v(x))|| \leq M||y(x)||.$$

If we join this with (5.9), we get

$$||y||_1 \leq (1 + M)||y(a)||e^{M(b-a)},$$

which allows us to make use of inequalities (5.10) with $||y||_1$ in place of $||y||_\infty$ and conclude in the same way. The corollary is thus completely proved.

We have seen that in order to study a given boundary value problem by this technique, it is necessary to associate a set U of linear problems with it in such a way that the hypotheses of the preceding theorem are satisfied; the corollary furnishes a practical criterion that is sufficiently convenient. For linear systems, we have a stronger result.

Theorem for Linear Equations. Let $A(x)$ be an $n \times n$ matrix whose entries are continuous functions on $[a,b]$, and let U be the set of all $L \in \mathscr{L}(C^1([a,b],R^n),R^n)$ such that the boundary value problem

$$y' = A(x)y, \quad L(y) = r$$

has a unique solution for every $r \in R^n$. Then U is open and dense in $\mathscr{L}(C^1([a,b],R^n),R^n)$.

Proof: We begin by proving that U is open. Let $\Gamma: R^n \to C^1([a,b],R^n)$ be the transformation that associates with every $c \in R^n$ the unique solution of the initial value problem

$$y' = A(x)y, \quad y(a) = c.$$

The results of Chapters I and II insure that Γ is linear and continuous. We may therefore define a transformation $\Phi: \mathscr{L}(C^1([a,b],R^n)R^n) \to \mathscr{L}(R^n,R^n)$ by composition:

$$\Phi(L) = L \circ \Gamma.$$

It is easy to verify that Φ is continuous. It follows from Chapter II that the set H of all injective linear operators from R^n to R^n (that is, of nondegenerate matrices) is an open subset of $\mathscr{L}(R^n,R^n)$. Thus $\Phi^{-1}(H)$ is an open subset of $\mathscr{L}(C^1([a,b],R^n),R^n)$. But the definition of U implies that $U = \Phi^{-1}(H)$, and so U is an open set as desired. We now pass on to show that U is dense in $\mathscr{L}(C^1([a,b],R^n),R^n)$. We must prove that for every $L \in \mathscr{L}(C^1([a,b],R^n),R^n)$ and every $\varepsilon > 0$ there is an $L_1 \in U$ such that $||L - L_1|| < \varepsilon$. Let $Y(x)$ be the fundamental matrix of $y' = A(x)y$. If we put $L_c(y) = y(a)$ and denote an element of $\mathscr{L}(R^n,R^n)$ and its associated matrix by the same symbol, we define $u: R \to R$ by

$$u(\lambda) = \det[(\lambda L_c + L)Y]. \tag{5.11}$$

The function u is analytic because the determinant on the right of (5.11) is. We now prove:

(*) There is a sequence $(\lambda_n)_{n=1}^{\infty}$ converging to 0 such that $\lambda_n \neq 0$ and $u(\lambda_n) \neq 0$ for each n.

If (*) is false, then $u(\lambda) = 0$ for each λ sufficiently small. Since u is analytic, this implies $u \equiv 0$. Thus

$$\det[(L_c + \tfrac{1}{\lambda}L)Y] = 0 \qquad (\lambda > 0). \tag{5.12}$$

But this is absurd; since the Cauchy problems for $y' = A(x)y$ have unique solutions, $\det(L_cY) \neq 0$, and thus (5.12) cannot be true for λ large enough. This contradiction implies that (*) is true. (*) means that all the problems

$$y' = A(x)y, \quad (\lambda_n L_c + L)(y) = r$$

have unique solutions, that is, that $\lambda_n L_c + L \in U$. Since $\lim||(\lambda_n L_c + L) - L|| = 0$, we have proven the theorem.

We observe that for nonlinear systems the conclusion of the last theorem is in general false, as was shown by a counterexample in Chow and Lasota [14].

Exercise 1. What must be changed in the corollary if we consider a norm in $\mathbb{R}^n$ different from the Euclidean norm? What must be changed in the examples in which the corollary was applied?

5.2. Nonlinear Functional Problems

In this section, we shall briefly treat nonlinear functional boundary value problems on a compact interval I and prove a uniqueness and existence theorem; in the statement of this theorem we shall use the concept of derivative for mappings between Banach spaces. Specifically, let X and Y be two Banach spaces, let $U \subseteq X$ be open, and let $f: U \to Y$. The derivatives of f at the point $x_0 \in U$ is a linear operator $u: X \to Y$ with the following property: for every $\varepsilon > 0$ there exists a $\delta > 0$ such that

$$||f(x) - f(x_0) - u(x-x_0)|| \le \varepsilon||x - x_0|| \qquad (||x - x_0|| < \delta).$$

It can be shown that u is unique when it exists; it is denoted by $f'(x_0)$. We say that f is of class C^1 if $f'(x)$ exists for every $x \in U$ and if $x \to f'(x)$ is a continuous mapping from U into the space $\mathscr{L}(X,Y)$ of linear operators.

Theorem. Let $f: [a,b] \times \mathbb{R}^n \to \mathbb{R}^n$ and $L: C([a,b],\mathbb{R}^n) \to \mathbb{R}^n$ be C^1-functions. If y_0 is a solution of the functional boundary value problem

$$y' = f(x,y), \quad L(y) = r_0,$$

and if the linear problem for the variational equation

$$z' = \frac{\partial}{\partial y} f(x,y_0(x))\cdot z, \quad L'(y_0)(z) = 0$$

admits $z \equiv 0$ for its unique solution, then there exists $\varepsilon > 0$ such that the problem

$$y' = f(x,y), \quad L(y) = r$$

have unique solutions for $||r - r_0|| \leq \varepsilon$.

The theorem can be expressed in a different way by saying that existence and uniqueness are preserved under small perturbations of the data r provided that the variational equation corresponding to the datum r_0 admits a unique solution for the linear problem $L'(y_0)(z) = 0$.

As an example of an application of this theorem, we examine the nonlinear problem of the type of Nicoletti

$$y'' = f(x,y), \quad y(0) = 0, \quad y'(1) = y^2(1). \tag{5.13}$$

We consider the first order system $z' = F(x,z)$ associated with $y'' = f(x,y)$ and define $L\colon C([0,1],\mathbb{R}^n) \to \mathbb{R}^n$ by

$$L(z) = (z_1(0), z_2(1) - z_1^2(1)),$$

where z_1 and z_2 are the two coordinate functions of z. Then problem (5.13) is equivalent to

$$z' = F(x,z), \quad L(z) = (0,0).$$

We can easily verify that the derivative of L at u is the linear operator $L'(u)$ defined by

$$L'(u)(z) = (z_1(0), z_1(1) - 2u_1(1)z_1(1)),$$

where z_i and u_i are the coordinate functions of z and u. For $u \equiv 0$, we get

$$L'(0)(z) = (z_1(0), z_2(1)).$$

Thus, if we suppose that f is of class C^1, we see that the problem for the variational equation corresponding to $u \equiv 0$ is

$$z' = \frac{\partial}{\partial z} F(x,0)z, \quad L'(0)(z) = (0,0)$$

which is equivalent (cf. Exercise 3) to the problem of Nicoletti

$$w'' = \frac{\partial}{\partial y} f(x,0)w, \quad w(0) = 0, \quad w'(1) = 0.$$

In agreement with the uniqueness theorems of Sec. 3, this problem will have a unique solution if $\partial f(x,0)/\partial y$ is always less than the first eigenvalue of

$$v'' + \lambda v = 0, \quad v(0) = 0, \quad v'(0) = 0$$

or always strictly in between two successive eigenvalues. If we observe that $y \equiv 0$ is a solution of (5.13) if $f(x,\cdot) \equiv 0$, we can use the theorem we have just stated to get existence and uniqueness of the solutions to the problem

$$y'' = f(x,y), \quad y(0) = \alpha, \quad y'(1) = y^2(1) + \beta$$

when α and β are sufficiently small.

Proof of the Theorem: To say that y is a solution of the functional boundary value problem

$$y' = f(x,y), \quad L(y) = r$$

is equivalent (cf. Exercise 1) to saying that y satisfies the identity

$$y(x) = y(a) + L(y) - r + \int_a^x f(s,y(s))ds.$$

We define $T\colon C([a,b],\mathbb{R}^n) \to C([a,b],\mathbb{R}^n)$ by

$$Ty(x) = y(x) - y(a) - L(y) - \int_a^x f(s,y(s))ds.$$

Then, to prove the theorem it suffices to show that T satisfies the following

> Local Inversion Theorem. If X and Y are Banach spaces, if $f\colon X \to Y$ is of class C^1, and if $f'(x_0)$ is invertible, then there is a neighborhood U of x_0 and a neighborhood V of $f(x_0)$ such that $f|_U$ is a homeomorphism of U onto V.

for $x_0 = y_0$ and $f = T$, since $-r = T(y_0)$. In the first place, we observe that it follows from the mean value theorem that the derivative $T'(y)$ exists and is defined by

$$T'(y)(u)(x) = u(x) - u(a) - L'(y)(u) - \int_a^x \frac{\partial}{\partial y} f(s,y(s))u(s)ds$$

and therefore T is of class C^1. To say that $T'(y_0)(u) = 0$ is equivalent to saying that u is a solution of the problem

$$z' = \frac{\partial}{\partial y} f(x,y_0(x))\cdot z, \quad L'(y_0)(u) = 0,$$

so $u \equiv 0$ by virtue of the hypothesis. This means that the kernel of $T'(y_0)$ contains only the origin. Since $T'(y_0)$ is the difference between the identity mapping and a compact linear mapping, the Fredholm alternative implies that $T'(y_0)$ is a bijection. This completes our proof.

Exercise 1. Prove that y is a solution of the boundary value problem

$$y' = f(x,y), \quad L(y) = r$$

if and only if

$$y(x) = y(a) + L(y) - r + \int_a^x f(s,y(s))ds.$$

Exercise 2. In the example analyzed in this section, we used without proof the fact that if (i) $y'' = f(x,y,y')$ is a second order equation, and if (ii) $z' = F(x,z)$ is the first order system associated with (i), then the first order system associated with the variational equation of (i) agrees with the variational system of (ii), that is, the system corresponding to

$$u'' = \frac{\partial}{\partial y} f(x,y(x),y'(x))u + \frac{\partial}{\partial y'} f(x,y(x),y'(x))u'$$

coincides with

$$w' = \frac{\partial}{\partial z} F(x,(y(x),y'(x)))w,$$

where we have used $\partial F/\partial z$ to indicate the Jacobian matrix and the vector (w_1,w_2) for the column matrix

$$\begin{pmatrix} w_1 \\ w_2 \end{pmatrix}.$$

6. BIBLIOGRAPHICAL NOTES

For an introduction to nonlinear functional analysis, we refer to the books of Miranda [74] and Prodi and Ambrosetti [89], while for an exposition of the theory of fixed points and the topological degree, see Amann [1]. It was pointed out in Section 1 that the topological degree can not be extended to all continuous functions on infinite dimensional Banach spaces. For a panorama of the various kinds of functions for which the degree can be defined, consult Browder [11], Gaines and Mawhin [32], Sadovskii [94], Lloyd [123], and the works they cite. It is interesting to observe that Amann and Weiss [3] have proved the uniqueness of the topological degree in the sense that all the various techniques used to introduce this concept lead to the same result. Since only the positive solutions are of interest in many physical problems, there has been a development of the theory of fixed points for ordered Banach spaces; see Amann [2] and the works in his bibliography. Among the most interesting results of this theory are an iterative method for approximating what could be called the minimum and maximum fixed points and an estimation of the number of solutions. This can also be done by means of the topological degree; see Cronin [21] and Nussbaum [81]. Various attempts have been made not only to give a proof of Brouwer's fixed point theorem that is simple and autonomous (in the sense of not relying on more or less complicated theories like that of the topological degree or of algebraic topology) but also to approximate numerically the fixed points of continuous functions. For the first question, consult Kuga [57] and Hönig [43] and the works they cite; the second, which is of great interest in numerical analysis and mathematical economics, was first solved by Scarf [96]. For more recent results, see Lütni [66] and Todd [109]; for an application of this algorithm to boundary problems of ordinary equations, see Chen [13].

For further applications of the results of Sec. 2 cf. Stampacchia [102],...,[105]. Ideas very similar to those in the proof of the existence theorem for geometric problems have been used by Muldowney and Willett [77]. For a recent contribution see Vidossich [117].

The idea of relating eigenvalues to existence and uniqueness theorems is due to Hammerstein [38], who, however, used only the first eigenvalue. Afterwards, Dolph [38] considered the case of the other eigenvalues. During the 1960's these results were extended to the periodic solutions of second order equations; see Mawhin [68] and the works in his bibliography. Toward the end of the 1960's Lazer, in collaboration with other mathematicians, began to study resonance for periodic solutions of ordinary differential equations and for the Dirichlet problem of elliptic equations (which corresponds to Picard's problem for ordinary equations of the second order). We may divide the great many works that have since appeared on this subject into three categories according to the methods used:

(i) Works that use classical methods, cf. Kazdan and Warner [52] and the works they cite and de Figueiredo and Gossez [23];

(ii) Works based on the methods of differential topology following the ideas of Caccioppoli, like Ambrosetti and Prodi [4], and Podolak [7] (Fučik [29] proved the result of Ambrosetti and Prodi using the theorem on contractions);

(iii) Works that use the theories of Morse and of Lusternik and Schnirelmann and their generalizations, cf. Fučik, Nečas, Souček and Souček [30].

These results have been applied by Kazdan and Warner in Riemannian geometry; cf. the works cited in [52]. Sturm-Liouville problems and their generalizations are used in Morse theory and in related areas; see Morse [75].

For a generalization of Hammerstein's theorem on the first eigenvalue, see Stampacchia [106, Sec. 10]. For a case in which the equation depends also on the first derivative, see Tippett [108]. The theory of eigenvalues can be used to estimate the number of solutions of the Picard problem for equations of order n, see Vidossich [118]. For applications to mechanics cf. Dickey [24].

Modern research on periodic solutions began with Poincaré's studies in celestial mechanics, studies which brought forth original ideas that lead to the development of dynamical systems and algebraic topology. We restricted ourselves to treating only a few special cases. For an exposition of the most recent results, cf. Mawhin [67] and Cronin [22]. For a treatment based on the methods of functional analysis, see the books of Rouche and Mawhin [93], Krasnoselskii [56] (where, in particular, there are numerous examples of guided functions) and Hale [37]. For the use of qualitative methods, cf. Halanay [36], Lefschetz [63], [64], Nemyckii and Stepanov [82], and the articles they cite. For equations of order n, see the book of Reissig, Sansone, and Conti [90]. For numerical methods, cf. Urabe [100]. For a necessary and sufficient condition for the existence of periodic solutions, see Becker and Vidossich [7], while for further applications of the ideas developed in Sec. 4.4, see Vidossich [113] and the works in his bibliography. Also consult Brock Fuller [10].

The study of linear functional problems was begun by Conti in a series of works which we can only cite here, [17] and [18]. It was then continued by Lasota and Opial [60], [61], Lasota and Olech [59] (which contains an interesting uniqueness theorem for the Nicoletti problem), Opial [85], and Antosiewicz [5]. Many proofs use multivalued operators and the nonlinear version of the Fredholm alternative due to Lasota [58]. The study of nonlinear functional problems is a very recent development; cf. McCandless [71], [72], Kartsatos [51] and Vidossich [119]. The results of Sec. 5.1 are connected with those of Chow and Lasota [14] , while those in

Sec. 5.2 are related to the cited works by McCandless.

The idea of using super and subsolutions was introduced by Perron for the Dirichlet problem (and thus for the Picard problem also). It was then progressively generalized to various boundary problems. For an historical panorama, see Jackson [47], while for an extension to Sturm-Liouville problems, see Kaplan, Lasota, and Yorke [50]. There are particularly simple proofs for the case of the Picard problem; for a constructive one that uses the iterative method of Cohen, cf. Sattinger [95]; for an nonconstructive one see Hess [42]. The theorem in Sec. 4.2 due to Knobloch [54] has been proved here using the iterative method of Cohen.

By virtue of the fact that it is often enough to verify the uniqueness of solutions than to establish their existence, the study of the relation between uniqueness and existence of solutions for boundary value problems has in the last few years begun to be developed for nonlinear equations. Above all, the case of the Picard problem for equations of order n has received special attention. Up till now it has been completely solved only for the cases when $n = 2$ or 3, the former by Lasota and Opial and the latter by Jackson and Schrader; for $n > 3$ only partial results are known. For a discussion and a bibliography, we refer to Jackson [49].

In the study of boundary value problems for second order equations, one of the fundamental questions is that of finding convenient a priori upper bounds for the first derivative of the solution. The most classic result is that of Nagumo [79], which is valid only on the real line; there are counterexamples against its extension to systems. Partial extensions of Nagumo's theorem to systems have been made by Hartman [40], Mawhin [69], and Schmitt [98], while a partial extension to equations of order n is due to Jackson [48] and Innes and Jackson [46]. See Rogers [91] also. For other types of a priori upper bounds for solutions cf. Cesari and Kannan [12], George and Sutton [34], Knobloch [55], Lasota and Yorke [62], Mawhin [70], Moyer [76], and Vidossich [114].

We did not treat a certain type of problem that is most important in fluid dynamics and other applied sciences, the so-called problem of singular perturbations. It concerns a family of boundary value problems P_ε for n-th order equations E_ε depending on a parameter ε such that the limit equation of the E_ε as $\varepsilon \to 0$ is an equation of order less than n. A typical example is the problem

$$\varepsilon y''_\varepsilon = f(x,y_\varepsilon,y'_\varepsilon,\varepsilon), \quad y_\varepsilon(0) = \alpha(\varepsilon), \quad y_\varepsilon(1) = \beta(\varepsilon). \qquad (P_\varepsilon)$$

As $\varepsilon \to 0$, the limiting equation is

$$0 = f(x, y_0, y_0', 0), \qquad (P_0)$$

which is a first order equation in implicit form. The interesting questions are 1) the relation between the solutions y_ε of P_ε and that y_0 of P_0, 2) how y_ε converges to y_0, 3) how to approximate y_ε conveniently, and 4) whether the existence of y_0 insures that of y_ε when ε is small. The matter is complicated by the fact that P_0 is an equation in implicit form (which is why we do not know if its solutions are defined on $[0,1]$ and if we can speak of uniform convergence of the y_ε to y_0) and by the fact that it is unclear for what boundary value problem we must find the solution y_0 of P_0. (The fact that P_0 is of the first order makes it very hard for $y_0(0) = \alpha(0)$ and $y_0(1) = \beta(0)$ to be satisfied; in concrete problems, the type of problem to be considered for P_0 is often suggested by physical laws.) For an exposition, see Ederlyi [28], Vasileva [111], Wasow [120], Dorr, Parter, and Shampine [26], and the books of Wasow [121] and O'Malley [83]. For applications, cf. Cole [16], and Nayfen [80] as well as the book by O'Malley already cited. For the most recent results, see Howes [45], Harris [39], and Rosenblat [92].

Another question of great physical and mathematical interest that we did not treat is that of bifurcation. For a general idea of bifurcation problems we refer to Prodi [88] and Prodi and Ambrosetti [89], while for recent results relative to Sturm-Liouville problems see Scheurle [97], Schmitt [124], and Amann [2] and the works they cite.

For a general construction of Green functions valid for various boundary value problems, cf. Hönig [44].

For special questions relative to various boundary value problems, see Bebernes and Gaines [6], Conti [19], Gamlen and Muldowney [33], Peterson [86], Schmitt [99], and Willett and Muldowney [122]. For boundary value problems with obstacles, cf. Vergara Caffarelli [112] and Kinderlehrer and Stampacchia [53].

Various boundary value problems are discussed in Conti [20] Bernfeld and Lakshmikantham [9], and Mawhin [125].

For an application of boundary value problems of ordinary differential equations to boundary value problems of partial differential equations, see the book of Lions cited in Chapter III above and Vidossich [115], [116].

[1] H. Amann, *Lectures on some fixed point theorems*, IMPA, Rio de Janeiro, 1974.

[2] H. Amann, Fixed point equations and nonlinear eigenvalue problems in ordered Banach spaces, *SIAM Rev.*, *18*(1976), 620-790.

[3] H. Amann and S. Weiss, On the uniqueness of the topological degree, *Math. A.*, *130*(1973), 39-54.

[4] A. Ambrosetti and G. Prodi, On the inversion of some differentiable mappings with singularities between Banach spaces, *Ann. Mat. Pura Appl.*, *93*(1972), 231-247.

[5] H. A. Antosiewicz, A general approach to linear problems for nonlinear ordinary differential equations, in L. Weiss (ed.): *Ordinary Differential Equations*, Academic Press, New York, 1972, pp. 3-10.

[6] J. N. Bebernes and R. Gaines, Dependence on boundary data and a generalized boundary value problem, *J. Diff. Eq.*, *4*(1968), 359-368.

[7] R. Becker and G. Vidossich, Some applications of a simple criterion for the existence of periodic solutions of ordinary differential equations, *J. Math. Anal. Appl.*, *48*(1974), 51-60.

[8] M. Berger and M. Berger, *Perspectives in Nonlinearity*, Benjamin, New York, 1968.

[9] S. R. Bernfeld and V. Lakshmikantham, *An Introduction to Nonlinear Boundary Value Problems*, Academic Press, New York, 1974.

[10] F. Brock Fuller, Bounds for the periods of periodic orbits, in Auslander and Gottschalk (eds.): *Topological Dynamics*, Benjamin, New York, 1968, pp. 205-215.

[11] F. E. Browder, *Nonlinear operators and nonlinear equations of evolution in Banach spaces*, Proc. Symp. Pure Math., vol. 18 (part 2), Amer. Math. Soc., Providence, R. I., 1975.

[12] L. Cesari and R. Kannan, Solutions in the large of Liénard systems with forcing terms, *Ann. Mat. Pura Appl.*, *111*(1976), 103-124.

[13] H. C. C. Chen, A constructive existence method for nonlinear boundary value problems, *J. Math. Anal. Appl.*, *59*(1977), 454-468.

[14] S. N. Chow and A. Lasota, On boundary value problems for ordinary differential equations, *J. Diff. Eq.*, *14*(1973), 326-337.

[15] E. A. Coddington and N. Levinson, *The Theory of Ordinary Differential Equations*, McGraw-Hill, New York, 1955.

[16] J. D. Cole, *Perturbation Methods in Applied Mathematics*, Giun-Blaisdell, Waltham, 1968.

[17] R. Conti, I problemi ai limiti lineari per i sistemi di equazioni differenziali ordinarie: teoremi di esistenza, *Ann. Mat. Pura Appl.*, *35*(1953), 155-182.

[18] R. Conti, Problèmes linéaires pour les équations différentielles ordinaires, *Math. Nach.*, *23*(1961), 161-178.

[19] R. Conti, On ordinary differential equations with interface conditions, *J. Diff. Eq.*, *4*(1968), 4-11.

[20] R. Conti, Recent trends in the theory of boundary problems for ordinary differential equations, *Boll. Un. Mat. Ital.*, *22*(1967), 135-178.

[21] J. Cronin, Using Leray-Schander degree, *J. Math. Anal. Appl.*, *25*(1969), 414-424.

[22] J. Cronin, Some mathematics of biological oscillations, *SIAM Rev.*, *19*(1977), 100-138.

[23] D. DeFigueiredo and J. P. Gossez, Perturbation non lineaire d'un problème elliptique linéaire pres de sa première valeur propre, *C. R. Acad. Sc. Paris*, *284*(1977), 163-166.

[24] R. W. Dickey, *Bifurcation Problems in Nonlinear Elasticity*, Pitman, London, 1976.

[25] C. L. Dolph, Nonlinear integral equations of the Hammerstein type, *Trans. Amer. Math. Soc.*, *66*(1949), 289-307.

[26] F. W. Dorr, S. V. Parter and L. F. Shampine, Applications of the maximum principle to singular perturbation problems, *SIAM Rev.*, *15*(1973), 43-88.

[27] J. Dugundji, *Topology*, Allyn and Bacon, Boston, 1966.

[28] A. Erdelyi, A case history in singular perturbations, in Antosiewicz (ed.): *International Conference on Differential Equations*, Academic Press, New York, 1975.

[29] S. Fučik, Remarks on a result by A. Ambrosetti and G. Prodi, *Boll. Un. Mat. Ital.*, *11*(1975), 259-267.

[30] S. Fučik, J. Nacas, J. Soucek and V. Soucek, *Sprectral Analysis of Nonlinear Operators*, LN in Math., 346, Springer-Verlag, Berlin, 1973.

[31] R. E. Gaines, A priori bounds for solutions to nonlinear two-point boundary value problems, *Applicable Anal.*, *3*(1973), 157-167.

[32] R. E. Gaines and J. Mawhin, *Coincidence Degree and Nonlinear Differential Equations*, Lecture Notes in Math., 568, Springer-Verlag, Berlin, 1977.

[33] J. L. Gamlen and J. S. Muldowney, An intermediate value property, *Proc. Amer. Math. Soc.*, *51*(1975), 413-420.

[34] J. H. George and W. G. Sutton, Application of Liapunov theory to boundary value problems, *Proc. Amer. Math. Soc.*, *25*(1970), 666-671.

[35] A. Granas, The theory of compact vector fields and some of its applications to topology of functional spaces (1), *Roszprawy Math.*, *30*(1962).

[36] A. Halanay, *Differential Equations: Stability, Oscillations Time Lags,* Academic Press, New York, 1966.

[37] J. K. Hale, *Oscillations in Nonlinear Systems,* McGraw-Hill, New York, 1963.

[38] A. Hammerstein, Nichtlineare Integralgleichungen nebst Anwendungen, *Acta Math., 54*(1930), 117-176.

[39] W. A. Harris, Jr., Applications of the method of differential inequalities in singular perturbation problems, in Eckhaus (ed.), *New Developments in Differential Equations,* North-Holland, Amsterdam, 1976, pp. 111-115.

[40] P. Hartman, *Ordinary Differential Equations,* Wiley, New York, 1964.

[41] E. Heinz, An elementary analytic theory of degree, *J. Math. Mech., 8*(1959), 231-247.

[42] P. Hess, On the solvability of nonlinear elliptic boundary value problems, *Indiana Univ. Math. J., 25*(1976), 461-466.

[43] C. S. Honig, *Applicacoes da Topologia e Analise,* IMPA, Rio de Janeiro, 1976.

[44] C. S. Honig, *The Abstract Riemann - Stieltjes Integral and Its Applications to Linear Differential Equations with Generalized Boundary Conditions,* Notas do Instituto de Matematica e Estatistica, Univ. de Sao Paulo, Sao Paulo, 1973.

[45] F. A. Howes, Singular perturbations and differential inequalities, *Memoirs Amer. Math. Soc., 168*(1976), 1-75.

[46] J. E. Innes and L. K. Jackson, Nagumo conditions for ordinary differential equations, in Antosiewciz, see citation in [28], pp. 385-398.

[47] L. K. Jackson, Subfunctions and second order differential inequalities, *Adv. Math., 2*(1968), 307-363.

[48] L. K. Jackson, A Nagumo condition for ordinary differential equations, *Proc. Amer. Math. Soc., 57*(1976), 93-96.

[49] L. K. Jackson, Uniqueness and existence of solutions of boundary value problems for ordinary differential equations, in Weiss (ed.), see citation in [5], pp. 137-149.

[50] J. L. Kaplan, A. Lasota and J. A. Yorke, An application of the Wazewski retract method to boundary value problems, *Zeszyty Nauk Uniw. Jagiello.*

[51] A. G. Kartsatos, The Hildebrandt-Graves theorem and the existence of solutions of boundary value problems in infinite intervals, *Math. Nach., 67*(1975), 91-100.

[52] J. L. Kazdan and F. W. Warner, Remarks on some quasilinear elliptic equations, *Comm. Pure Appl. Math., 28*(1975), 567-597.

[53] D. Kinderlehrer and G. Stampacchia, *An Introduction to Variational Inequalities and Their Applications*, Academic Press, New York, 1980.

[54] H. W. Knobloch, Comparison theorems for nonlinear second order differential equations, *J. Diff. Eq.*, *1*(1965), 1-26.

[55] H. W. Knobloch, On the existence of periodic solutions for second order vector differential equations, *J. Diff. Eq.*, *9*(1971), 67-85.

[56] M. A. Krasnoselskii, *Translation along Trajectories of Differential Equations*, Amer. Math. Soc., Providence, R. I., 1968.

[57] K. Kuga, Brouwer's fixed-point theorem: an alternative proof, *SIAM J. Math. Anal.*, *5*(1974), 893-897.

[58] A. Lasota, Une généralization du premier théorème de Fredholm et ses applications à la theorie des équations differentielles ordinaires, *Ann. Polon. Math.*, *18*(1966), 65-77.

[59] A. Lasota and C. Olech, An optimal solution of Nicoletti's boundary value problem, *Ann. Polon. Math.*, *18*(1966), 131-139.

[60] A. Lasota and Z. Opial, L'existence e l'unicité des solutions du problème d'interpolation pour l'equation differentielle ordinaire d'ordre n, *Ann. Polon. Math.*, *15*(1964), 253-278.

[61] A. Lasota and Z. Opial, Sur la dependance continue des solutions de des équations differentielles ordinaires de leurs second membres des conditions aux limites, *Ann. Polon. Math.*, *19*(1967), 13-36.

[62] A. Lasota and J. A. Yorke, Existence of solutions of two-point boundary value problems for nonlinear systems, *J. Diff. Eq.*, *11*(1972), 509-518.

[63] S. Lefschetz, *Differential Equations: Geometric Theory*, Interscience, New York, 1957.

[64] S. Lefschetz, *Contributions to the Theory of Nonlinear Oscillations*, Princeton Univ. Press.

[65] J. Leray and J. Schauder, Topologie et equations fonctionelles, *Ann. Sci. Ecole Norm. Sup.*, *51*(1934), 45-78.

[66] H. J. Lüthi, *Komplementaritäts-und Fixpunktalgorithmen in der mathematischen Programmierung, Spieltheorie und Ökonomie*, Lecture Notes in Economics and Math. Sy. 129, Springer-Verlag, Berlin, 1976.

[67] J. Mawhin, Recent results on periodic solutions of differential equations, in Antosiewicz (ed.), see citation in [28], pp. 537-556.

[68] J. Mawhin, Contractive mappings and periodically perturbed conservative systems, *Archivum Math.*, *12*(1976), 67-74.

[69] J. Mawhin, Boundary value problems for nonlinear second-order vector differential equations, *J. Diff. Eq.*, *16*(1974), 257-269.

[70] J. Mawhin, L_2-estimates and periodic solutions of some nonlinear differential equations, *Boll. Univ. Mat. Ital.*, *10*(1974), 341-352.

[71] W. L. McCandless, Existence theorems for nonlinear boundary value problems, *Canad. J. Math.*, *26*(1974), 884-892.

[72] W. L. McCandless, Newton's method and nonlinear boundary value problems, *J. Math. Anal. Appl.*, *48*(1974), 434-445.

[73] J. Milnor, *Topology from the Differentiable Viewpoint*, Univ. Press of Virginia, Charlottesville, 1965.

[74] C. Miranda, *Problemi di esistenza in analisi funzionale*, Quaderni mat., Scuola Normale Superiore, Pisa, 1975.

[75] M. Morse, *Variational Analysis*, Wiley, New York, 1973.

[76] R. D. Moyer, Second order differential equations of monotonic type, *J. Diff. Eq.*, *2*(1966), 281-292.

[77] J. S. Muldowney and D. Willett, An elementary proof of the existence of solutions to second order nonlinear boundary value problems, *SIAM J. Math. Anal.*, *5*(1974), 701-707.

[78] M. Nagumo, A theory of degree of mappings based on infinitesimal analysis, *Amer. J. Math.*, *73*(1951), 485-496.

[79] M. Nagumo, Ueber die Differentialgleichung $y'' = f(x,y,y')$, *Proc. Phys. Math. Soc. Japan*, *19*(1937), 861-866.

[80] A. H. Nayfeh, *Perturbation Methods*, Wiley, New York, 1973.

[81] R. Nussbaum, Estimates for the number of solutions of operator equations, *Applicable Anal.*, *1*(1971), 183-200.

[82] V. V. Nemyckii and V. V. Stepanov, *Qualitative Theory of Differential Equations*, Princeton Univ. Press, 1960.

[83] R. E. O'Malley, Jr., *Introduction to Singular Perturbations*, Academic Press, New York, 1974.

[84] T. O'Neil and J. W. Thomas, The calculation of the topological degree by quadrature, *SIAM J. Numer. Anal.*, *12*(1975), 673-680.

[85] Z. Opial, Linear problems for systems of nonlinear differential equations, *J. Diff. Eq.*, *3*(1967), 580-594.

[86] A. C. Peterson, Comparison theorems and existence theorems for ordinary differential equations, *J. Math. Anal. Appl.*, *55*(1976), 773-784.

[87] E. Podolak, On the range of operators equations with an asymptotically nonlinear term, *Indiana Univ. Math. J.*, *25*(1976), 1122-1137.

[88] G. Prodi, Problemi di diramazione per equazioni funzionali, *Boll. Un. Mat. Ital.*, *22*(1967), 413-433.

[89] G. Prodi and A. Ambrosetti, *Analisi nonlineare, I*, Quaderni mat., Scuola Normale Superiore, Pisa, 1973.

[90] R. Reissig, G. Sansone and R. Conti, *Nonlinear Differential Equations of Higher Order,* Nordhoff, Leyden, 1974.

[91] T. Rogers, On Nagumo's condition, *Canad. Math. Bull., 15*(1972), 609-611.

[92] S. Rosenblat, Asymptotically equivalent singular perturbation problems, *Studies Appl. Math., 55*(1976), 249-280.

[93] N. Rouche and J. Mawhin, *Equations Différentielles Ordinaires, Vol. II,* Masson, Paris, 1973.

[94] B. N. Sadovskii, Limit-compact and condensing operators, *Russian Math. Surveys, 27*(1972), 85-155.

[95] D. H. Sattinger, *Topics in Stability and Bifurcation Theory,* Lecture Notes in Math., 309, Springer-Verlag, Berlin, 1973.

[96] H. Scarf, The approximation of fixed points of a continuous mapping, *SIAM J. Appl. Math., 15*(1967), 1328-1343.

[97] J. Scheurle, Selective iteration and application, *J. Math. Anal. Appl., 59*(1977), 596-616.

[98] K. Schmitt, Randwertaufgaben für gewohnliche Differentialgleichungen, Proc. Steiermark Math. Symp., Graz, 1973.

[99] K. Schmitt, Boundary value problems and comparison theorems for ordinary differential equations, *SIAM J. Appl. Math., 26*(1974), 670-678.

[100] J. T. Schwartz, *Nonlinear Functional Analysis,* Gordon and Breach, New York, 1968.

[101] E. H. Spanier, *Algebraic Topology,* McGraw-Hill, New York, 1966.

[102] G. Stampacchia, Sulle condizioni che determinano gli integrali dei sistemi di equazioni differenziali ordinarie del primo ordine, *Giorn. Mat. Battaglini, 77*(1947), 55-60.

[103] G. Stampacchia, Un'osservazione su un problema ai limiti per l'equazione $y^{(n)} = \lambda f(x,y,y',\ldots,y^{(n-1)})$, *Boll. Un. Mat. Ital., 4*(1949), 235-239.

[104] G. Stampacchia, Sopra una generalizzazione dei problemi ai limiti per i sistemi di equazioni differenziali ordinarie, *Ricerche Mat., 3*(1954), 76-94.

[105] G. Stampacchia, Problemi ai limiti per i sistemi di equazioni differziali ordinari, *Le Matematiche (Catania), 11*(1950), 121-134.

[106] G. Stampacchia, On some regular multiple integral problems in the calculus of variations, *Comm. Pure Appl. Math., 16*(1963), 383-421.

[107] R. B. Thompson, A unified approach to local and global fixed point indices, *Adv. Math., 3*(1969), 1-71.

[108] J. Tippett, An existence-uniqueness theorem for two point boundary value problems, *SIAM J. Math. Anal., 5*(1974), 153-157.

[109] M. J. Todd, *The computation of Fixed Points and Applications,* Lecture Notes in Economics and Math. Sy. 124, Springer-Verlag, Berlin, 1976.

[110] M. Urabe, *Nonlinear Oscillations,* Academic Press, New York, 1967.

[111] A. B. Vasileva, Asymptotic behavior of solutions of the problems for ordinary nonlinear differential equations with a small parameter multuplying the highest derivative, *Russian Math. Surveys,* *18*(1963), 13-84.

[112] G. Vergara Caffarelli, Su un problema al contorno con vincoli per operatori differenziali ordinari, *Boll. Un. Mat. Ital.,* *3*(1970), 566-584.

[113] G. Vidossich, On the structure of periodic solutions of differential equations, *J. Diff. Eq.,* *21*(1976), 263-278.

[114] G. Vidossich, The two-point boundary value problem from the Cauchy problem.

[115] G. Vidossich, Dirichlet problem for nonlinear elliptic equations using ordinary differential equations.

[116] G. Vidossich, Periodic solutions of nonlinear hyperbolic equations using ordinary differential equations.

[117] G. Vidossich, On the geometric boundary value problem of Stampacchia.

[118] G. Vidossich, Boundary value problems for n-th order differential equations using eigenvalues of second order equations.

[119] G. Vidossich, Solving Kartsatos problem on nonlinear boundary value problems for ordinary differential equations.

[120] W. Wasow, The capriciousness of singular perturbations, *Neiuw. Arch Wisk.,* *18*(1970), 190-210.

[121] W. Wasow, *Asymptotic Expansions for Ordinary Differential Equations,* Wiley, New York, 1965.

[122] D. Willett and J. S. Muldowney, An intermediate value property of operators with applications to integral and differential equations, *Canad. J. Math.,* *26*(1974), 27-41.

[123] N. G. Lloyd, *Degree Theory,* Cambridge University Press, Cambridge, 1978.

[124] K. Schmitt: A study of eigenvalue and bifurcation problems for nonlinear elliptic partial differential equations via topological continuation methods, Notes de Seminaire 1982, Institut de Mathematique Université Catholique de Louvain, Louvain-la-Neuve (Belgium).

[125] J. Mawhin, *Topological degree methods in nonlinear boundary value problems,* CBMS Regional Confer. Series, n. 40, AMS, Providence, 1979.

Chapter V
Questions of Stability

This chapter is devoted to presenting a panorama of the problems and methods of the theory of stability for ordinary differential equations. Roughly speaking, one could say that stability is the continuous dependence of the solutions as functions of the data on *infinite intervals;* the reason it is necessary to start a new chapter is that the theorems on continuous dependence in Chapter III are only valid on compact intervals. Keeping in mind that in general, physical phenomena develop over infinite intervals of time, the following observations may convince the reader of the necessity of undertaking a study of stability.

(i) In the description of physical phenomena by means of differential equations, it is necessary to impose additional conditions like, for example, initial values, in order to determine the solution uniquely. These initial conditions are obtained by experiment and are subject to experimental error. We are therefore led to the study of the effect on the solutions of a small change in the initial values. The concept of stability is essentially the requirement that a small change in the initial values produce only a small change in the solution.

(ii) Let us consider the solution of a system of ordinary differential equations that describes a phenomenon. Suppose there is a perturbation of short duration in the course of the physical phenomenon and that this perturbation cannot be known exactly and therefore cannot be taken into consideration in the mathematical description of the phenomenon. Once the perturbation is over, the phenomenon continues to be described by the same system of equations even though at the instant t_0 at which the action of the perturbation ends the value of the new solution will be different from that given by the solution originally considered. In other words, the effect of brief perturbations consists in the passing from one solution with a certain initial value at t_0 to a second with another initial value at t_0. Since the perturbation is not known in its real form, it is necessary, in order to produce an accurate description of the phenomenon that does not change the physical meaning, that the equation of the mathematical model have the property that a small change in the initial values produces a small change in the solution.

There are various definitions of the concept of stability. This is due in part to the personal views of the various physicists of what really should remain stable and in part to the technical requirements of mathematicians. One can find in Stoker [32] an interesting discussion of the difficulties involved in giving a definition of stability that is reasonable from a physical point of view and at the same time consistent from the mathematical point of view.

In order to illustrate the practical significance of stability in the experimental sciences, we shall examine certain equations that appear in biology, chemistry, and control theory. We shall even show the connection between stability and problems of numerical integration by illustrating the method of Runge and Kutta.

1. STABILITY OF THE SOLUTIONS OF LINEAR SYSTEMS

In this section, we study the concept of the stability of the solutions of a system of linear differential equations. As we shall prove in Section 1.1, one can use the results of the linear case when studying stability problems for nonlinear systems. In Section 1.2 we solve the problem for autonomous linear systems, in Section 1.3 we treat the case of autonomous linear systems of order two in detail, and finally, in Section 1.4, we present certain stability theorems for nonautonomous linear systems.

1.1. Definition of Stability

We consider the system of differential equations

$$z' = f(t,z). \tag{1.1}$$

Suppose that the integral $z(t)$ of the system (1.1) satisfying the initial condition $z(t_0) = z_0$ is defined on the half line $t \geq t_0$ and that the same is true for the integrals $y(t)$ passing through the points y_0 of a neighborhood of z_0. One may then pose the question of determining whether the continuous dependence on the initial data is uniform on all the half line $t \geq t_0$, that is, whether

$$\lim_{y_0 \to z_0} [y(t) - z(t)] = 0 \tag{1.2}$$

uniformly as t *varies in* $[t_0, \infty[$.

The answer is in general no, as we see from the simple example $y' = y$. Whenever, though, (1.2) does hold, we say that *the given solution* $z(t)$ *is a stable solution of equation* (1.1).

If we set $y(t) = z(t) + w(t)$, then, since $y(t)$ is a solution of (1.1), we have

$$z' + w' = f(t, z + w). \tag{1.3}$$

Suppose that the function f on the right of (1.1) is differentiable with respect to the variables $(z_1, z_2, \ldots, z_n)$; then we may write

$$f(t, z + w) = f(t,z) + A(t,z)w + f_1(t,z,w), \tag{1.4}$$

where

$$A(t,z) = \begin{pmatrix} \dfrac{\partial f^{(1)}}{\partial z_1} & \cdots & \dfrac{\partial f^{(1)}}{\partial z_n} \\ \\ \dfrac{\partial f^{(n)}}{\partial z_1} & \cdots & \dfrac{\partial f^{(n)}}{\partial z_n} \end{pmatrix}$$

and

$$\lim_{||w|| \to 0} \frac{||f_1(t,z,w)||}{||w||} = 0.$$

By (1.4), since $z(t)$ is a solution of (1.1), it follows from (1.3) that

$$w' = A(t,z)w + f_1(t,z,w),$$

and the problem of determining the stability of the solution $z(t)$ for Eq. (1.1) is reduced to the study of the stability of the solution $w(t)$ for this last equation.

We may, therefore, without loss of generality, restrict ourselves to the study of the stability of the solution $z(t) \equiv 0$ for systems of the type

$$z' = A(t)z + f(t,z), \tag{1.5}$$

where $A(t)$ is an assigned matrix of order n and $f(t,z)$ satisfies the condition

$$\lim_{||z||\to 0} \frac{||f(t,z)||}{||z||} = 0.$$

Then one has to see if it is possible to associate with every $\varepsilon > 0$ some $\delta > 0$ such that if z_0 is any vector satisfying $||z_0|| < \delta$, then the solutions $z(t)$ of Eq. (1.5) determined by the initial conditions $z(t_0) = z_0$ satisfies the condition $||z(t)|| < \varepsilon$ for every $t \geq t_0$.

1.2. Stability for Autonomous Linear Systems

We first of all study the stability of autonomous linear homogeneous systems with constant coefficients

$$y' = Ay, \tag{1.6}$$

where A is a matrix of order n with constant entries.

The results we shall use are contained in Chapter II, Sections 1.1, 1.3, and 3.2. The results of Section 3.2 in particular imply the stability theorems we now state.

Theorem 1.1. If the roots $\lambda_1,\lambda_2,\ldots,\lambda_n$ of the characteristic equation $\det(\lambda I - A) = 0$ are pairwise distinct, a necessary and sufficient condition for all the solutions of the system (1.6) to be infinitesimal as $t \to \infty$ is that the real parts of the roots $\lambda_1,\lambda_2,\ldots,\lambda_n$ be all negative. A necessary and sufficient condition for all the solutions of the system (1.6) to be bounded for $t \geq t_0$ is that the real parts of the roots $\lambda_1,\lambda_2,\ldots,\lambda_n$ be all negative or zero.

Theorem 1.2. If the roots $\lambda_1,\lambda_2,\ldots,\lambda_n$ of the equation $\det(\lambda I - A) = 0$ are pairwise distinct, a necessary and sufficient condition for the stability of the system (1.6) is that the real parts of the roots $\lambda_1,\lambda_2,\ldots,\lambda_n$ be all negative or zero.

If we give up the hypothesis that the roots be all distinct, then we have the following theorems.

Theorem 1.3. A necessary and sufficient condition for all the integrals of the equation (1.6) to be infinitesimal as $t \to \infty$ is that the real parts of the roots $\lambda_1,\lambda_2,\ldots,\lambda_n$ be all negative.

Theorem 1.4. A sufficient condition for the stability of the solutions of the equation (1.6) is that the real parts of the roots $\lambda_1,\lambda_2,\ldots,\lambda_n$ of the characteristic equation be all negative.

The proofs are all obvious consequences of the results of Section 3.2 in Chapter II. Section 1.3 contains applications of these theorems.

It is worthwhile to recall a method that allows us to give the signature of the roots of an algebraic equation from an analysis of the coefficients. We present this method, Routh's algorithm, but omit its long proof.

<u>Routh's Algorithm</u>. We wish to determine the signature of the real part of the roots of the equation

$$a_n x^n + a_{n-1} x^{n-1} + \dots + a_1 x + a_0 = 0,$$

where $a_n > 0$, $a_0 \neq 0$. (This does not impose any restriction on the applicability of the method.) We construct a sequence of $n + 1$ lines of decreasing length, whose elements will be denoted by $b_{i,j}$, with $i = n, n-1, \dots, 1, 0$ counting from the top.

<u>Initial Phase</u>. In the two top lines (n and $n-1$) we arrange the coefficients of the equation as follows:

$$\begin{array}{lll} n & b_{n,0} = a_n & b_{n,1} = a_{n-2} \\ n-1 & b_{n-1,0} = a_{n-1} & b_{n-1,1} = a_{n-3} \\ \vdots & & \\ 1 & & \\ 0 & & \end{array}$$

If n is even, the line $n - 1$ may have all entries equal to zero. In this case, pass directly to Phase B.

<u>Phase A</u>. New lines are constructed according to the following rule:

$$b_{i,j} = -\mathrm{sgn}(b_{i+1,0}) \det \begin{pmatrix} b_{i+2,0} & b_{i+2,j+1} \\ b_{i+1,0} & b_{i+1,j+1} \end{pmatrix}, \quad i = n-2, n-3, \dots, 0.$$

Note that $\mathrm{sgn}(0)$ is either $+1$ or -1, with no effect on the final result. One may also multiply the whole line by a positive number without changing the results.

If one gets a line with all 0-entries, then one proceeds to Phase B. Note that the lines become shorter and that the 0 line has only one element.

<u>Analysis of Phase A</u>. Two cases may occur:

1. Phase A terminates at the last line. Consider the sequence of $n+1$ terms

$$\mathrm{sgn}(b_{n,0}),\ \mathrm{sgn}(b_{n-1,0}),\ldots,\mathrm{sgn}(b_{1,0}),\ \mathrm{sgn}(b_{0,0});$$

to each change in sign there corresponds a root with positive real part; to each permanence of sign there corresponds a root with negative real part.

2. Phase A terminates with the 2h-1st line having 0 entries. (Note that this happens only at odd lines.) Then the sequence

$$\mathrm{sgn}(b_{n,0}),\ \mathrm{sgn}(b_{n-1,0}),\ldots,\mathrm{sgn}(b_{2h,0})$$

gives information about the first $n-2h$ roots according to the rule in 1. The remaining roots exhibit the same behavior as the roots of the equation

$$b_{2h,0}x^{2h} + b_{2h,1}x^{2(h-1)} + \cdots + b_{2h,h} = 0.$$

This behavior can also be investigated according to Phase B.

Phase B. Replace the 0 entries of line $2h-1$ by

$$b_{2h-1,j} = 2(h-j)b_{2h,j}, \qquad j = 0,\ldots,h-1$$

Then construct the new lines as in Phase A.

Analysis of Phase B. Consider the sequence

$$\mathrm{sgn}(b_{2h,0}),\ \mathrm{sgn}(b_{2h-1,0}),\ldots,\mathrm{sgn}(b_{0,0}).$$

To each change in the sequence there corresponds a pair of opposite roots, one of which has positive real part and the other of which has negative real part. In case the number of changes is $k < h$, then there are also $2(h-k)$ roots with vanishing real part.

We give now some examples:

Only A. $x^4 - 5x^3 + 11x^2 - 10x + 4 = 0$ (roots: $1, 2, 1+i, 1-i$)

4	1	11	4
3	-5	-10	(0)
2	45		20
1	-350		(0)
0	7000		

Sequence: + - + - +, four changes.

Both A and B: $x^4 - 2x^3 + 2x - 1 = 0$ (roots: 1,1,1,-1)

4	1	0	-1	Phase A.
3	-2	2	(0)	
2	2	-2		
1	0			

The sequence is + - +; there are two changes, hence two roots with positive real parts.

2	2	-2	Phase B.
1	4	(0)	
0	-8		

The sequence is + + -; there is one change, hence one root with positive real part and one with negative real part. There are no other roots; hence no purely imaginary roots.

Both A and B: $x^4 + 1 = 0$ (roots $\sqrt{2}/2 \pm i\sqrt{2}/2$, $-\sqrt{2}/2 \pm i\sqrt{2}/2$)

4	1	0	1	Phase A, no sequence
3	0	0		

4	1	0	1	Phase B.
3	4	0	(0)	
2	+0	4		
1	-16	(0)		
0	64			

The sequence is + + + - +; there are two changes, hence two roots with positive real part and two with negative real part. There are no other roots.

Both A and B: $x^4 - 1 = 0$ (roots ± 1, $\pm i$)

4	1	0	-1	Phase A, no sequence
3	0	0		

4	1	0	-1	Phase B.
3	4	0	(0)	
2	+0	-4		
1	16	(0)		
0	-64			

The sequence is + + + + 0; there is one change, hence one root with positive real part and one with negative real part. The two remaining roots are purely imaginary.

1.3. Autonomous Linear Systems of the Second Order

As an application of the theorems of Section 1.2 we shall examine in detail the case of systems of two equations. This case is particularly interesting because it is possible to give an effective graphical interpretation connected with first order homogeneous differential equations. The orbits of the solutions permit a detailed classification connected with the structure of the matrix and its eigenvalues. Let A be the 2×2 matrix with constant entries that appears on the right of the system

$$y' = Ay. \tag{1.7}$$

Then we can distinguish three general cases for the eigenvalues λ_1 and λ_2 of A:

1. They are distinct real numbers with the same sign.
2. They are real but of different sign.
3. They are complex conjugates.

Aside from these general cases, it is useful to distinguish explicitly the following particular cases:

4. One eigenvalue is 0, and the other different from 0.
5. The eigenvalues are equal and different from 0, and the matrix is diagonalizable.
6. The eigenvalues are equal and different from 0, and the matrix is not diagonalizable.
7. The eigenvalues are purely imaginary numbers.
8. The eigenvalues are equal to 0, and the matrix is not diagonalizable.

We do not count the trivial case $A = 0$.

First of all, we consider the cases in which the eigenvalues are real. There is, as we saw in Chapter II, a real nonsingular matrix T such that $A_1 = T^{-1}AT$ is a Jordan matrix. We note that the solutions of (1.7) are of the form $T^{-1}y_1$, where y_1 is a solution of $y_1' = A_1y_1$. We therefore study this last equation first.

In Cases 1 and 2 we have $A = \begin{pmatrix} \lambda_1 & 0 \\ 0 & \lambda_2 \end{pmatrix}$, so the solutions of the equation are

$$\begin{cases} u(t) = \exp(\lambda_1 t)\cdot u_0 \\ v(t) = \exp(\lambda_2 t)\cdot v_0 \end{cases}. \tag{1.8}$$

Since λ_1 and $\lambda_2 \neq 0$, we get the equation of the trajectories from (1.8): if $u_0 = 0$, then $u = 0$; if $u_0 \neq 0$, then

$$v(t) = c_0|u(t)|^{\lambda_2/\lambda_1}, \quad c_0 = v_0\cdot|u_0|^{-\lambda_2/\lambda_1}. \tag{1.9}$$

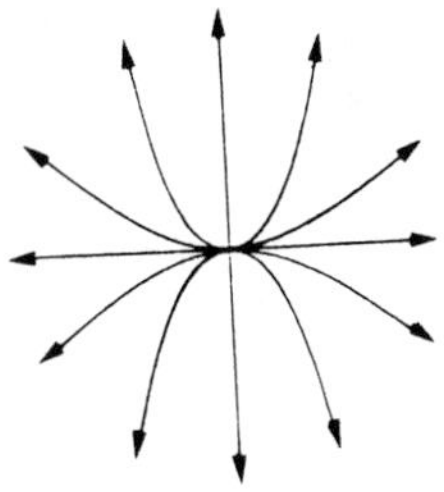

Unstable

Fig. 1a

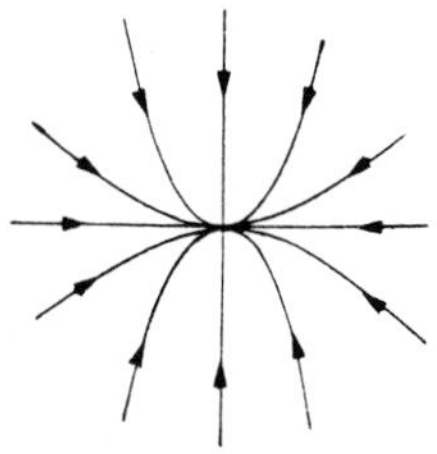

Stable

Fig. 1b

Thus, in case 1, with $\lambda_1 > \lambda_2 > 0$, we have the situation in Fig. 1a, while if $\lambda_2 < \lambda_1 < 0$, we have that in Fig. 1b.

In case 2, we have the situation pictured in Fig. 2. (In the figure, $\lambda_2 > -\lambda_1 > 0$.)

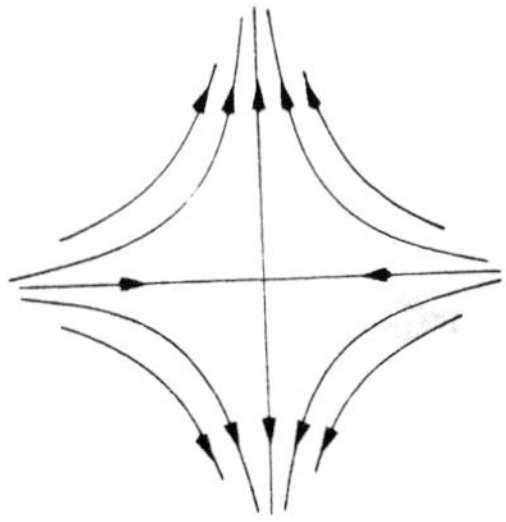

Unstable

Fig. 2

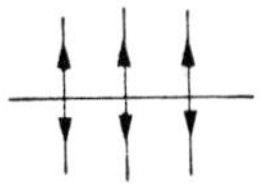

Unstable

Fig. 4a

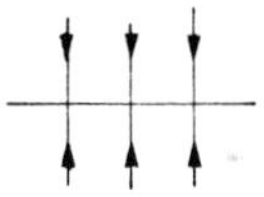

Stable

Fig. 4b

Case 4 is a limiting case between the two preceding ones; let $A_1 = \begin{pmatrix} 0 & 0 \\ 0 & \lambda \end{pmatrix}$. The solutions of the equation are given by

$$\begin{cases} u(t) = u_0 \\ v(t) = \exp(\lambda t)v_0 \end{cases} .$$

The trajectories are half lines; note that in this case all the points for which $v_0 = 0$ are stationary points. There is instability (Fig. 4a) for $\lambda > 0$ and stability (Fig. 4b) for $\lambda < 0$.

Case 5 is associated with the matrix $A_1 = \lambda I$. The trajectories are therefore half lines going out from the origin; there is instability for

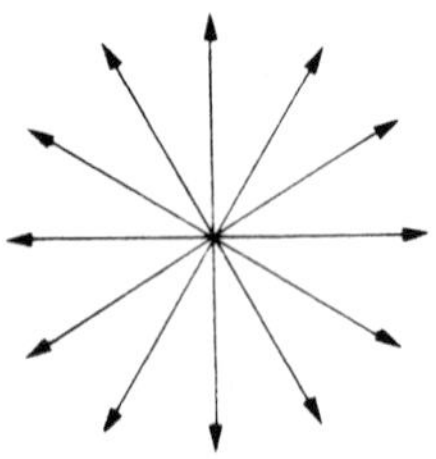

Unstable

Fig. 5a

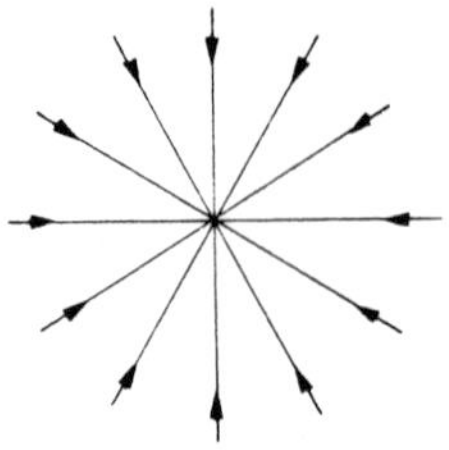

Stable

Fig. 5b

$\lambda > 0$ and stability for $\lambda < 0$ (see Fig. 5a and 5b). Note that this is a special case of Case 1.

In Case 6, the matrix is $A_1 = \begin{pmatrix} \lambda & 1 \\ 0 & \lambda \end{pmatrix}$; the corresponding solution is

$$\begin{cases} u(t) = \exp(\lambda t)u_0 + t\exp(\lambda t)v_0 \\ v(t) = \exp(\lambda t)v_0 \end{cases}. \tag{1.10}$$

The trajectories are given by $v = 0$ if $v_0 = 0$ and by

$$u = c_0 v + \lambda^{-1} v\, \ell g|v|, \quad c_0 = \frac{u_0}{v_0} - \lambda^{-1}\, \ell g|v_0| \tag{1.11}$$

if $v_0 \neq 0$. There is instability if $\lambda > 0$ and stability if $\lambda < 0$. See Fig. 6a and 6b.

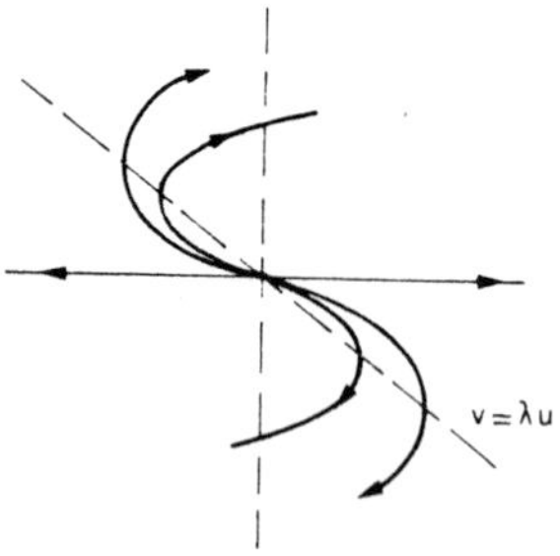

Unstable

Fig. 6a

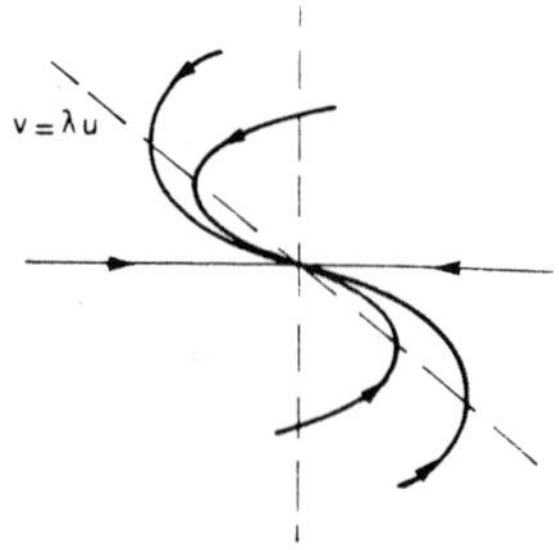

Stable

Fig. 6b

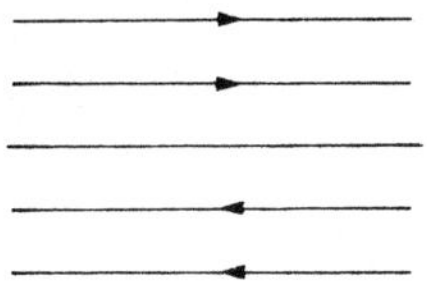

Unstable

Fig. 8

Case 8 is a limiting case of Case 6. The matrix is $A_1 = \begin{pmatrix} 0 & 1 \\ 0 & 0 \end{pmatrix}$ with the solution

$$\begin{cases} u = u_0 + tv_0 \\ v = v_0 \end{cases},$$

and the trajectories are lines. The case is unstable. Note that the points for which $v_0 = 0$ are stationary. See Fig. 8.

In studying the case of complex eigenvalues, it is convenient to use a similarity real matrix T. The canonical form to which A can be reduced is given by

$$A_1 = \begin{pmatrix} \alpha & -\omega \\ \omega & \alpha \end{pmatrix}, \quad \alpha = \mathrm{Re}\lambda_i, \quad \omega = |\mathrm{Im}\ \lambda_i|.$$

It is convenient to express the system in polar coordinates and to transform it into

$$\begin{cases} \theta' = \dfrac{-u'v + v'u}{u^2 + v^2} = \omega \\ \rho' = \dfrac{uu' + vv'}{\sqrt{u^2 + v^2}} = \alpha\rho. \end{cases} \tag{1.12}$$

Then one has the following solution for Case 3:

$$\begin{cases} \theta(t) = \theta_0 + \omega t \\ \rho(t) = \rho_0 \exp(\alpha t), \end{cases} \tag{1.13}$$

and the trajectories are logarithimic spirals with equation

$$\rho = c_0 \exp\left(\frac{\alpha}{\omega}\theta\right), \quad c_0 = \rho_0 \exp\left(-\frac{\alpha}{\omega}\theta_0\right).$$

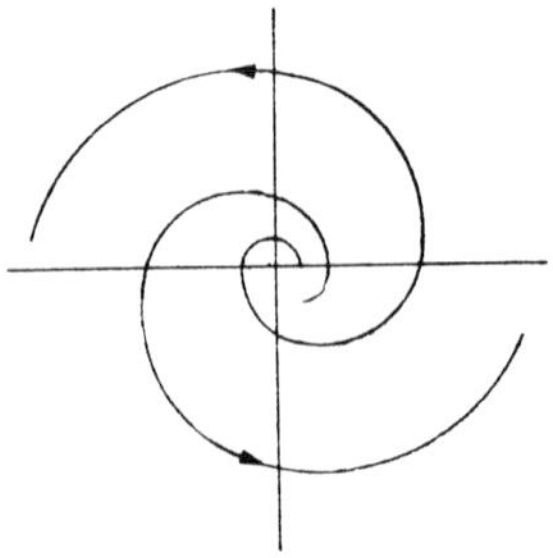

Unstable

Fig. 3a

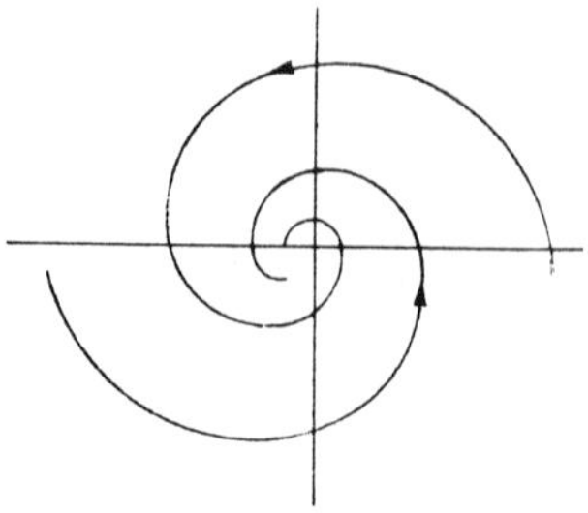

Stable

Fig. 3b

There is instability for $\alpha = \text{Re}\ \lambda_i > 0$ and stability for $\alpha = \text{Re}\ \lambda_i <$ See Fig. 3a and 3b. Note that as $\omega \to 0$, we have the limiting Case 5.

Finally, the limiting Case 7 is the only one that has nontrivial periodic solutions; in this case, $\alpha = 0$, and the solutions are given by

$$\begin{cases} \theta(t) = \theta_0 + \omega t \\ \rho(t) = \rho_0 \end{cases}.$$

The orbits are circles (Fig. 7). The solutions are periodic with period $2\pi/\omega$; the case is stable.

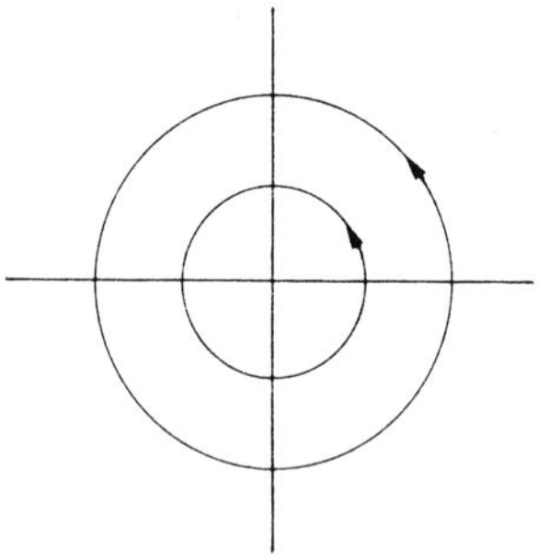

Stable

Fig. 7

The results considered up to now are results relative to the matrix A_1 in canonical form. To obtain the solutions for general A, it is necessary, as we have said, to apply a nonsingular linear transformation. The trajectories in general turn out to be of the type presented in the following figures.

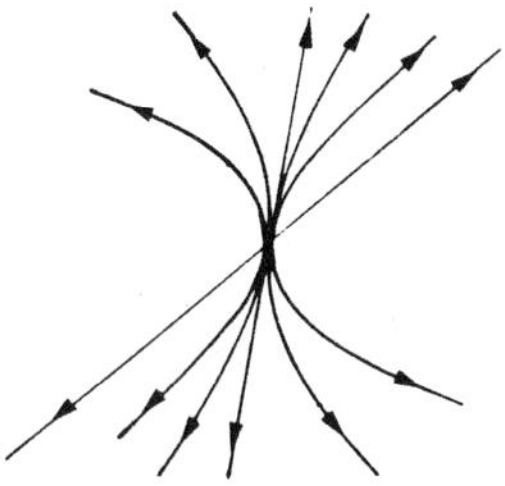

Case 1a

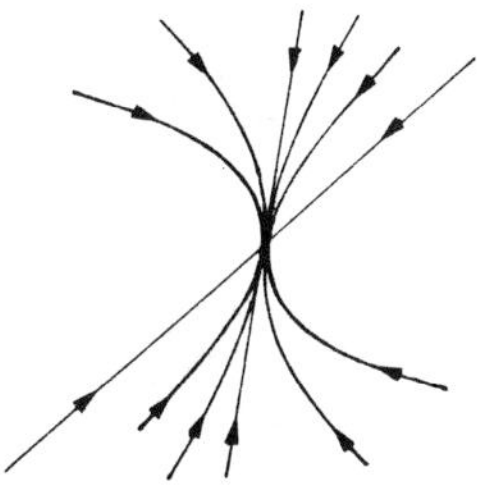

Case 1b

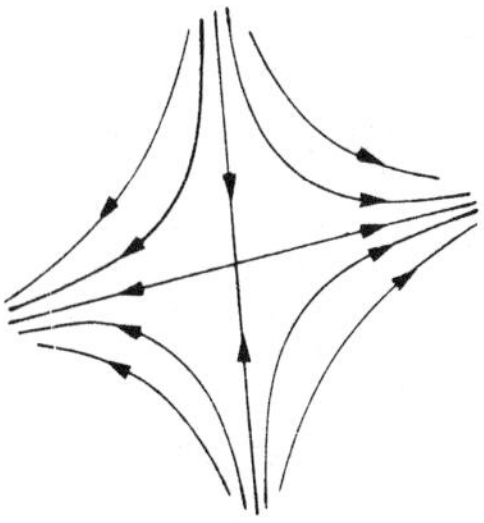

Case 2

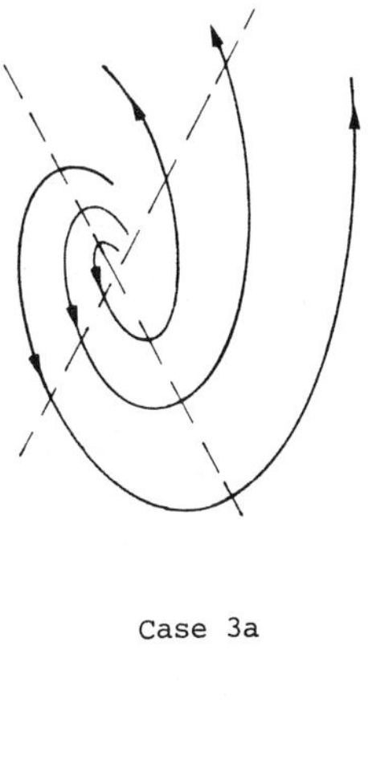

Case 3a

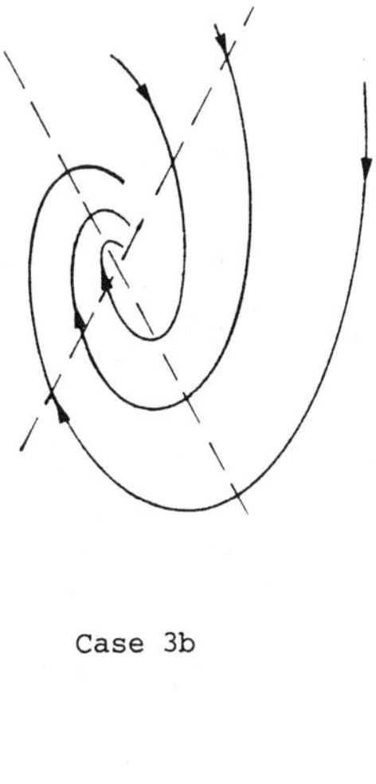

Case 3b

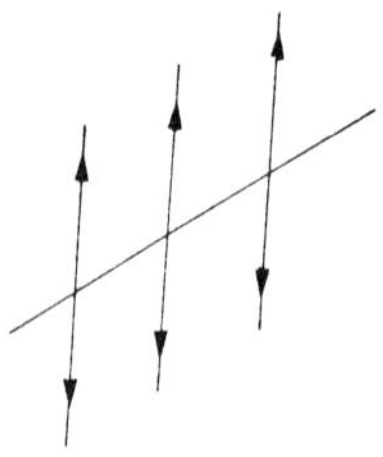

Case 4a

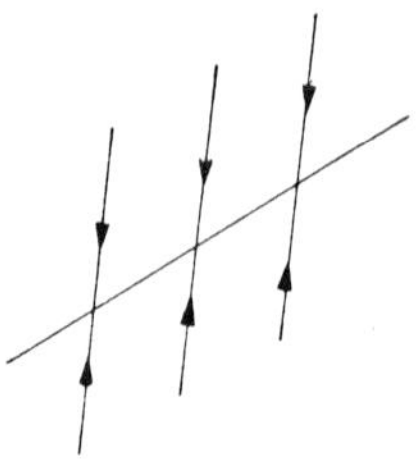

Case 4b

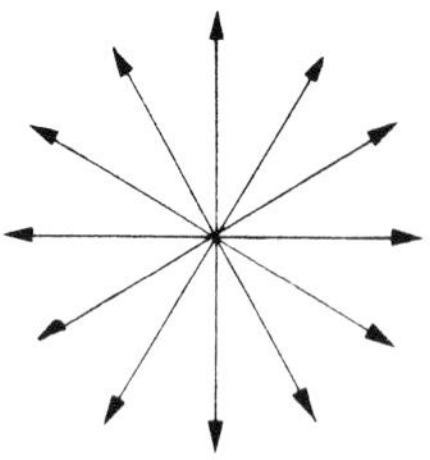

Case 5a

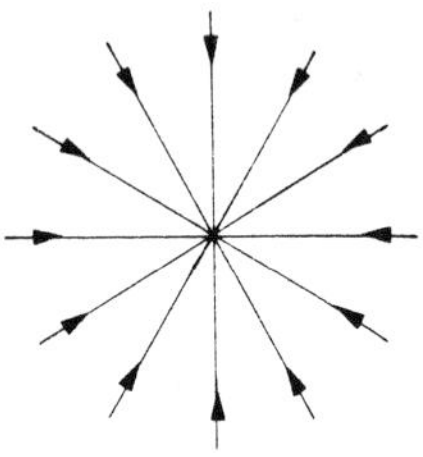

Case 5b

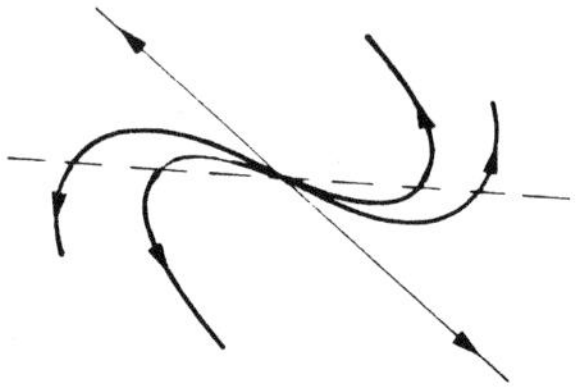

Case 6a

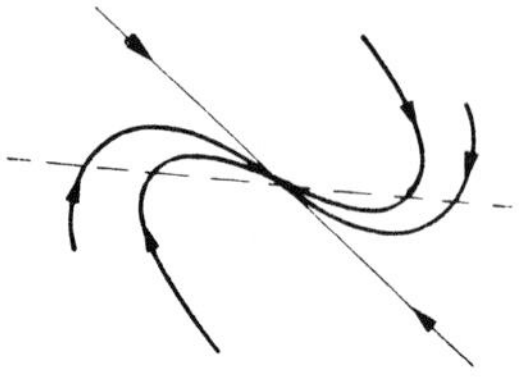

Case 6b

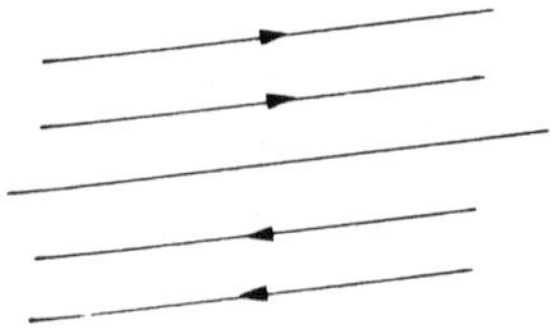

Case 7

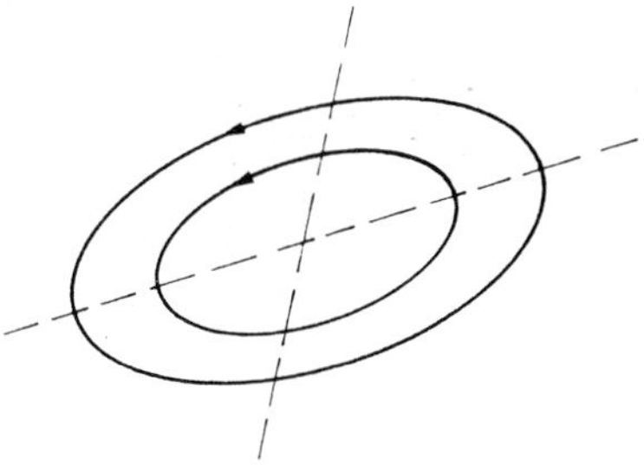

Case 8

It is sometimes useful to keep in mind that there are rectilinear trajectories if and only if the eigenvalues are real.

We recall, finally, that the trajectories, being arcs of a curve $v = v(u)$, are locally solutions of the first order equation

$$\frac{dv}{du} = \frac{cu + dv}{au + bv}\ .$$

The trajectories $u(t)$ and $v(t)$ constitute a parametric representation of the solutions of the equation.

1.4. Certain Stability Problems for Nonautonomous Linear Systems

The simplest case is that of a single first order equation

$$y' = a(t)y, \quad a(t) \text{ a continuous function.} \tag{1.14}$$

Since the general solution of (1.14) is given by

$$y(t) = y(t_0)\ \exp\left[\int_{t_0}^{t} a(\tau)d\tau\right], \tag{1.15}$$

a sufficient condition for stability is

$$\limsup_{t\uparrow+\infty} t^{\alpha}a(t) \leq c, \quad \alpha > 1, \tag{1.16}$$

while a sufficient condition that the solutions be infinitesimal at infinity is given by

$$\limsup_{t\uparrow+\infty} a(t) < 0. \tag{1.17}$$

Let us now look at some other condition. If

$$\limsup_{t\to+\infty} \int_{t_0}^{t} a(\tau)d\tau \leq K < +\infty, \tag{1.18}$$

then the solutions of (1.14) are stable; if we have

$$\limsup_{t\to+\infty} \int_{t_0}^{t} a(\tau)d\tau = -\infty, \tag{1.19}$$

the solutions of (1.14) are also infinitesimal at infinity. In the special case when $a(t)$ is periodic with period T, the solutions of (1.14) are stable if and only if

$$\int_0^T a(\tau)d\tau \le 0 \tag{1.20}$$

and are infinitesimal if and only if

$$\int_0^T a(\tau)d\tau < 0. \tag{1.21}$$

The proofs of these statements are obvious consequences of (1.15).

The following theorem is a result for systems, but it is only valid under quite restrictive hypotheses.

Theorem 1.5. Consider the linear system

$$y' = A(t)y, \tag{1.22}$$

where $A(t)$ is a selfadjoint matrix (Hermitian, if the entries are complex) whose entries are continuous. (It would be enough for the entries to be in L^1.) If $\lambda_1(t),\lambda_2(t),\ldots,\lambda_n(t)$ are the eigenvalues of $A(t)$, suppose

$$\limsup_{t\uparrow+\infty} \lambda_i(t) < 0 \qquad (i = 1,2,\ldots,n). \tag{1.23}$$

Then the solutions of (1.22) are stable and infinitesimal as $t \to \infty$.

Proof: Let $\max_i \limsup_{t\uparrow+\infty} \lambda_i(t) = -2M < 0$; there exists t^* such that if $t > t^*$, then

$$\lambda_i(t) \le -M \qquad (i = 1,2,\ldots,n). \tag{1.24}$$

Since the system is linear with continuous coefficients, there is a constant K such that

$$|y(t^*)| = |\mathop{E}_{t_0}^{t^*} A(\tau)y_0| \le K|y_0|. \tag{1.25}$$

Starting with t^*, consider the following system with piecewise constant coefficients

$$\begin{cases} \dfrac{dy_k^{(s)}}{dt} = A(t_{s,k})y_k^{(s)} & t_{s,k} < t < t_{s+1,k} \\ y_k^{(s)}(t_{s,k}) = y_k^{(s-1)}(t_{s,k});\ y_k^{(0)}(t_{0,k}) = y(t^*) & \end{cases} \tag{1.26}$$

where $t_{s,k} = t^* + s2^{-k}$. Since the eigenvalues of $A(t_{s,k})$ satisfy (1.24) and the matrices are selfadjoint, the eigenvalues μ_i of $E_{s,k}(t) = \exp[A(t_{s,k})(t-t_{s,k})]$ satisfy the relation

$$\mu_i(t) \leq e^{-M(t-t_{s,k})}, \qquad t_{s,k} \leq t < t_{s+1,k} \tag{1.27}$$

and we have, in addition, $|E_{s,k}(t)y| \leq \max_i \mu_i(t)\cdot|y| \leq e^{-M(t-t_{s,k})}\cdot|y|$.

We denote by $y_k(t)$ the function obtained by joining the $y_k^{(s)}$ together as s varies; this function satisfies the inequality

$$|y_k(t)| \leq |y(t^*)|\exp[-M(t-t^*)]. \tag{1.28}$$

(1.28) implies that the $y_k(t)$ are equibounded functions on $[t^*,\infty[$; it then follows from (1.26) and the continuity of $A(t)$ that they are equicontinuous on every bounded interval. In addition, for every $\varepsilon > 0$, there is a t_ε such that for $t > t_\varepsilon$,

$$|y_k(t^*)|\cdot\exp[-M(t-t^*)] < \varepsilon \tag{1.29}$$

for every k. By virtue of Ascoli's theorem, we can extract a subsequence $y_{k,1}$ that converges uniformly in the interval $[t^*,t^*+1]$. We may then extract a sequence $(y_{k,2})_k$ that converges uniformly in $[t^*,t^*+2]$, and so on. The sequence $(y_{s,s})_s$ converges pointwise to a limit function $y(t)$ in $[t^*,\infty[$ and converges uniformly there on every bounded interval. On the other hand, for every $\varepsilon > 0$, (1.29) and (1.28) guarantee that for each s and for every $t > t_\varepsilon$, we also have $|y_{s,s}(t) - y(t)| < 2\varepsilon$. The sequence $(y_{s,s})_s$ thus converges uniformly to y on $[t^*,\infty[$. Kamke's theorem (see Chapter I) then assures us that on every interval, $y(t)$ is the solution of (1.22) that takes on the value $y(t^*)$ at t^*. (1.25) and (1.28) then imply the inequality

$$|y(t)| \leq K\exp[-M(t-t^*)]\cdot|y_0| \tag{1.30}$$

which, in turn, implies the theorem. In the special case when, for each t, $\lambda_i(t) \leq -M$, (1.30) can be strengthened to $|y(t)| \leq \exp[-M(t-t_0)]|y_0|$.

Another case which is valid only under very strict hypotheses is described by the following

Theorem 1.6. Consider the system

$$y' = f_t(A)\cdot y, \tag{1.31}$$

where for each t, $f_t(A)$ is an analytic function of a fixed matrix A and depends continuously on t. Suppose that the eigenvalues of A are all distinct. (This hypothesis may be replaced by the assumption that A is diagonalizable.) If

$$\limsup_{t\to+\infty} \int_0^t \mathrm{Re}[f_t(\lambda_i)]dt \le K < +\infty \qquad (i = 1,2,\ldots,n) \tag{1.32}$$

then the solutions are stable. If we also have

$$\lim_{t\to+\infty} \int_0^t \mathrm{Re}[f_t(\lambda_i)]dt = -\infty \qquad (i = 1,2,\ldots,n) \tag{1.33}$$

then the solutions are infinitesimal as $t \to \infty$.

Proof: The argument is based on Theorems 3.5 and 1.9 of Chapter II. Since the $f_t(A)$ are analytic functions of the same matrix for all t, they have the same eigenvectors as A. Furthermore, the eigenvalues of A are distinct, so that it is possible to diagonalize all the matrices $f_t(A)$ by means of the same transformation $T^{-1}AT$ that diagonalizes A. In this case, the system reduces to

$$\begin{cases} z_1' = f_t(\lambda_1)z_1 \\ \ldots \\ z_n' = f_t(\lambda_n)z_n \end{cases}$$

and we are back to the elementary case of stability which we examined above for first order equations.

The preceding theorem illustrates that for nonsymmetric matrices, the positions of the eigenvectors as well as the real parts of the eigenvalues are relevant to the determination of stability. A complete treatment of the problem is beyond the scope of this book. We conclude this section with two examples which, though quite simple, are characteristic for the kind of problems that arise.

Example 1. Consider the following system with piecewise constant coefficients that are periodic with period 2

$$x' = A(t)x \tag{1.34}$$

with

$$A(t) = \begin{cases} \begin{pmatrix} -\lambda_1 & 0 \\ 0 & -\lambda_2 \end{pmatrix} & \text{for } 2n < t < 2n+1. \\ \begin{pmatrix} a & b \\ c & d \end{pmatrix} & \text{for } 2n-1 < t < 2n. \end{cases}$$

Suppose that λ_1 and λ_2 are strictly positive and that the eigenvalues of $\begin{pmatrix} a & b \\ c & d \end{pmatrix}$ are $-\lambda_1$ and $-\lambda_2$. We set

$$E_1(t) = \exp\left[\begin{pmatrix} -\lambda_1 & 0 \\ 0 & -\lambda_2 \end{pmatrix} t\right], \quad E_2(t) = \exp\left[\begin{pmatrix} a & b \\ c & d \end{pmatrix} t\right].$$

To establish the nature of the solutions, it is sufficient, since $A(t)$ is periodic, to study $\underset{0}{\overset{2}{E}} A(\tau)$ because for each t,

$$\underset{0}{\overset{t}{E}} A(\tau) = \underset{0}{\overset{t-[t/2]}{E}} A(\tau)\cdot\left(\underset{0}{\overset{2}{E}} A(\tau)\right)^{[t/2]},$$

where $[\frac{1}{2}t]$ is the greatest integer less or equal to $\frac{1}{2}t$. We obviously have $\underset{0}{\overset{2}{E}} A(\tau) = E_2(1)E_1(1)$. To study the behavior of $E_2(1)E_1(1)$, it us useful to give a new representation of the matrix $\begin{pmatrix} a & b \\ c & d \end{pmatrix}$ that takes into account the position of its eigenvectors. With a little loss of generality which can be eliminated by passing to the limit, we shall suppose that $(1,\varepsilon)$ is an eigenvector associated with λ_1 and $(\eta,1)$ an eigenvector associated with λ_2. We then have

$$\begin{pmatrix} a & b \\ c & d \end{pmatrix} = \frac{1}{1-\varepsilon\eta}\begin{pmatrix} 1 & \eta \\ \varepsilon & 1 \end{pmatrix}\begin{pmatrix} -\lambda_1 & 0 \\ 0 & -\lambda_2 \end{pmatrix}\begin{pmatrix} 1 & -\eta \\ -\varepsilon & 1 \end{pmatrix}$$

$$= \frac{1}{1-\varepsilon\eta}\begin{pmatrix} -\lambda_1 + \varepsilon\eta\lambda_2 & \eta(\lambda_1 - \lambda_2) \\ \varepsilon(\lambda_2 - \lambda_1) & -\lambda_2 + \varepsilon\eta\lambda_1 \end{pmatrix}.$$

It follows (see Chapter II)

$$E_2(1) = \frac{1}{1-\varepsilon\eta}\begin{pmatrix} e^{-\lambda_1} - \varepsilon\eta e^{-\lambda_2} & -\eta(e^{-\lambda_1} - e^{-\lambda_2}) \\ \varepsilon(e^{-\lambda_1} - e^{-\lambda_2}) & -\varepsilon\eta e^{-\lambda_1} + e^{-\lambda_2} \end{pmatrix},$$

while clearly

$$E_1(1) = \begin{pmatrix} e^{-\lambda_1} & 0 \\ 0 & e^{-\lambda_2} \end{pmatrix}.$$

We therefore obtain

$$\overset{2}{\underset{0}{E}} A(\tau) = E_2(1)E_1(1) = \frac{1}{1-\varepsilon\eta}\begin{pmatrix} e^{-2\lambda_1} - \varepsilon\eta e^{-(\lambda_1+\lambda_2)} & -\eta(e^{-(\lambda_1+\lambda_2)} - e^{-2\lambda_2}) \\ \varepsilon(e^{-2\lambda_1} - e^{-(\lambda_1+\lambda_2)}) & -\varepsilon\eta e^{-(\lambda_1+\lambda_2)} + e^{-2\lambda_2} \end{pmatrix}$$

In particular,

$$\operatorname{tr}\left(\overset{2}{\underset{0}{E}} A(\tau)\right) = \frac{1}{1-\varepsilon\eta}\left[e^{-2\lambda_1} + e^{-2\lambda_2} - 2\varepsilon\eta e^{-(\lambda_1+\lambda_2)}\right]$$

and

$$\det\left(\overset{2}{\underset{0}{E}} A(\tau)\right) = \det(E_2(1))\det(E_1(1)) = e^{-2(\lambda_1+\lambda_2)}.$$

The characteristic polynomial is

$$\xi^2 - \frac{1}{1-\alpha}\left(e^{-2\lambda_1} + e^{-2\lambda_2} - 2\alpha e^{-(\lambda_1+\lambda_2)}\right)\xi + e^{-2(\lambda_1+\lambda_2)} = 0$$

where $\alpha = \varepsilon\eta$.

A sufficient condition for stability is that the eigenvalues of $\overset{2}{\underset{0}{E}} A(\tau)$ be strictly less than 1 in modulus; this follows from the fact that in order to have stability for the equation $y' = \left[\ell g \overset{2}{\underset{0}{E}} A(\tau)\right] y$ with constant coefficients, it is sufficient that the eigenvalues have real part negative. A condition for $|\xi_1|, |\xi_2| < 1$ is that

$$1 - \operatorname{tr}\left(\overset{2}{\underset{0}{E}} A(\tau)\right) + \det\left(\overset{2}{\underset{0}{E}} A(\tau)\right) > 0,$$

or that one of the following two relations hold:

$$\alpha < \left(1 - e^{-2\lambda_1}\right)\left(1 - e^{-2\lambda_2}\right)\left(1 - e^{-(\lambda_1+\lambda_2)}\right)^{-2} \quad (\le 1) \tag{1.35}$$

$$\alpha > 1. \tag{1.36}$$

The roots are real if $\alpha \le (e^{-\lambda_1} + e^{-\lambda_2})^2 e^{\lambda_1} e^{\lambda_2}/4$; otherwise they are complex conjugates and we have $|\xi_1| = |\xi_2| = e^{-(\lambda_1+\lambda_2)}$. To interpret (1.35) and (1.36), observe that if θ_1 is the angle that the vector $(1,\varepsilon)$ corresponding to λ_1 makes with the positive u axis and if θ_2 is the angle that the vector $(\eta,1)$ corresponding to λ_2 makes with the positive v axis, then $\alpha = -\tan\theta_1 \cdot \tan\theta_2$.

Example 2. Consider the equation $y'' + a(t)y = 0$, where $a(t) = a^2$ if $2n < t < 2n + 1$ and $a(t) = b^2$ if $2n - 1 < t < 2n$; we shall find the relation between a^2 and b^2 that gives stability of the solutions. Writing the associated system, we have

$$\begin{cases} y' = v \\ v' = -a(t)y \end{cases}, \quad \text{or} \quad z' = A(t)z.$$

As in the preceding example, it is sufficient to study the eigenvalues of $E_0^2 A(\tau)$. We get

$$E_0^2 A(\tau) = \begin{pmatrix} \cos b & \frac{1}{b}\sin b \\ -b\sin b & \cos b \end{pmatrix}\begin{pmatrix} \cos a & \frac{1}{a}\sin a \\ -a\sin a & \cos a \end{pmatrix}$$

whence

$$\det\left(E_0^2 A(\tau)\right) = 1, \quad \operatorname{tr}\left(E_0^2 A(\tau)\right) = 2\cos a\cos b - \left(\frac{b}{a} + \frac{a}{b}\right)\sin a \sin b.$$

If we have $|\operatorname{tr}(E_0^2 A(\tau))| < 2$, the roots of the characteristic equation are complex conjugates and therefore have modulus 1. If, however, we have $|\operatorname{tr}(E_0^2 A(\tau))| \geq 2$, there are real roots and we have instability. The conditions on the trace insure that for every fixed a, there is a neighborhood in which b can vary without loss of stability. Outside of this neighborhood, there may be instability. As an example, take $a = \frac{1}{2}\pi$, $b = 3\pi/2$; here we have $\operatorname{tr}(E_0^2 A(\tau)) = 10/3 > 2$.

Exercise 1. Are the results of Example 1 valid if we do not require that λ_1 and $\lambda_2 > 0$?

Exercise 2. Generalize the results of Example 1 to the case when the matrices are piecewise constant in two arbitrary intervals.

Exercise 3. Repeat Exercise 2 for Example 2.

Exercise 4. In the example of instability with $a = \frac{1}{2}\pi$, $b = 3\pi/2$, find a solution $y_1(t)$ such that $\limsup_{t\to\infty} y_1(t) = \infty$ and a solution $y_2(t)$, not identically 0, such that $\lim_{t\to\infty} y_2(t) = 0$.

2. SOME METHODS FOR THE DETERMINATION OF THE STABILITY OF NONLINEAR SYSTEMS

In this section we consider certain methods for the study of the stability of systems of nonlinear equations, viz., Liapunov's method, the fixed points method, Olech's method, the method of the logarithmic derivative, and the method of invariant sets.

We shall examine in particular depth the system

$$y' = A(x)y + f(x,y)$$

obtained by perturbing the linear system $z' = A(x)z$ by a nonlinear term f. As we already pointed out in Sec. 1.1 and as we shall see in detail below, the results of this type can be applied to the nonlinear systems

$$y' = F(x,y),$$

when F is of class C^1, by examining the associated variational equation.

2.1. Definitions

We consider the system of ordinary equations

$$y' = f(x,y) \tag{2.1}$$

written in vector form, where $f\colon [a,\infty[\times U \to R^n$ is continuous and $U \subseteq R^n$ is a neighborhood of the origin. Let y_0 be a solution of (2.1) defined on $[a,\infty[$.

We have already had occasion to observe that there are various definitions of stability for the solution y_0 of (2.1). We shall now examine four of them. They are related in that they represent successive developments of the same idea of the concept of stability. We begin by stating the two definitions introduced by Liapunov.

Definition 1. The solution y_0 of (2.1) is *stable* (or, more precisely, *stable on the interval* $[a,\infty[$) if, for every $\varepsilon > 0$ and for every $x_0 \geq a$, there is $\delta = \delta(\varepsilon,x_0) > 0$ such that, for every solution y of (2.1), $||y_0(x_0) - y(x_0)|| < \delta$ implies $||y_0(x) - y(x)|| < \varepsilon$ for $x \geq x_0$.

In other words, we are simply dealing with the continuous dependence on the initial data at x_0 with respect to the topology of uniform convergence on the whole interval $[x_0,\infty[$. It is thus a different situation from that of the continuous dependence of the solutions on the data studied in Chapters I and III where the intervals were required to be compact. The results on continuous dependence in Chapter III show that if we assume that the solutions for the initial value problems relative to (2.1) are unique, then y_0 is stable if there exists $\bar{x} \geq a$ with the property that for every $\varepsilon > 0$, there exists $\delta = \delta(\varepsilon) > 0$ such that for any solution y of (2.1), $||y_0(\bar{x}) - y(\bar{x})|| < \delta$ implies that $||y_0(x) - y(x)|| < \varepsilon$ for $x \geq \bar{x}$. Thus, the definition of stability takes on a simpler form when we have the uniqueness of solutions for initial value problems.

Definition 2. The solution y_0 of (2.1) is *asymptotically stable* (on the interval $[a,\infty[$) if it is stable and, furthermore, for every $x_0 \geq a$ there is $\eta = \eta(x_0) > 0$ such that $||y_0(x_0) - y(x_0)|| < \eta$ implies that $\lim_{x\to\infty}||y_0(x) - y(x)|| = 0$ for every solution y of (2.1).

These stability properties are too weak to be preserved after small perturbations of the second member f of (2.1). For this reason, we require stronger properties. This is the objective of the next definitions.

Definition 3. The solution y_0 of (2.1) is *uniformly stable* (on the interval $[a,\infty[$) if for every $\varepsilon > 0$ there exists a $\delta = \delta(\varepsilon) > 0$ such that for each $x_0 \geq a$ and each solution y of (2.1), $||y_0(x_0) - y(x_0)|| < \delta$ implies $||y_0(x) - y(x)|| < \varepsilon$ for $x \geq x_0$.

The difference between this and Definition 1 lies in the fact that δ is now determined independently of x_0. The corresponding concept of uniform asymptotic stability is a bit more complicated.

Definition 4. The solution y_0 of (2.1) is *uniformly asymptotically stable* (on the interval $[a,\infty[$) if it is uniformly stable and, furthermore, there exists an $\eta > 0$ such that to each $\varepsilon > 0$ there corresponds $T = T(\varepsilon) > 0$ such that for every $x_0 \geq a$ and for every solution y of (2.1), $||y_0(x_0) - y(x_0)|| < \eta$ implies $||y_0(x) - y(x)|| < \varepsilon$ for $x \geq x_0 + T$.

This means not only that $||y_0(x) - y(x)||$ tends to 0, but also that the required interval in which $||y_0(x) - y(x)||$ must be small is independent of the initial point:

$$\lim_{(x-x_0)\uparrow+\infty} ||y_0(x) - y(x)|| = 0.$$

There are two remarks which we make to illustrate the definitions just given.

The first is that we did not specify the domain of existence of the solution y because the results of Chapters I and III concerning global existence insure that y exists on the whole interval $[x_0,\infty[$ when a relation of the type

$$||y_0(x) - y(x)|| < \varepsilon$$

holds in the whole domain of y.

The other remark is that it suffices to study the stability of the null solution. To see this, define the function $g\colon [a,\infty[\times A \to \mathbb{R}^n$ by

$$g(x,z) = f(x,z + y_0(x)) - f(x,y_0(x))$$

where A is a suitable neighborhood of the origin. Then the solution y_0 satisfies one of the four definitions above relative to Eq. (2.1) if and only if the solution $z \equiv 0$ of

$$z' = g(x,z) \tag{2.2}$$

satisfies the corresponding definition for Eq. (2.2). This follows from the fact that y is a solution of (2.1) if and only if $z = y - y_0$ is a solution of (2.2). *We can consequently restrict ourselves to studying the stability of the null solution under the hypothesis that*

$$f(x,0) = 0 \qquad (x \geq a).$$

This implies a technical simplification which will become clear if one rewrites the four definitions above for the case $y_0 = 0$.

Before proceeding to prove the mutual independence of the four concepts of stability that we have just introduced, we need to state the following characterization for linear systems.

Theorem on Stability for Linear Systems. Let $A(x)$ be an $n \times n$ matrix continuous on $[a,\infty[$, and let $Y(x)$ be a fundamental matrix of the linear system

$$y' = A(x)y. \tag{2.3}$$

We have the following characterizations:

(i) The solution $y = 0$ of (2.3) is stable if and only if $Y(x)$ is uniformly bounded, that is, if and only if there exists $M > 0$ such that

$$||Y(x)|| \leq M \quad (x \geq a);$$

(ii) The solution $y = 0$ of (2.3) is asymptotically stable if and only if

$$\lim_{x\uparrow\infty} ||Y(x)|| = 0;$$

(iii) The solution $y = 0$ of (2.3) is uniformly stable if and only if there exist $x_0 \geq 0$ and $M > 0$ such that

$$||Y(x)Y^{-1}(x_0)|| \leq M \qquad (x \geq x_0 \geq a);$$

(iv) The solution $y = 0$ of (2.3) is uniformly asymptotically stable if and only if there are $x_0 \geq a$, $M > 0$ and $\alpha > 0$ such that

$$||Y(x)Y^{-1}(x_0)|| \leq Me^{-\alpha(x-x_0)} \qquad (x \geq x_0 \geq a).$$

Proof: We only prove the necessity of (iv) since the other implications are proved similarly or are very easy to demonstrate. We therefore suppose that $y = 0$ is uniformly asymptotically stable. We know that the unique solution of (2.3) with initial point y_0 at x_0 can be represented by the formula

$$y(x) = Y(x)Y^{-1}(x_0)y_0 \qquad (x \geq a).$$

By virtue of Definition 4, there exists an $\eta > 0$ such that corresponding to every $\varepsilon > 0$ there is a $T = T(\varepsilon)$ such that for $||y_0|| < \eta$ we have

$$||Y(x)Y^{-1}(x_0)y_0|| < \varepsilon \qquad (x \geq x_0 + T;\ x_0 \geq a).$$

Since this relation is true for all vectors y_0 whose norm is less than η, we may deduce (via, for example, the definition of the norm of the linear operator associated with the matrix $Y(x)Y^{-1}(x_0)$) that

$$||Y(x)Y^{-1}(x_0)|| < \frac{\varepsilon}{\eta} \qquad (x \geq x_0 + T;\ x_0 \geq a). \tag{2.4}$$

We now fix $\varepsilon_0 < \eta$ and put $\theta = \varepsilon_0/\eta$. It follows from (2.4) that

$$||Y(x + T)Y(x)|| < \theta \qquad (x \geq a). \tag{2.5}$$

Since $y = 0$ is also uniformly stable, (iii) implies the existence of a constant $K > 0$ such that

$$||Y(x + h)Y^{-1}(x)|| \leq K \qquad (x \geq a;\ 0 \leq h \leq T). \tag{2.6}$$

If we fix $x_0 \geq a$, there corresponds to each $x \geq x_0$ an integer n such that $x_0 + nT \leq x < x_0 + (n+1)T$. Then we have

$$\begin{aligned} ||Y(x)Y^{-1}(x_0)|| &\leq ||Y(x)Y^{-1}(x_0+nT)||\ ||Y(x_0+nT)Y^{-1}(x_0+(n-1)T||\ldots \\ &\quad \ldots||Y(x_0+T)Y^{-1}(x_0)|| \\ &\leq K\theta^n \quad \text{(by virtue of (2.6) and (2.5))}. \end{aligned} \tag{2.7}$$

Set $\sigma = -(\ell g\ \theta)/T$. Since $\theta < 1$, $\sigma > 0$. (2.7) implies that

$$||Y(x)Y^{-1}(x_0)|| \leq Ke^{-n\sigma T} = \frac{K}{\theta}e^{-\sigma(n+1)T}$$
$$\leq \frac{K}{\theta}e^{-\sigma(x-x_0)}.$$

This completes the proof of the theorem.

In the case of scalar equations, we have the following

Corollary. Let $h\colon [a,\infty[\to \mathbb{R}$ be continuous. The following holds for the equation

$$y' = h(x)y.$$

(i) The solution $y = 0$ is stable if and only if

$$\int_{x_0}^{x} h(t)dt \leq M(x_0) \qquad (x \geq x_0),$$

where $M(x_0)$ is finite for every $x_0 \geq a$.

(ii) $y = 0$ is uniformly stable if and only if

$$\int_{x_0}^{x} h(t)dt \leq M < +\infty \qquad (x \geq x_0 \geq a):$$

(iii) $y = 0$ is asymptotically stable if and only if

$$\lim_{x\uparrow\infty} \int_{x_0}^{x} h(t)dt = -\infty;$$

(iv) $y = 0$ is uniformly asymptotically stable if and only if there exists $\alpha > 0$ such that

$$\int_{x_0}^{x} h(t)dt \leq -\alpha(x - x_0) \qquad (x \geq x_0 \geq a).$$

We leave it to the reader to supply the proof of the corollary, which by the way is similar to that of the results of Sec. 1. Since no two of the conditions (i),...,(iv) of the corollary (and of the theorem) are equivalent the four types of stability that we have introduced are all distinct. We have the following diagram to illustrate their mutual relationship:

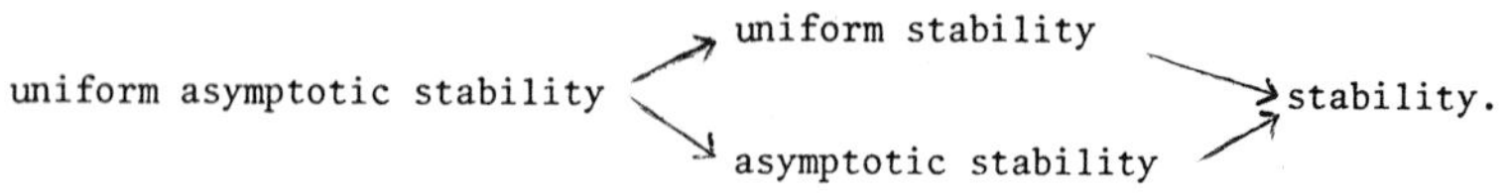

No arrow can be inverted; in each case the opposite implication is false. Furthermore, neither one of uniform stability and asymptotic stability implies the other. To see this, note that condition (ii) of the corollary is satisfied by $h(x) = \cos x$, while (iii) is not. Thus, uniform stability does not imply asymptotic stability. If we set

$$h(x) = \sin \ell g\, x + \cos \ell g\, x - \alpha$$

with $1 < \alpha < \sqrt{2}$, we have

$$\int_1^x h(t)dt = x \sin \ell g\, x - \alpha(x-1)$$

and (iii) is satisfied. Nevertheless, (ii) is not satisfied, for there exists an interval $[\theta_1,\theta_2]$ such that $\theta_1 < \frac{1}{4}\pi < \theta_2$ and $\sin x + \cos x \geq \beta > \alpha$ for $\theta_1 \leq x \leq \theta_2$ and suitable β. Thus, for $a_n = e^{2n\pi+\theta_1}$, $b_n = e^{2n\pi+\theta_2}$ we have

$$\int_{a_n}^{b_n} h(t)dt \geq (\beta-\alpha)(e^{\theta_2} - e^{\theta_1})e^{2n\pi},$$

which tends to ∞ as $n \to \infty$. This shows that asymptotic stability does not imply uniform stability.

Exercise 1. Prove that for a constant solution of an autonomous equation $y' = f(y)$, stability is equivalent to uniform stability and asymptotic stability is equivalent to uniform asymptotic stability.

Exercise 2. Let A be a *constant* matrix of order n. Prove that $y = 0$ is uniformly asymptotically stable for $y' = Ay$ if and only if the real parts of all the eigenvalues of A are less than 0.

2.2. Liapunov's Method

Liapunov studied the stability of the null solution for the system

$$y' = f(x,y) \tag{2.1}$$

by associating a C^1-function $V\colon [a,\infty[\times U \to \mathbb{R}$ with it in such a way that there exist two functions $\alpha,\beta\colon \mathbb{R}^+ \to \mathbb{R}^+$ that are continuous and strictly increasing, take the value 0 at the origin, and satisfy

$$\alpha(||y||) \leq V(x,y) \leq \beta(||y||)$$

$$\frac{\partial}{\partial x} V(x,y) + \left(\frac{\partial}{\partial y} V(x,y) \,\middle|\, f(x,y)\right) < 0$$

where $(\cdot|\cdot)$ represents, as usual, the scalar product in R^n. Functions V of this type are called *Liapunov functions*.

It is traditional to denote $\frac{\partial}{\partial x} V(x,y) + \left(\frac{\partial}{\partial y} V(x,y) \middle| f(x,y)\right)$ by $\dot{V}(x,y)$, but we shall not do so since we shall be dealing with a situation more general than that which Liapunov considered and moreover the notation $\dot{V}$ does not make clear that the function depends on the second number f of Eq. (2.1).

The following result allows us to study the stability of a given system by examining the stability of a scalar equation.

Comparison Theorem for Stability. Let $U \subseteq R^n$ be a neighborhood of the origin and let $f\colon [a,\infty[\times U \to R^n$ and $\omega\colon [a,\infty[\times]-\rho,\rho[\to R$ be continuous functions such that $f(x,0) = \omega(x,0) \equiv 0$. Let $V\colon [a,\infty[\times U \to R^+$ be a C^1-function with $V(x,0) = 0$ satisfying one of the two conditions

(A) $\inf_{y\in U, ||y||=\varepsilon, x\geq a} V(x,y) > 0$ for every $\varepsilon > 0$;

(B) $\inf_{y\in U, ||y||\geq\varepsilon, x\geq a} V(x,y) > 0$ for every $\varepsilon > 0$.

If the differential inequality

$$\frac{\partial}{\partial x} V(x,y) + \left(\frac{\partial}{\partial y} V(x,y) \middle| f(x,y)\right) \leq \omega(x,V(x,y))$$

holds, then we have the following results:

(i) (A) and the stability of $u = 0$ for the equation $u' = \omega(x,u)$ imply the stability of $y = 0$ for the equation $y' = f(x,y)$;

(ii) (A), $\lim_{y\to 0} V(x,y) = 0$ uniformly in x and the uniform stability of $u = 0$ for $u' = \omega(x,u)$ imply the uniform stability of $y = 0$ for $y' = f(x,y)$;

(iii) (B) and the asymptotic stability of $u = 0$ for $u' = \omega(x,u)$ imply the asymptotic stability of $y = 0$ for $y' = f(x,y)$;

(iv) (B), $\lim_{y\to 0} V(x,y) = 0$ uniformly in x, and the uniform asymptotic stability of $u = 0$ for $u' = \omega(x,u)$ imply the uniform asymptotic stability of $y = 0$ for $y' = f(x,y)$.

Before proving the theorem, we examine a few examples to illustrate how the auxiliary functions V are found and used.

We consider first of all the equation of the pendulum

$$y'' + h(x)y' + \sin y = 0$$

where h is a real continuous function with nonnegative values. Let us consider the first order system associated with the given equation:

$$y' = z$$

$$z' = -\sin y - h(x)z,$$

i.e., $u' = F(x,u)$ with $u = (y,z)$ and $F(x,u) = (z,-\sin y - h(x)z)$. The function $V: \mathbb{R} \times \mathbb{R}^2 \to \mathbb{R}$ defined by

$$V(x,(y,z)) = \frac{1}{2} z^2 - \cos y + 1$$

is a C^1-function that is positive in a neighborhood of the origin. Moreover, we have

$$\frac{\partial}{\partial x} V(x,u) + \left(\frac{\partial}{\partial u} V(x,u) | F(x,u)\right) = -h(x)z^2 \leq 0,$$

so that the theorem is applicable with $\omega \equiv 0$.

One of the major difficulties in the application of this method is to find an appropriate auxiliary function V. In many cases it is convenient to choose

$$V(x,y) = \alpha||y||^2$$

where $\alpha > 0$ is a suitable constant. When $\alpha = 1$, the inequality $\frac{\partial}{\partial x} V + \left(\frac{\partial}{\partial y} V|f\right) \leq \omega(x,V)$ simplifies to

$$(f(x,y)|y) \leq 2\omega(x,||y||^2)$$

so that the comparison theorem for stability assumes a form similar to the global existence theorem of Chapter III. For an example, we consider the Van der Pol equation

$$y'' + \varepsilon(1 - y^2)y' + y = 0$$

with $\varepsilon > 0$. An equivalent first order system is

$$y' = z + \varepsilon(\tfrac{1}{3} y^3 - y)$$

$$z' = -y.$$

For $u = (y,z)$, we have $u' = F(u)$ with $F(u) = \left(z + \varepsilon(\frac{1}{3} y^3 - y),-y\right)$. We define $V(x,u) = \frac{1}{2} ||u||^2$. Then

$$\frac{\partial}{\partial x} V(x,u) + \left(\frac{\partial}{\partial u} V(x,u)|F(u)\right) = \frac{-1}{3} \varepsilon y^2(3 - y^2)$$

which is ≤ 0 for $||u||^2 \leq 3$. Therefore, upon applying the theorem with $\omega \equiv 0$, we see that the null solution for the system under consideration is stable and so the null solution for the given equation is stable, too.

Proof of the Comparison Theorem for Stability: (i) Fix $\varepsilon > 0$ and $x_0 \geq a$. Let $\varepsilon_0 = \inf_{y \in U,\ ||y||=\varepsilon,\ x \geq a} V(x,y)$. By virtue of the stability of $v = 0$ for

$$u' = \omega(x,u), \tag{2.8}$$

there is $\delta_0 > 0$ such that $|u(x_0)| < \delta_0$ implies

$$|u(x)| < \varepsilon_0 \qquad (x \geq x_0)$$

for every solution u of (2.8). Since V is continuous, there is $\delta > 0$ such that

$$V(x_0,y) < \delta_0 \qquad (||y|| < \delta).$$

Let us consider a solution y of (2.1) with $||y(x_0)|| < \delta$. To prove that $||y(x)|| < \varepsilon$ for $x \geq x_0$, we suppose that there exists $x_1 \geq x_0$ such that $||y(x_1)|| = \varepsilon$ and deduce a contradiction. Put $u_0 = V(x_0, y(x_0))$ and let u be the maximal solution of (2.8) satisfying the initial condition $u(x_0) = u_0$. The theorem on differential inequalities of Sec. 2.4 of Chapter III implies that

$$V(x,y(x)) \leq u(x) \qquad (x \geq x_0)$$

by virtue of the fact that

$$\frac{\partial}{\partial x} V(x,y(x)) = \frac{\partial}{\partial x} V(x,y(x)) + \left(\frac{\partial}{\partial y} V(x,y(x)) \middle| f(x,y(x))\right) \leq \omega(x,V(x,y(x)).$$

But $u_0 < \delta_0$, and we therefore have $u(x) < \varepsilon_0$ for $x \geq x_0$. We thus arrive at

$$\varepsilon_0 \leq V(x_1,y(x_1)) \leq u(x_1) < \varepsilon_0,$$

which is a contradiction. This completes the proof of (i); the proof of (ii) is similar.

(iii) By virtue of (i), it remains to show that $\lim_{x\to\infty} y(x) = 0$ if $y(x_0)$ is sufficiently small. Fix $x_0 \geq a$. Since $v = 0$ is an asymptotically stable solution of (2.8), $\eta_0 = \eta_0(x_0) > 0$ exists such that for every solution u of (2.8), $|u(x_0)| < \eta_0$ implies that $\lim_{x\to\infty} u(x) = 0$. Let $\eta > 0$ be such that

$$V(x_0,y) < \eta_0 \qquad (||y|| < \eta).$$

Let y be a solution of (2.1) such that $||y(x_0)|| < \eta$. To prove that $\lim_{x\to\infty} y(x) = 0$, we suppose that x_n exists tending to infinity such that $\lim_n y(x_n) = y_0 \neq 0$ and arrive at a contradiction. Let $0 < \varepsilon' < ||y_0||$, $\varepsilon_0 = \inf_{z\in U,\ ||z||\geq\varepsilon,\ x\geq a} V(x,z)$ and let u be the maximal solution of

$$u' = \omega(x,u), \qquad u(x_0) = V(x_0,y(x_0)).$$

Since $V(x_0,y(x_0)) < \eta_0$, we have $\lim_{x\to\infty} u(x) = 0$. If we use the theorem on differential inequalities as in (i), we get

$$V(x,y(x)) \leq u(x) \quad (x \geq x_0).$$

There is n_0 such that $y(x_n) > \varepsilon'$ for $n \geq n_0$. Then we have the contradiction

$$0 < \varepsilon' \leq \lim_n V(x_n,y(x_n)) \leq \lim_n u(x_n) = 0,$$

and (iii) is proved.

(iv) By virtue of (ii), it is sufficient to show that the following is true:

(*) There is $\eta > 0$ with the property that for every $\varepsilon > 0$ there is a $T = T(\varepsilon) > 0$ such that for every $x_0 \geq a$ and every solution y of (2.1), $||y(x_0)|| < \eta$ implies that $||y(x)|| < \varepsilon$ for all $x \geq x_0+T$.

Since this same condition (*) holds for Eq. (2.8), there exists $\eta_0 > 0$ such that for each $\varepsilon > 0$, there is a $T_0 = T_0(\varepsilon) > 0$ with the property that for every $x_0 \geq a$ and every solution u of (2.8), $|u(x_0)| < \eta_0$ implies that $|u(x)| < \varepsilon$ for $x \geq x_0 + T_0$. Let $\eta > 0$ be such that

$$V(x,y) < \eta_0 \qquad (||y|| < \eta;\ x \geq a).$$

We shall show that (*) is satisfied for this η. If this were not the case, then there would be an $\varepsilon_0 > 0$ such that for each $T > 0$ there would be $x_T \geq a$, $x'_T \geq x_T + T$ and a solution y_T of (2.1) such that $||y_T(x_T)|| < \eta$ and $||y_T(x'_T)|| \geq \varepsilon$. Now choose ε' so that the relation

$$0 < \varepsilon' \leq \inf_{z\in U,\ ||z||\geq\varepsilon_0,\ x\geq a} V(x,z)$$

holds and that the solutions of (2.8) with initial point of absolute value less than ε' exist in the large and are equibounded; this is possible because of the stability of the null solution of (2.8). For every $T \geq 0$, let u_T be the maximal solution of (2.8) such that $u_T(x_T) = V(x_T, y_T(x_T))$. If we use the theorem on differential inequalities of Chapter III as we did in (i), we get

$$V(x,y(x)) \leq u_T(x) \qquad (x \geq x_T).$$

In particular, we have

$$\varepsilon' \leq V(x'_T, y_T(x'_T)) \leq u_T(x'_T) \tag{2.9}$$

for every $T \geq 0$. Since $u_T(x_T) < \eta_0$ there is a $T' > 0$ corresponding to ε' such that $u_T(x) < \varepsilon'$ for each $x \geq x_T + T'$. If we take $T = T'$ in (2.9), then we get the contradiction

$$\varepsilon' \leq V(x'_{T'}, y_{T'}(x'_{T'})) \leq u_{T'}(x'_{T'}) < \varepsilon',$$

and our result is established. This completes our proof.

As an application of the preceding theorem, we now prove a result about the stability of linear systems under suitable small perturbations.

Theorem on Stability in the First Approximation. Let A be a symmetric, $n \times n$ constant matrix, $U \subseteq R^n$ a neighborhood of the origin, and $f\colon [a,\infty[\times U \to R^n$ a continuous function such that $f(x,0) \equiv 0$. If the real parts of all the eigenvalues of A are negative and if

$$\lim_{y \to 0} \frac{(f(x,y)|y)}{||y||^2} = 0$$

uniformly for $x \geq a$, then $y = 0$ is uniformly asymptotically stable for $y' = Ay + f(x,y)$.

In particular, the condition

$$\lim_{y \to 0} \frac{(f(x,y)|y)}{||y||^2} = 0$$

is true if

$$\lim_{y \to 0} \frac{||f(x,y)||}{||y||} = 0$$

uniformly for $x \geq a$, as follows immediately from the Cauchy-Schwarz inequality.

The theorem we have just stated permits us to study the stability of the nonlinear autonomous systems

$$y' = F(y)$$

when $F(0) = 0$ and F is of class C^1 in a neighborhood U of the origin of R^n, for the given equation is equivalent to

$$y' = \frac{\partial}{\partial y} F(0)y + \left[F(y) - \frac{\partial}{\partial y} F(0)y\right]$$

where $\partial F(0)/\partial y$ is the Jacobian matrix of F at the origin. Since the function $f(y) = F(y) - \frac{\partial F(0)}{\partial y} y$ satisfies

$$\lim_{y \to 0} \frac{||f(y)||}{||y||} = 0,$$

the Theorem on Stability in the First Approximation can be applied when the real parts of the eigenvalues of $\partial F(0)/\partial y$ are all negative and the matrix $\partial F(0)/\partial y$ is symmetric. The nonlinear equation $y' = F(y)$ and the linear equation $y' = \frac{\partial F(0)y}{\partial y}$ (which is its variational equation) thus behave in exactly the same way with respect to stability.

Proof of the Theorem on Stability in the First Approximation: The fact that A is symmetric and that the real parts of all its eigenvalues are negative implies that there is a constant $\alpha < 0$ such that the function

$$H(y) = (y|Ay)$$

satisfies the inequality

$$H(y) \leq \alpha||y||^2.$$

Let us consider the following auxiliary function

$$V(x,y) = \frac{1}{2} ||y||^2.$$

We have

$$\frac{\partial}{\partial x} V(x,y) + \left(\frac{\partial}{\partial y} V(x,y)|Ay + f(x,y)\right) = (y|Ay + f(x,y))$$

$$= (y|Ay) + (y|f(x,y)) \leq \alpha||y||^2 + (y|f(x,y)). \qquad (2.10)$$

Let $\beta > 0$ be such that $\alpha + \beta < 0$. By virtue of the hypothesis on f, there exists $\delta < 0$ such that

$$(y|f(x,y)) < \beta||y||^2 \qquad (||y|| < \delta).$$

This means that (2.10) implies

$$\frac{\partial}{\partial x} V(x,y) + \left(\frac{\partial}{\partial y} V(x,y)|Ay + f(x,y)\right) < (\alpha+\beta)||y||^2 \qquad (||y|| < \delta;\ x \geq a).$$

We are now in a position to apply (iv) of the Comparison Theorem for Stability to the restriction of V and $Ay + f$ to the set $[a,\infty[\times B$, where B is the ball of radius δ with center at the origin; we take $\omega(x,u) = (\alpha+\beta)u$. The theorem is then proved.

Exercise 1. Let $U \subseteq \mathbb{R}^n$ be a neighborhood of the origin, $f\colon [a,\infty[\times U \to \mathbb{R}^n$ a continuous function, and $V\colon [a,\infty[\times U \to \mathbb{R}^+$ a C^1-function. V is called *positive definite* if $V(x,0) \equiv 0$ and if a continuous and strictly increasing function $a\colon \mathbb{R}^+ \to \mathbb{R}^+$ exists such that $a(0) = 0$ and $a(||y||) \leq V(x,y)$ for $x \geq a$, $y \in U$. V is called *negative definite* if $-V$ is positive definite. V is *decreasing* if there is a continuous strictly increasing function $b\colon \mathbb{R}^+ \to \mathbb{R}^+$ such that $b(0) = 0$ and $V(x,y) \leq b(||y||)$ for $x \geq a$, $y \in U$. Prove the following results.

a. (Liapunov's first theorem on stability) If V is positive definite and

$$\frac{\partial}{\partial x} V + \left(\frac{\partial}{\partial y} V \middle| f\right) \leq 0,$$

then $y = 0$ is stable for (2.1).

b. (Persidski's theorem on uniform stability) If V is positive definite and decreasing and

$$\frac{\partial}{\partial x} V + \left(\frac{\partial}{\partial y} V \middle| f\right) \leq 0,$$

then $y = 0$ is uniformly stable for (2.1).

c. (Liapunov's second theorem on stability) If V is positive definite and decreasing and

$$\frac{\partial}{\partial x} V + \left(\frac{\partial}{\partial y} V \middle| f\right)$$

is negative definite, then $y = 0$ is uniformly asymptotically stable for (2.1).

Exercise 2. Determine conditions for h and g that insure that the comparison theorem on stability can be applied using the function $\omega(x,u) = h(x)g(u)$.

Exercise 3. Using the formula for the variation of constants, prove that if A is an $n \times n$ constant matrix where eigenvalues have negative real parts, if $B(x)$ is an $n \times n$ continuous matrix such that $\lim_{x\to\infty} ||B(x)|| = 0$, and if $\lim_{y\to 0} \frac{||f(x,y)||}{||y||} = 0$ uniformly for $x \geq a$, then

$y = 0$ is asymptotically stable for $y' = (A + B(x))y + f(x,y)$.

Exercise 4. Study the stability of the system $y'' = f(y,y')$, where f is continuous and such that $f(0,0) = 0$, by considering the associated first order system and using the Liapunov function

$$V(x,z) = z_1^2 - \int_0^{z_1} f(u,0)du$$

with $z = (z_1,z_2)$.

Exercise 5. Consider the system

$$y_1' = 2y_1^3y_2^2 - y_1$$
$$y_2' = -y_2.$$

Prove that the origin is asymptotically stable by using the Liapunov function $V(y) = ||y||^2$.

Exercise 6. Consider the equation of the simple pendulum

$$\theta'' + \sin\theta = 0$$
$$\theta(0) = \theta_0$$
$$\theta'(0) = \omega_0$$

and determine the values of θ_0 and ω_0 for which the solution is stable. Hint: Consider the relation

$$\theta'^2 = \omega_0^2 + \cos\theta(t) - \cos\theta_0$$

and get stability for $|\cos\theta_0 - \omega_0^2| < 1$.

Exercise 7. In the preceding exercise, prove that the stable solutions are periodic and give the values of the periods.

Exercise 8. Study the behavior of the autonomous system

$$\begin{cases} 2y' = y(y^2 + z^2 - 3) + \dfrac{2(y+z)}{y^2+z^2} \\ 2z' = z(y^2 + z^2 - 3) + \dfrac{2(y-z)}{y^2+z^2} \end{cases}$$

with particular attention to stability. Hint: Multiply the first equation by z and the second by y. Put $y^2 + z^2 = \rho^2$ and get

$$(\rho^2)' = \rho^2(\rho^2 - 3) + 2$$

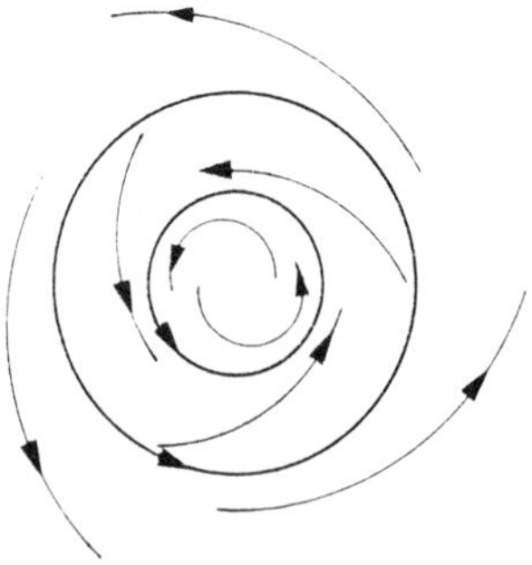

Figure 5.1

whence $\rho^2 = (Ke^t-2)/(Ke^t-1)$ with $K = (2-\rho_0^2)/(1-\rho_0^2)$ if $\rho_0^2 \neq 1$, $\rho_0^2 \neq 2$; with $\rho^2 = 1$ if $\rho_0^2 = 1$; with $\rho^2 = 2$ if $\rho_0^2 = 2$. Correspondingly we have

$$\theta' = \frac{y'z - z'y}{y^2 + z^2} = \frac{1}{\rho^2} .$$

The solutions are thus stable for $\rho_0 \leq 2$. See the above figure for the behavior of the solution.

2.3. The Fixed Point Method: Asymptotic Equivalence

Two systems of ordinary differential equations are *asymptotically equivalent* if there is a correspondence between the solutions of the two systems such that the difference between two corresponding solutions tends to zero as x tends to infinity.

In this section, we shall study asymptotic equivalence between a linear system

$$z' = A(x)z \tag{2.11}$$

and the nonlinear system

$$y' = A(x)y + f(x,y) \tag{2.12}$$

obtained by perturbing the linear system. We shall also see how asymptotic equivalence can lead to stability.

Theorem on Asymptotic Equivalence for Bounded Solutions. Let $A(x)$ be an $n \times n$ matrix continuous for $x \geq a$, $Z(x)$ a fundamental matrix of the system (2.11) and $f\colon [a,\infty[\times \mathbb{R}^n \to \mathbb{R}^n$ a continuous function such that

$$\int_a^{+\infty} ||f(t,0)||dt < +\infty \quad \text{and} \quad ||f(x,y) - f(x,\bar{y})|| \leq h(x)||y - \bar{y}||$$

with $\int_a^\infty h(t)dt < \infty$. If there are two supplementary projections P_1 and P_2 of $\mathbb{R}^n$ and a constant $K > 0$ such that

$$||Z(x)P_1Z^{-1}(t)|| \leq K \quad \text{for} \quad a \leq t \leq x,$$

$$||Z(x)P_2Z^{-1}(t)|| \leq K \quad \text{for} \quad a \leq x \leq t,$$

$$\lim_{x\uparrow+\infty} Z(x)P_1 = 0.$$

then there is a 1-1 correspondence S between the set of bounded solutions of (2.11) and the set of bounded solutions of (2.12); this correspondence is a homeomorphism in the topology of uniform convergence on $[a,\infty[$ and satisfies the relations

$$\lim_{x\uparrow\infty} ||z(x) - Sz(x)|| = 0, \quad z(c) - Sz(c) = \int_c^{+\infty} Z(c)P_2Z^{-1}(t)f(t,Sz(t))dt$$

for every bounded solution z of (2.11), with $c \geq a$ sufficiently large.

The proof of Corollary 1 provides an example for the production of the projections P_1 and P_2 while Corollary 2 gives an application of the theorem to stability. Observe that the initial points of the two corresponding solutions z and Sz are the same when $P_2 \equiv 0$.

Proof of the Theorem on Asymptotic Equivalence for Bounded Solutions: Let $c \geq a$ be such that

$$2K \int_c^{+\infty} h(t)dt < 1.$$

By virtue of the uniqueness and continuous dependence of solutions on compact intervals in the Lipschitz case, it is enough to prove the theorem for solutions on the interval $[c,\infty[$. We therefore consider the Banach space X of bounded and continuous functions from $[c,\infty[$ into $\mathbb{R}^n$ with the sup norm $||\cdot||_\infty$. Let z be a bounded solution of (2.11). Corresponding to z we define a transformation T_z on X by

$$T_zy(x) = z(x) + \int_c^x Z(x)P_1Z^{-1}(t)f(t,y(t))dt$$
$$- \int_x^{+\infty} Z(x)P_2Z^{-1}(t)f(t,y(t))dt.$$

The hypotheses of the theorem insure that the last integral converges and that T_z maps X into itself, $T_z(X) \subseteq X$. For every pair $y_1, y_2 \in X$, it follows from the definition of T_z that

$$\begin{aligned}||T_zy_1(x) - T_zy_2(x)|| &\leq \int_c^x ||Z(x)P_1Z^{-1}(t)||\ ||f(t,y_1(t))-f(t,y_2(t))||dt \\ &\quad + \int_x^{+\infty} ||Z(x)P_2Z^{-1}(t)||\ ||f(t,y_1(t))-f(t,y_2(t))||dt \\ &\leq \int_c^x Kh(t)||y_1(t)-y_2(t)||dt + \int_x^{+\infty} Kh(t)||y_1(t)-y_2(t)||dt \\ &\leq \int_c^x Kh(t)||y_1-y_2||_\infty dt + \int_x^{+\infty} Kh(t)||y_1-y_2||_\infty dt \\ &\leq 2K||y_1-y_2||_\infty \int_c^{+\infty} h(t)dt.\end{aligned}$$

This proves that T_z is a contraction on the space X. T_z has a unique fixed point y_z. We now define S by putting $Sz = y_z$. Let y be a bounded solution of (2.12). We define z by

$$\begin{aligned}z(x) = y(x) &- \int_c^x Z(x)P_1Z^{-1}(t)f(t,y(t))dt \\ &+ \int_x^{+\infty} Z(x)P_2Z^{-1}(t)f(t,y(t))dt.\end{aligned} \tag{2.13}$$

Upon differentiating, we see at once that z is a solution of (2.11). Moreover, we have

$$\begin{aligned}||z(x)|| &\leq ||y(x)|| + \int_c^x ||Z(x)P_1Z^{-1}(t)||\ ||f(t,y(t)||dt \\ &\quad + \int_x^{+\infty} ||Z(x)P_2Z^{-1}(t)f(t,y(t))||dt \\ &\leq ||y(x)|| + \int_c^x K(||f(t,y(t)) - f(t,0)|| + ||f(t,0)||)dt \\ &\quad + \int_x^{+\infty} K(||f(t,y(t)) - f(t,0)|| + ||f(t,0)||)dt \\ &\leq ||y(x)|| + \int_c^x Kh(t)||y(t)||dt + \int_c^x K||f(t,0)||dt \\ &\quad + \int_x^{+\infty} Kh(t)||y(t)||dt + \int_x^{+\infty} Kh(t)||f(t,0)||dt.\end{aligned}$$

If we now take $\sup_x$ we see that z is bounded. From (2.13) and the uniqueness of the fixed point of T_z, it follows that $Sz = y$. We have thus proved that S is a 1-1 correspondence between the set of bounded solutions of (2.11) and the set of bounded solutions of (2.12) and, moreover, that the formula

$$z(c) - Sz(c) = \int_c^{+\infty} Z(c)P_2Z^{-1}(t)f(t,Sz(t))dt$$

is valid. From the fact that T_z is a contraction with constant independent of z, we easily see that S is a homeomorphism. To complete the proof, it only remains to show that $\lim_{x\to\infty} ||z(x) - Sz(x)|| = 0$. From (2.13), for $x \geq x_1 \geq c$, we have

$$\begin{aligned}||z(x) - Sz(x)|| &\leq \int_c^x ||Z(x)P_1Z^{-1}(t)f(t,Sz(t))||dt \\ &\quad + \int_x^{+\infty} ||Z(x)P_2Z^{-1}(t)f(t,Sz(t)||dt \\ &= \int_c^{x_1} ||Z(x)P_1Z^{-1}(t)f(t,Sz(t))||dt \\ &\quad + \int_{x_1}^{x} ||Z(x)P_1Z^{-1}(t)f(t,Sz(t))||dt \\ &\quad + \int_x^{+\infty} ||Z(x)P_2Z^{-1}(t)f(t,Sz(t))||dt \\ &\leq ||Z(x)P_1|| \int_c^{x_1} ||Z^{-1}(t)f(t,Sz(t)||dt \\ &\quad + 2\int_{x_1}^{+\infty} K||f(t,Sz(t))||dt.\end{aligned}$$

We fix $\varepsilon > 0$ and determine $x_1 \geq c$ such that the second integral in the right-hand side of this inequality is less than $\varepsilon/4$. With x_1 thus fixed, the first term on the right-hand side of the inequality above tends to zero as x tends to infinity by virtue of the hypothesis $\lim_{x\to\infty} Z(x)P_1 = 0$. We may therefore conclude that $\lim_{x\to\infty} ||z(x) - Sz(x)|| = 0$, and the theorem is completely proved.

Corollary 1. Let A be a constant $n \times n$ matrix whose eigenvalues have nonpositive real part; suppose that the eigenvalues with real part zero are simple. Let $f\colon [a,\infty[\times \mathbb{R}^n \to \mathbb{R}^n$ be continuous and such that

$$\int_a^{+\infty} ||f(t,0)||dt < +\infty \quad \text{and} \quad ||f(x,y) - f(x,\bar{y})|| \leq h(x)||y - \bar{y}||$$

with $\int_a^\infty h(t)dt < \infty$. Then the systems $z' = Az$ and $y' = Ay + f(x,y)$ are asymptotically equivalent by means of a homeomorphism between the sets of their respective solutions where the topology is that of uniform convergence on $[a,\infty[$.

Proof: The matrix A is similar to a matrix of the form

$$\begin{pmatrix} A_1 & 0 \\ 0 & A_2 \end{pmatrix}$$

where A_1 and A_2 are square matrices such that the real parts of the eigenvalues of A_1 are negative while the real parts of those of A_2 are zero. Corresponding to a matrix of this form, there is a fundamental matrix $Z(x)$ of (2.11) with $Z(a)$ equal to the identity and

$$Z(x) = \begin{pmatrix} Z_1(x) & 0 \\ 0 & Z_2(x) \end{pmatrix}$$

where $Z_1(x)$ consists of decreasing exponential functions and $Z_2(x)$ contains constants and complex exponentials. To this decomposition there correspond two supplementary projections P_1 and P_2 of $\mathbb{R}^n$ such that

$$Z(x)P_1 = \begin{pmatrix} Z_1(x) & 0 \\ 0 & 0 \end{pmatrix} \quad \text{and} \quad Z(x)P_2 = \begin{pmatrix} 0 & 0 \\ 0 & Z_2(x) \end{pmatrix}.$$

It then follows that a constant $K > 0$ exists such that

$$||Z(x)P_1Z^{-1}(t)|| \le K \quad \text{for} \quad 0 \le t \le x,$$

$$||Z(x)P_2Z^{-1}(t)|| \le K \quad \text{for} \quad 0 \le x \le t.$$

Moreover,

$$\lim_{x\uparrow+\infty} Z(x)P_1 = 0.$$

We are now in a position to apply the theorem on asymptotic equivalence for bounded solutions. Since all the solutions of the two systems are bounded (see Exercise 1), it follows that the two given systems are asymptotically equivalent, and the corollary is proved.

Corollary 2. Let A be a constant $n \times n$ matrix whose eigenvalues are all negative, and let $f: [a,\infty[\times \mathbb{R}^n \to \mathbb{R}^n$ be a continuous function such that $f(t,0) \equiv 0$ and

$$||f(x,y) - f(x,\bar{y})|| \leq h(x)||y - \bar{y}||$$

with $\int_a^\infty h(t)dt < \infty$. Then $y = 0$ is stable for $y' = Ay + f(x,y)$

Proof: Since the solution of our equations depend continuously over compact intervals on the data, it suffices to prove the theorem on an interval $[c,\infty[$ with $c \geq a$. As we see from the proof of Corollary 1, the conditions of the theorem on asymptotic equivalence for bounded solutions are satisfied with $P_2 = 0$. Let K be the constant determined in the proof of Corollary 1, and let $c \geq a$ be such that

$$2K \int_c^{+\infty} h(t)dt < 1.$$

Let $x_0 \geq c$. If we apply Corollary 1 on the interval $[x_0,\infty[$, we obtain a homeomorphism S from the set of solutions of (2.11) onto the set of solutions of (2.12) with the topology is that of uniform convergence on $[x_0,\infty[$. Since $P_2 = 0$, we have

$$z(x_0) = Sz(x_0)$$

by virtue of the relation between initial values of two corresponding solutions stated by the theorem on asymptotic equivalence for bounded solutions. Finally, $z = 0$ is stable for (2.11) under the present hypotheses. These three facts clearly imply the statement of the corollary, which is thus proven.

Exercise 1. Prove that under the hypotheses of Corollary 1, all the solutions of the two systems (2.11) and (2.12) are bounded.

Exercise 2. Study the asymptotic equivalence between $y' = F(x,y)$ and the variational equation $z' = \frac{\partial}{\partial y} F(x,0)z$, where F is of class C^1 and $\frac{\partial}{\partial y} F(x,0)$ is the Jacobian matrix.

Exercise 3. Those who know Schauder's fixed point theorem should extend both the theorem on asymptotic equivalence for bounded solutions and Corollary 1 by assuming that

$$||f(x,y)|| \leq \omega(x,||y||)$$

rather than that f satisfies a Lipschitz condition, where $\omega: [a,\infty[\times R^+ \to R^+$ is continuous, $\omega(x,\cdot)$ is increasing, and for every $c \geq 0$, $\int_a^\infty \omega(x,c)dx < \infty$. One can now no longer claim that the transformation S is a homeomorphism, but only that it is surjective.

2.4. Olech's Method

We illustrate this method by studying the following question. Consider the autonomous system

$$y' = f(y) \tag{2.14}$$

in $\mathbb{R}^2$, that is, $y = (y_1, y_2)$ and $f = (f_1, f_2)$. We suppose that $f(0) = 0$, that f is of class C^1, and that the Jacobian matrix $J_f(y)$ of f is always symmetric and that the real parts of its eigenvalues are negative. We then know from the Theorem on Stability on the First Approximation that $y = 0$ is asymptotically stable; thus, all the solutions of (2.14) that pass through a sufficiently small neighborhood of the origin tend to 0 as x tends to infinity. We propose to see if a stronger property holds; is it true that *all* the solutions of (2.14) tend to 0 as $x \to \infty$, independently of the initial point? When this happens, we say that $y = 0$ is globally asymptotically stable. More precisely, $y = 0$ is *globally asymptotically stable* if it is asymptotically stable and, furthermore, every solution of the given system tends to zero as $x \to \infty$.

We have the following result.

Olech's Theorem. Let $f: \mathbb{R}^2 \to \mathbb{R}^2$ be a C^1-function such that $f(0) = 0$, and suppose that the Jacobian matrix $J_f(y)$ is symmetric and that the real parts of all its eigenvalues are negative, no matter what $y \in \mathbb{R}^2$ is. If there are two positive constants ρ and r such that

$$||f(y)|| \geq \rho \quad \text{for} \quad ||y|| \geq r,$$

then $y = 0$ is globally asymptotically stable for (2.14).

We have the following interesting consequence.

Corollary. If $f: \mathbb{R}^2 \to \mathbb{R}^2$ is a C^1-function such that $f(0) = 0$, if, for all $y \in \mathbb{R}^2$, the Jacobian matrix $J_f(y)$ is symmetric and the real parts of all its eigenvalues are negative, and if f is injective, then $y = 0$ is globally asymptotically stable.

Olech proved [24] that if f is of class C^1, if $f(0) = 0$, if the Jacobian matrix $J_f(y)$ is symmetric, and if the real parts of all its eigenvalues are negative, then f is injective when one of the following two conditions holds in all of $\mathbb{R}^2$:

$$\frac{\partial}{\partial y_1} f_1 \cdot \frac{\partial}{\partial y_2} f_2 \neq 0 \quad \text{or} \quad \frac{\partial}{\partial y_2} f_1 \cdot \frac{\partial}{\partial y_1} f_2 \neq 0.$$

These conditions are in particular satisfied in the hypotheses of certain theorems of Hartman and Markus-Yamabe.

We shall need two lemmas to prove Olech's theorem. First we introduce some notation.

We shall denote by $y(x,P)$ the solution of (2.14) with initial point P; $y(0,P) = P$. We shall denote by $I(P)$ the set of all points $y(x,P)$, where x varies in the greatest interval in which $y(x,P)$ is defined. We call $I(P)$ the *orbit* of (2.14) passing through P. If P_1 and P_2 belong to $I(P)$, then

$$P_1P_2$$

will denote the set $\{y(x,P) | x_1 \leq x \leq x_2\}$, where $y(x_i,P) = P_i$ for $i = 1,2$. The set P_1P_2 is called an *orbit segment* of (2.14). Along with system (2.14), let us consider the system

$$y' = g(y) \tag{2.14)*$$

where $g = (g_1,g_2)$, $g_1 = -f_2$, $g_2 = f_1$. This is the system of the orthogonal trajectories to the orbits of (2.14). The symbols $y^*(x,P)$ and $I^*(P)$ will have the same meaning for the system (2.14)* as $y(x,P)$ and $I(P)$ have for (2.14).

Let PQ be an orbit segment of (2.14)*. We denote by $L(P,Q)$ the function

$$L(P,Q) = \left| \int_{PQ} ||f(y^*(t,P))||^2 dt \right| = \int_0^{s_0} ||f(y^*(t(s),P))|| ds$$

where $ds = \left| ||f(y^*(t,P)|| dt \right|$ is the arc element and s_0 is the length of the segment PQ.

Lemma 1. Let G be a region in the plane whose boundary is the curve $P_1P_2Q_2Q_1$ where P_1P_2, Q_1Q_2, and P_1Q_1, P_2Q_2 are orbit segments of (2.14) and (2.14)* respectively. Suppose that for every $P \in P_1Q_1$ the solution $y(x,P)$ enters into G, that is, there exists $\varepsilon > 0$ such that

$$y(x,P) \in \overline{G} \quad \text{for} \quad P \in P_1Q_1 \quad \text{and} \quad 0 < x < \varepsilon,$$

and that for every $P \in P_2Q_2$, the solution $y(x,P)$ leaves $\overline{G}$, that is, there exists $\varepsilon > 0$ such that

$$y(x,P) \in \overline{G} \quad \text{for} \quad P \in P_2Q_2 \quad \text{and} \quad -\varepsilon < x < 0.$$

If f is as in the statement of Olech's theorem, then $L(P_2,Q_2) < L(P_1,Q_1)$.

Proof: We observe that the hypotheses we have imposed on the eigenvalues of the Jacobian matrix $J_f(y)$ imply that

$$\operatorname{tr} J_f(y) = \frac{\partial}{\partial y_1} f_1(y) + \frac{\partial}{\partial y_2} f_2(y) < 0 \quad \text{for} \quad y \in \mathbf{R}^2. \tag{2.15}$$

Since the given equation is autonomous, we may without loss of generality assume that

$$P_iQ_i = \{y \mid y = y^*(x,P_i), \quad 0 \leq x \leq x_i\} \quad \text{for} \quad i = 1,2.$$

By Green's formula, we have

$$\oint (f_1(y)dy_2 - f_2(y)dy_1) = \iint_G \operatorname{tr} J_f(y)dy_1dy_2 < 0. \tag{2.16}$$

Note that the orientation of the boundary in Green's formula is such that the vector $(dy_2,-dy_1)$ is normal to the boundary of G and is directed towards the exterior of G. The curvilinear integral in (2.16) is therefore positive along P_2Q_2 because of the hypothesis of the lemma that $y(x,P)$ leaves $\overline{G}$; it is negative along P_1Q_1 because of the hypothesis that $y(x,P)$ enters G, and it is null along P_1P_2 and Q_1Q_2 since they are segments of curves that are solutions of (2.14). It is thus easy to see that

$$\begin{aligned}\oint (f_1(y)dy_2 - f_2(y)dy_1) &= \int_{P_2Q_2} ||f(y^*(t,P_2))||^2 dt \\ &\quad - \int_{P_1Q_1} ||f(y^*(t,P_1))||^2 dt \\ &= L(P_2,Q_2) - L(P_1,Q_1).\end{aligned} \tag{2.17}$$

The inequality $L(P_2,Q_2) < L(P_1,Q_1)$ immediately follows from relations (2.16) and (2.17). The lemma is thus proved.

Before stating the next lemma, we introduce some more notation. Let $y(x,P)$ be a nonconstant solution of (2.14). If $\varepsilon > 0$, we denote by $N(P,\varepsilon)$ the set of all Q in $\mathbf{R}^2$ for which there exists $x_0 \geq 0$ such that $y(x_0,P) \in I^*(Q)$ and $L(y(x_0,P),Q) < \varepsilon$. We denote by $O(P,\varepsilon)$ the "neighborhood" with center $\{y(x,P): x \geq 0\}$ and radius ε, that is

$$O(P,\varepsilon) = \bigcup_{x \geq 0} \{Q \in \mathbf{R}^2 \mid ||Q - y(x,P)|| < \varepsilon\}.$$

Lemma 2. Let f be as in Olech's theorem, and let $y(x,P)$ be a nonconstant solution of (2.14). If there exist $\eta > 0$ and $d > 0$ such that

$$||f(y)|| \geq d \quad \text{for} \quad y \in O(P,\eta),$$

then for every $0 < \varepsilon < \eta d$ and every $Q \in N(P,\varepsilon)$, we have

$$y(x,Q) \in N(P,\varepsilon)$$

for each $x \geq 0$ in the domain of definition of $y(x,Q)$.

Proof: We begin by establishing the relation

$$N(P,\varepsilon) \subseteq O(P,\eta) \quad \text{if} \quad \varepsilon < \eta d.$$

We shall prove this relation by demonstrating a more general fact: if $Q \in N(P,\varepsilon)$ and $\varepsilon < \eta d$, then

$$||Q - y(x_0,P)|| < \eta \tag{2.18}$$

where $y(x_0,P)$ is the point of $I(P)$ that can be joined to Q by a curve that is a solution of (2.14)*; this is possible because of the definition of $N(P,\varepsilon)$. Suppose that (2.18) is false. Then the hypothesis of the lemma that $||f(y)|| \geq d$ for $y \in O(P,\eta)$ implies that $L(y(x_0,p),Q) \geq \eta d$, which contradicts the definition of $N(P,\varepsilon)$. (2.18) is thus valid.

We now take $Q \in N(P,\varepsilon)$ and let $\varepsilon < \eta d$. Let $\tau(x)$ be a continuous and increasing function for $x \geq 0$ such that $\tau(0) = 0$ and such that the points

$$Q_x = y(x,P) \quad \text{and} \quad P_x = y(x_0 + \tau(x),P)$$

belong to the same orbit of (2.14)*. Because of the continuous dependence of the solutions on the initial data, the function τ exists and is defined in a maximal interval $[0,\omega[$ with $\omega \leq \infty$. Let us prove that

$$y(x,Q) \in N(P,\varepsilon) \quad \text{for} \quad 0 \leq x < \omega. \tag{2.19}$$

We consider the function

$$\phi(x) = L(P_x,Q_x),$$

which is continuous in $[0,\omega[$. By virtue of Lemma 1, we can prove that ϕ is locally decreasing. To do so, note that for every $x_0 \in [0,\omega[$ there exists $\delta > 0$ such that, for any $X_0 < X < X_0 + \delta$, the orbit segments $P_{x_0} P_x$, $Q_{x_0} Q_x$ and $P_{x_0} Q_{x_0}$, $P_x Q_x$ of (2.14) and (2.14)* define a region G in which Lemma 1 can be applied. Thus $\phi(x) < \phi(x_0)$ for $x_0 < x < x_0 + \delta$ which, together with the continuity of ϕ, implies that ϕ is decreasing for $0 \leq x < \omega$. Since $\phi(0) < \varepsilon$, we therefore have that

$$\phi(x) \leq \phi(0) < \varepsilon \quad \text{for} \quad 0 \leq x < \omega. \tag{2.20}$$

This inequality and the definitions of ϕ and $N(P,\varepsilon)$ imply (2.19). In order to complete the proof of the lemma, it suffices at this point to show that ω is the least upper bound of the domain of $y(x,P)$. To do this, we begin by observing that if we use the same argument that allowed us to establish (2.18), then from (2.20), the definition of ϕ, and the hypothesis of the lemma that $||f(y)|| \geq d$ for $y \in O(P,\eta)$, it follows that

$$||Q_x - P_x|| \leq \frac{\phi(0)}{d} < \eta \quad \text{for} \quad 0 \leq x < \omega. \tag{2.21}$$

We now suppose that ω is less than the least upper bound b of the domain of $y(x,P)$ and arrive at a contradiction. We have

$$\lim_{x\to\omega} Q_x = Q_\omega$$

where $Q_x = y(x,Q)$ and $Q_\omega = y(\omega,Q)$. Hence (2.21) implies that $\delta > 0$ exists such that

$$||Q_\omega - P_x|| < \eta \quad \text{for} \quad \omega - \delta < x < \omega.$$

This inequality, together with the hypothesis $||f(y)|| \geq d$ for $x \in O(P,\eta)$, proves that $\lim_{x\to\omega} P_x$ exists and $\lim_{x\to\omega} (x_0 + \tau(x)) < b$. It is now easy to see that $y^*(x,P_\omega)$ must pass through Q_ω. This means that $[0,\omega[$ is not the maximal interval of definition of τ. Thus $\omega = b$, which completes the proof of the lemma.

Proof of Olech's Theorem: Let

$$\Omega = \{P \in R^2 \,|\, \lim_{x\uparrow+\infty} y(x,P) = 0\}.$$

We must show that $\Omega = R^2$. Suppose that $\Omega \neq R^2$. Since Ω contains the origin and is therefore not empty, its boundary $\partial\Omega$ is not empty. Because of the continuous dependence on the initial data and the fact that $y(x_0,P) \in \Omega$ for some $x_0 \geq 0$ implies that $\lim_{x\uparrow+\infty} y(x,P) = 0$, we have that $y(x,P) \in \partial\Omega$ for each $x \geq 0$ if $P \in \partial\Omega$. Since all the constant solutions are stable as a result of the theorem on stability in the first approximation, Ω cannot contain any singular point of (2.14). (By *singular* or *critical points* of (2.14), we mean points where f takes on the value 0; thus Q is a critical point of (2.14) if and only if $\{Q\}$ is an orbit of (2.14).) As a result, $y(x,P)$ cannot tend to any singular point of

(2.14) for $P \in \partial\Omega$. We now fix $P \in \partial\Omega$. What we have just established shows that $y(x,P)$ is not constant and cannot converge towards any critical point of (2.14). Therefore, for every critical point Q of (2.14) there is a ball $B(Q,\varepsilon_Q)$ with center Q and radius $\varepsilon_Q > 0$ such that $y(x,P)$ is outside of $B(Q,\varepsilon_Q)$. Since by hypothesis $||f(y)|| \geq \rho$ for $||y|| \geq r$, the set of critical points of (2.14) is compact and can be covered by finitely many balls $B(Q_1,\varepsilon_{Q_1}/2),\ldots,B(Q_n,\varepsilon_{Q_n}/2)$. If $\eta = \min_{i=1,\ldots,n} \varepsilon_{Q_i}/2$, there exists $d > 0$ such that

$$||f(y)|| \geq d \quad \text{for } y \in O(P,\eta). \tag{2.22}$$

To see this, suppose on the contrary that such a d does not exist. Then $y_n \in O(P,\eta)$ exists such that $\lim_n f(y_n) = 0$. Since by hypothesis, $||f(y)|| \geq \rho$ for $||y|| \geq r$, a subsequence $(y_{n_k})_{k=1}^{\infty}$ exists such that $||y_{n_k}|| \leq r$. Then a subsequence $(y_{n_{k_i}})_{i=1}^{\infty}$ exists which converges to an element y_0, and we have $f(y_0) = 0$. Since y_0 is a critical point of (2.14), we must have $||y_0 - y(x,Q)|| \geq 2\eta$ for each x in the domain of $y(x,P)$. But this contradicts $y_n \in O(P,\eta)$, and (2.22) holds. We are thus in a position to apply Lemma 2, which contradicts the fact that $P \in \partial\Omega$. $\partial\Omega$ is therefore empty. This completes the proof.

Proof of the Corollary: It is enough to show that there exist ρ, $r > 0$ such that

$$||f(y)|| \geq \rho \quad \text{for} \quad ||y|| > r,$$

since we can apply Olech's theorem. If no such ρ and r exist, then there would be a sequence $(y_n)_{n=1}^{\infty}$ of points of R^2 such that

$$\lim_n ||y_n|| = +\infty \quad \text{and} \quad \lim_n f(y_n) = 0.$$

By virtue of the implicit function theorem, there exist α and $\beta > 0$ such that the balls $B(0,\alpha)$ and $B(0,\beta)$ satisfy the relation $f(B(0,\alpha)) \supseteq B(0,\beta)$. Thus, for sufficiently large n, we have $y_n \notin B(0,\alpha)$ and $||f(y_n)|| < \beta$. This contradicts the fact that f is injective, since there must then be an element in $B(0,\alpha)$ at which f assumes the same value as it does at y_n. The corollary is therefore completely proved.

Remark. The theorem of stability in the first approximation holds without assuming that the matrix A is symmetric if we make the less general assumption on f:

$$\lim_{x \to 0} \frac{||f(x,y)||}{||y||} = 0.$$

As a consequence, the results of this section are valid without the hypothesis that the Jacobian matrix $J_f(y)$ is symmetric.

2.5. The Method of the Logarithmic Norm

The *logarithmic norm* of a matrix A is defined by the relation

$$\mu(A) = \lim_{h \downarrow 0} \frac{||I + hA|| - 1}{h}$$

where I is the identity matrix.

It is an immediate consequence of the definition that

$$\mu(\alpha A) = \alpha\mu(A) \quad \text{for } \alpha \geq 0$$

$$|\mu(A)| \leq ||A||$$

$$\mu(A + B) \leq \mu(A) + \mu(B).$$

These last inequalities in turn imply

$$|\mu(A) - \mu(B)| \leq ||A - B||.$$

The value $\mu(A)$ depends on the particular norm used for vectors and matrices. For example, if $||y||$ represents the Euclidean norm, then $\mu(A)$ is the greatest eigenvalue of $\frac{1}{2}(A + A^*)$, where A^* is the transpose of A, and the corresponding norm $||A||$ of the matrix A is the square root of the greatest eigenvalue of A^*A. If, on the other hand, $||y|| = \sum_{i=1}^{n} |y_i|$ and $||A|| = \sup_k \sum_{i=1}^{n} |a_{ik}|$, then

$$\mu(A) = \sup_k \left(\text{Re } a_{kk} + \sum_{i=1,\ldots,n,i \neq k} |a_{ik}|\right).$$

Finally, we observe that the real part of each eigenvalue of A is $\leq \mu(A)$.

We prove the following result in order to illustrate the use of the logarithmic derivative in the study of stability

Brauer's Theorem. Let U be a neighborhood of the origin in $\mathbb{R}^n$, and let $f: [a,\infty[\times U \to \mathbb{R}^n$ be a continuous function such that $f(x,0) \equiv 0$. Suppose furthermore that $\frac{\partial}{\partial y} f(x,0)$ exists and is continuous, and denote by $f_y(x,0)$ the value at the origin of the Jacobian matrix of $f(x,\cdot)$. If the relation

$$\limsup_{x\to+\infty} \frac{1}{x-a}\int_a^x \mu(f_y(t,0))dt < 0$$

holds, then $y \equiv 0$ is asymptotically stable for $y' = f(x,y)$.

Proof: Fix $\varepsilon > 0$ and $x_0 \geq a$. It is easy to see that

$$\limsup_{x\to+\infty} \frac{1}{x-x_0}\int_{x_0}^x \mu(f_y(t,0))dt = \alpha < 0. \tag{2.23}$$

Since $f(x,0) \equiv 0$, $\delta = \delta(\varepsilon) > 0$ exists such that $\delta \leq \varepsilon$ and, if F be defined by

$$f(x,y) = f_y(x,0)y + F(x,y), \tag{2.24}$$

such that

$$||F(x,y)|| \leq \varepsilon||y|| \quad \text{for} \quad ||y|| \leq \delta \quad \text{uniformly in} \quad x. \tag{2.25}$$

It follows from (2.23) that if x is sufficiently large, then

$$\int_{x_0}^x \mu(f_y(t,0))dt \leq \frac{\alpha}{2}(x - x_0).$$

Thus, for $\varepsilon < |\frac{1}{2}\alpha|$, we have

$$\lim_{x\uparrow+\infty} \exp\left\{\varepsilon(x - x_0) + \int_{x_0}^x \mu(f_y(t,0))dt\right\} = 0. \tag{2.26}$$

Then $K > 0$ exists such that

$$K > \sup_{x\geq x_0} \exp\left\{\varepsilon(x - x_0) + \int_{x_0}^x \mu(f_y(t,0))dt\right\}.$$

Let $\delta_1 > 0$ be such that

$$K\delta_1 < \delta.$$

We shall show that the following property holds for the solutions y of $y' = f(x,y)$: if $||y(x_0)|| < \delta_1$, then $||y(x)|| < \delta$ for each $x \geq x_0$. If this were not true, then there would be an $x_1 > x_0$ and a solution y of the given equation such that

$$||y(x_1)|| = \delta, \; ||y(x)|| \leq \delta \quad \text{for} \quad x_0 \leq x \leq x_1.$$

For every $h > 0$, we have, by virtue of (2.24),

$$\begin{aligned}||y(x) + hy'(x)|| &= ||y(x) + hf(x,y(x))|| \\ &\leq ||I + hf_y(x,0)|| \; ||y(x)|| + h||F(x,y(x))||,\end{aligned}$$

where I is the identity matrix. For x varying in $[x_0,x_1]$, it follows from (2.25) that

$$||y(x) + hy'(x)|| \leq (||I + hf_y(x,0)|| + \varepsilon h)||y(x)||$$

and so

$$||y(x) + hf(x,y(x))|| - ||y(x)|| \leq (||I + hf_y(x,0)|| - 1+\varepsilon h)||y(x)||.$$

We now divide by $h > 0$ and take the limit as $h \downarrow 0$. By virtue of the exercise at the end of this section, the limit of the quantity on the left of the inequality is $||y'(x)||$. Hence we have

$$||y'(x)|| \leq (\mu(f_y(x,0)) + \varepsilon)||y(x)|| \qquad (x_0 \leq x \leq x_1).$$

Using this inequality, we obtain

$$\begin{aligned} \frac{d}{dx}||y(x)||^2 &= 2(y(x)|y'(x)) \leq 2||y(x)||\ ||y'(x)|| \\ &\leq 2(\mu(f_y(x,0)) + \varepsilon)||y(x)||^2 . \end{aligned}$$

Now we can apply the theorem on differential inequalities (Sec. 2.4 in Chapter III) with $\omega(x,u) = 2(\mu(f_y(x,0)) + \varepsilon)u$ and obtain

$$||y(x)||^2 \leq ||y(x_0)||^2 \exp 2\left\{\varepsilon(x-x_0) + \int_{x_0}^{x} \mu(f_y(t,0))dt\right\},$$

that is,

$$||y(x)|| \leq ||y(x_0)||\exp\left\{\varepsilon(x-x_0) + \int_{x_0}^{x} \mu(f_y(t,0))dt\right\} \tag{2.27}$$

for $x_0 \leq x \leq x_1$. This result immediately leads us to a contradiction:

$$\begin{aligned} \delta = ||y(x_1)|| &= ||y(x_0)||\exp\left\{\varepsilon(x-x_0) + \int_{x_0}^{x} \mu(f_y(t,0))dt\right\} \\ &< \delta_1 K \leq \delta. \end{aligned}$$

This shows that $||y(x_0)|| < \delta_1$ implies that $||y(x)|| < \delta \leq \varepsilon$ for $x \geq x_0$ and every solution y of the given equation. $y = 0$ is therefore stable. It remains to prove that the solutions y tend to zero if $y(x_0)$ is sufficiently small. We have seen that if ε is sufficiently small, then (2.26) holds. Moreover, (2.27) holds on the whole domain of the solution y if $||y(x_0)|| < \delta_1$, where δ_1 corresponds to ε as above, for then $||y(x)|| < \delta$ for all points x in the domain by virtue of the preceding argument. Thus $\lim_{x\uparrow\infty} y(x) = 0$ if $y(x_0)$ is sufficiently small. This completes the proof of the theorem.

Exercise. Prove that if $f\colon I \to \mathbb{R}^n$, $I \subseteq \mathbb{R}^n$ an interval, has a right derivative at $t = t_0$, and this derivative has value u, then $||f(\cdot)||$ also has a right derivative at $t = t_0$ which is equal to

$$\lim_{h \downarrow 0} \frac{||f(t_0) + hu|| - ||f(t_0)||}{h}.$$

Hint: First prove that $\lim_{h\downarrow 0}(||x+hu||-||x||)/h$ exists for every $x,u \in \mathbb{R}^n$, and then show that

$$\lim_{h\downarrow 0} \frac{1}{h}\{[||f(t_0+h)|| - ||f(t_0)||] - [||f(t_0) + hu|| - ||f(t_0)||]\} = 0.$$

2.6. Invariant Sets

Let $U \subseteq \mathbb{R}^n$ be open and let $f\colon U \to \mathbb{R}^n$ satisfy a Lipschitz condition locally. Let us consider the autonomous system

$$y' = f(y). \tag{2.28}$$

In addition to the concept of orbit introduced in Sec. 2.4, we shall consider that of the *positive semi-orbit* starting at the point P. It is the set

$$I^+(P)$$

of all points of the form $y(x,P)$, where x is nonnegative and in the domain of $y(x,P)$ and $y(x,P)$ is the unique solution of (2.28) such that $y(0,P) = P$.

A *positive limit point* associated with P is any point $Q \in \mathbb{R}^n$ for which there exists a sequence $(x_n)_{n=1}^{\infty}$ tending to the least upper bound of the domain of $y(x,P)$ and such that $\lim_n y(x_n,P) = Q$. It is clear that if $Q \in I^+(P)$, then every positive limit point associated with P is also associated with Q and vice versa. We may therefore speak unambiguously of a positive limit point associated with an orbit. The *positive limit set* of P is the set

$$\Lambda^+(P)$$

of all positive limit points associated with P.

A set is *invariant* if it is a union of orbits. Thus, if P is a member of an invariant set, then so is $y(x,P)$ for any x. A set is *positive invariant* if $y(x,P)$ belongs to it whenever P does, for every $x \geq 0$.

Finally, we shall say that $y(x,P)$ *tends to the set* A and write

$y(x,P) \to A,$

if $\lim d(y(x,P),A) = 0$ as x tends to the least upper bound of the domain of $y(x,P)$.

We have the following result.

Theorem. Let $U \subseteq R^n$ be an open set and $f: U \to R^n$ satisfy a Lipschitz condition locally. Then the following results hold:

(i) $\overline{I^+(P)} = I^+(P) \cup \Lambda^+(P)$.

(ii) $\Lambda^+(P) = \bigcap_{Q \in I(P)} \overline{I^+(Q)}$.

(iii) If $\Lambda^+(x_0)$ is nonempty and bounded, then $y(x,P) \to \Lambda^+(P)$.

(iv) If $I^+(P)$ is bounded, then $\Lambda^+(P)$ is nonempty and compact.

(v) $\Lambda^+(P) \cap U$ is invariant.

Proof: We shall prove only (v), leaving the others as exercises for the reader. Let $Q \in U \cap \Lambda^+(P)$. There exists a sequence $(x_n)_{n=1}^{\infty}$ convergi[ng] to the least upper bound b of the domain of $y(x,P)$ and such that $y(x_n,P)$ converges to Q. b must necessarily be ∞ for otherwise, since r is an interior point of U, $y(x,P)$ would extend beyond b. It remains to show that for every $x \in [0,\infty[$, $y(x,P) \in \Lambda^+(P)$. To do this, fix $x \geq 0$. If n is sufficiently large, $x + x_n \geq 0$. We furthermore have the equality

$$y(x + x_n,P) = y(x,y(x_n,P)).$$

Thus, the continuous dependence on the data implies

$$\lim_n y(x + x_n,P) = y(x,Q)$$

since $\lim_n y(x_n,P) = Q$. This identity shows that $y(x,Q) = \Lambda^+(P)$, which completes the proof of the theorem.

LaSalle's Theorem. Let $U \subseteq R^n$ be an open set, let $f: U \to R^n$ satisfy a Lipschitz condition locally, and let $V: U \to R$ be a C^1-function. Let $K \subseteq U$ be a compact set such that

$$(\text{grad } V(y)|f(y)) \leq 0 \qquad (y \in K).$$

Let $E = \{y \in K|(\text{grad } V(y)|f(y)) = 0\}$, and let M be the greatest invariant subset of E. Then for every P such that $I^+(P) \subseteq K$, $y(x,P) \to M$.

Proof: We begin by proving that V is constant on $\Lambda^+(P)$. If y_0 and y_1 are two points of $\Lambda^+(P) \cap U$, then there exist two sequences $(x_n)_{n=1}^{\infty}$ and $(x'_n)_{n=1}^{\infty}$ tending to the least upper bound b of the domain of

$y(x,P)$ and furthermore satisfying

$$\lim_n y(x_n,P) = y_0, \qquad \lim_n y(x_n',P) = y_1.$$

The continuity of V then implies that

$$V(y_0) = \lim_n V(y(x_n,P)), \qquad V(y_1) = \lim_n V(y(x_n',P)).$$

But $V(y(x,P))$ is a decreasing function of x since its derivative is nonpositive. Then $\lim_{x\uparrow b} V(y(x,P))$ exists, and the uniqueness of the limit implies that $V(y_0) = V(y_1)$, which is exactly what we wanted to show. Since $V(x)$ is constant on $\Lambda^+(P)$, and since this set is invariant, the derivative of the function $V(y(x,P))$ of x must have value 0. We therefore obtain

$$(\operatorname{grad} V(y) | f(y)) = 0 \qquad \text{for} \quad y \in \Lambda^+(P).$$

Thus, $\Lambda^+(P) \subseteq M$. Since K is compact, its subset $\Lambda^+(P)$ is bounded and (iii) of the preceding theorem insures that $y(x,P) \to \Lambda^+(P)$, and so we certainly have $y(x,P) \to M$ since M is a superset of $\Lambda^+(P)$. The theorem is thus completely proven.

Corollary. Let $f\colon \mathbb{R}^n \to \mathbb{R}^n$ satisfy a Lipschitz condition locally, and let $V\colon \mathbb{R}^n \to \mathbb{R}^n$ be a C^1-function that is bounded below with $\lim_{||x||\to\infty} V(x) = \infty$ and

$$(\operatorname{grad} V(y) | f(y)) \leq 0 \qquad \text{for} \quad y \in \mathbb{R}^n.$$

Let $E = \{y \in \mathbb{R}^n | (\operatorname{grad} V(y)|f(y)) = 0\}$, and let M be the greatest invariant subset of E. Then all the solutions are bounded on $[0,\infty[$ and tend to M.

Proof: We pick $P \in \mathbb{R}^n$ and put $V(P) = a$. The set $K = \{y \in \mathbb{R}^n | V(y) \leq a\}$ is closed and bounded and therefore compact. It is also positive invariant. (See the exercise at the end of the section.) Let

$$E' = \{y \in K | (\operatorname{grad} V(y)|f(y)) = 0\}$$

and let M' be the greatest invariant subset of E'. LaSalle's theorem implies that $y(x,P) \to M'$, so $y(x,P) \to M$ since $M' \subseteq M$. The corollary is thus established.

For an example, consider the equation $my'' + hy' + ky = 0$, where m, h, and k are positive constants. The associated first order system is

$$z' = y$$
$$y' = -\frac{k}{m}z - \frac{h}{m}y.$$

Consider the auxiliary function

$$V(z,y) = \frac{1}{2}kz^2 + \frac{1}{2}my^2.$$

We have

$$(\operatorname{grad} V(u)|F(u)) = -hy^2$$

for $u = (z,y)$ and $F(z,y) = (y, -\frac{k}{m}z - \frac{h}{m}y)$. We may now apply the corollary to LaSalle's theorem; the set E reduces to the z axis and the set M to the origin. All the solutions are therefore bounded and tend to the origin as $x \to \infty$.

In conclusion, we present the following result, where the second derivative of the auxiliary function V is used.

Yorke's Theorem. Let $U \subseteq \mathbb{R}^n$ be open, let $f: U \to \mathbb{R}^n$ be a C^1-function, and let $V: U \to \mathbb{R}$ be a C^2-function. Let $K \subseteq \Omega$ be a compact invariant set (perhaps empty) such that $U \setminus K$ is simply connected. Put $\dot{V}(y) = (\operatorname{grad} V(y)|f(y))$ and $\ddot{V}(y) = (\operatorname{grad} \dot{V}(y)|f(y))$ and suppose that

$$\dot{V}(y) \neq 0 \quad \text{or} \quad \ddot{V}(y) \neq 0 \quad \text{for every} \quad y \in U \setminus K.$$

Then $\Lambda^+(P) \subseteq K$ for every $P \in U$.

We refer to the original work of Yorke [36] for the proof, which is based on the Mayer-Vietoris sequence of homology groups. We observe that if $n \neq 2$, $U = \mathbb{R}^n$, and $K = \{0\}$, the hypotheses of the theorem imply that either

$$\lim_{x \uparrow b} ||y(x,P)|| = +\infty$$

or

$$\lim_{x \uparrow b} y(x,P) = 0$$

must hold, where b is the least upper bound of the domain of $y(x,P)$.

Exercise. Consider the hypotheses of LaSalle's theorem. Prove that 1) if K is positive invariant, all the solutions starting in K tend to M as $x \to \infty$ and 2) if there is an $a \in \mathbb{R}$ such that $K = \{y \in U | V(y) \leq a\}$, then K is positive invariant.

3. SOME APPLICATIONS

The purpose of this section is to present some concrete examples both on the utility of the stability concept and on the way it is used in the experimental and applied sciences. We have chosen quite simple examples in order to focus attention on the underlying ideas.

3.1. Problems In Biology and Chemistry

In this section, we give some examples of nonlinear equations and systems arising in biology and chemistry; we pay particular attention to the stability of the equilibrium solutions. Since the independent variable here will be time, it will be indicated by t.

The simplest equation is the so-called logistic equation

$$x' = ax(b - x), \quad a > 0,\ b > 0, \quad t \geq 0,\ x \geq 0$$

where $x(t)$ represents the size of the population, a is a growth parameter, and b a parameter connected with external environmental conditions. The equation is integrable, and we get

$$x(t) = \frac{bx(0)\exp[bat]}{b + x(0)[\exp(bat) - 1]}$$

whence we see that 1) all the solutions with $x(0) > 0$ are stable (in particular the equilibrium solution $x(t) = b$) and 2) the equilibrium solution $x(t) \equiv 0$ is unstable. Note, however, that the zero solution is stable from a biological point of view, where it corresponds to the absence of the species. Also, when $b = 0$ the equation has solution $x(t) = x(0)/(1 + ax(0)t)$, which tends to 0, so in this case $x(t) \equiv 0$ is a stable equilibrium solution and the species tends to extinction.

As we shall see in the following examples, ecological systems are usually characterized by one or more parameters $\rho_1, \rho_2, \ldots$ that determine whether there is survival or extinction. In the case just examined, there is a unique parameter $\rho = b$ whose critical value is $\rho = 0$. The second example we consider involves two species and is known as the predator-prey model of Lotka and Volterra; x is the population of the prey, for which a variant of the logistic equation holds, and y is the population of the predator, whose possibilities for development are limited by the available prey. We have

$$x' = x(a - bx - cy)$$

$$y' = y(-e + fx), \quad a,b,c,e,f > 0; \quad x \geq 0,\ y \geq 0,\ t \geq 0.$$

The possible equilibrium points are $P_I = (0,0)$, $P_{II} = \left(\frac{a}{b}, 0\right)$, $P_{III} = \left(\frac{e}{f}, \frac{1}{c}(a - b\,\frac{e}{f})\right)$. P_I corresponds to the absence of both species, P_{II} to the absence of the predator, and P_{III}, provided that $\rho = af - be > 0$, to the presence of both species; P_{III} is thus the only nontrivial equilibrium point. We now examine the stability of P_{III}. The equation can be written as

$$x' = -b\bar{x}(x - \bar{x}) - c\bar{x}(y - \bar{y}) - b(x - \bar{x})^2 - c(x - \bar{x})(y - \bar{y})$$

$$y' = f\bar{y}(x - \bar{x}) + f(x - \bar{x})(y - \bar{y})$$

where $\bar{x} = \frac{e}{f}$, $\bar{y} = \frac{1}{c}\left(a - \frac{be}{f}\right)$; if we put

$$X = x - \bar{x},\ Y = y - \bar{y},\ Z = \begin{pmatrix} X \\ Y \end{pmatrix},\ A = \begin{pmatrix} -b\bar{x} & -c\bar{x} \\ f\bar{y} & 0 \end{pmatrix},$$

$$F = \begin{pmatrix} -bX^2 - cXY \\ fXY \end{pmatrix}$$

it becomes $Z' = AZ + F(Z)$, with $\lim_{|z|\to 0} \frac{F(z)}{|z|} = 0$. Therefore, for P_{III} to be stable, it is sufficient that the real parts of the eigenvalues of A be negative. The characteristic equation of A is $\xi^2 + b\bar{x}\xi + cf\bar{x}\bar{y} = 0$; a, b, c, e, and f are positive, so that if $\rho > 0$, then $\bar{x}$ and $\bar{y}$ are positive, and the roots are positive real numbers if b and $(\frac{b}{f} + 4) - 4af$ are ≥ 0; otherwise, they are complex conjugates whose real parts are negative. In each of these two cases there is a nontrivial stable equilibrium position. If $\rho < 0$, P_{II} is a stable equilibrium position, for if we proceed as before with $\bar{x} = a/b$ and $\bar{y} = 0$, we have

$$X = x - \bar{x},\quad Y = y,\quad Z = \begin{pmatrix} X \\ Y \end{pmatrix},\quad A = \begin{pmatrix} -b\bar{x} & -c\bar{x} \\ 0 & -e + f\bar{x} \end{pmatrix},$$

$$F = \begin{pmatrix} -bX^2 - cXY \\ fXY \end{pmatrix}$$

and hence the equation $Z' = AZ + F$, with $\lim_{|z|\to 0} \frac{F(z)}{|z|} = 0$. The characteristic equation of A is $\xi^2 + (b\bar{x}-f\bar{x}+e)\xi + b\bar{x}(e-f\bar{x}) = 0$, and we have $b\bar{x}(e - f\bar{x}) = \frac{a}{b}(eb - af) = -\rho\frac{a}{b} > 0$ and $b\bar{x} - f\bar{x} + e = a - \frac{\rho}{b} > 0$. Finally, $\xi_1 = \frac{\rho}{b} < 0$, $\xi_2 = -a < 0$. Real negative roots therefore exist, so the solution $X = 0$, $Y = 0$ is stable. We say that ρ is the survival-extinction parameter; its critical value is 0. If $\rho < 0$, we have the extinction of the predator species.

Exercise. What is the biological significance of the constants a, b, e, and f that appear in the formula for the parameter ρ? Why does the constant c not appear in the formula for ρ?

The third example is the quadratic model of competition between two species. Let $x(t)$ and $y(t)$ be the populations of the two competing species. The system governing the dynamics is

$$x' = x(a - bx - cy)$$

$$y' = y(e - fx - gy)$$

where $x,y,t \geq 0$; $a,b,c,e,f,g > 0$. The equilibrium points are

$$P_I = (0,0),\quad P_{II} = (0,eg^{-1}),\quad P_{III} = (ab^{-1},0),\quad P_{IV}\ \left(\frac{ag-ce}{bg-cf}, \frac{be-af}{bg-cf}\right)$$

where $bg - cf \neq 0$. (The case when $bg = cf$ is left for an exercise.) The point P_{IV} is a nontrivial equilibrium point since it involves the presence of both species; the points P_{II} and P_{IV} correspond to the presence of only one of the two species. We now examine the conditions under which P_{IV} exists and is stable.

A condition for existence is that

$$\begin{cases} ag - ce > 0 \\ be - af > 0 \end{cases} \quad \text{(this in fact also implies that } bg - cf > 0\text{)},$$

or

$$\begin{cases} ag - ce < 0 \\ be - af < 0 \end{cases} \quad \text{(which implies that } bg - cf < 0\text{)}.$$

The linear system associated with P_{IV} is

$$X' = -b\bar{x}X - c\bar{x}Y$$

$$Y' = -f\bar{y}X - g\bar{y}Y,$$

with

$$\bar{x} = \frac{ag - ce}{bg - cf},\quad \bar{y} = \frac{be - af}{bg - cf},\quad X = x - \bar{x},\quad Y = y - \bar{y}.$$

A condition for stability is that the real parts of the roots of the equation $\xi^2 + (bx + gy)\xi + (bg - cf)xy = 0$ be negative. If we put $\rho_1 = ag - ce$ and $\rho_2 = be - af$, the equilibrium point exists if $\rho_1\rho_2 > 0$, hence $b\bar{x} - g\bar{y} > 0$. The known term of the equation is positive only if $\rho_1 > 0$ and $\rho_2 > 0$ and therefore $bg - cf > 0$. We therefore have stability if $\rho_1 > 0$ and $\rho_2 > 0$ and instability if $\rho_1 < 0$ and $\rho_2 < 0$. There are no spiral orbits in either of the two cases since

$\Delta = (b\bar{x} - g\bar{y})^2 + 4cf\bar{x}\bar{y} > 0.$

We now consider P_{II}. The associated linear system is

$$X' = \left(a - \frac{ce}{g}\right)X$$

$$Y' = -\frac{fe}{g}X - eY,$$

whose characteristic equation is $\xi^2 + \left(\frac{1}{g}(ce - ag) + e\right)\xi - \frac{e}{g}(ag - ce) = 0.$ A necessary and sufficient condition for the real parts of the roots to be negative is that $\rho_1 = ag - ce < 0$. Thus P_{II} is stable if $\rho_1 < 0$. Similarly, P_{III} is stable if $\rho_2 < 0$. P_I is unstable, as is obvious, no matter what ρ_1 and ρ_2 are.

To summarize, if ρ_1 and $\rho_2 > 0$, the nontrivial equilibrium position P_{IV} is stable and P_I, P_{II}, and P_{III} are unstable. Note that this is equivalent to the fact that P_{IV} lies above the line between P_{II} and P_{III}.

If $\rho_1 > 0$ and $\rho_2 < 0$, P_{IV} does not exist, P_{III} is stable, and P_{II} is unstable. The species y becomes extinct.

If $\rho_1 < 0$ and $\rho_2 > 0$, we have the same situation as above, but with the species x becoming extinct.

Finally, if $\rho_1 < 0$ and $\rho_2 < 0$, P_{IV} is an unstable equilibrium point; P_{II} and P_{III} are both stable. In general, one of the two species becomes extinct; which one dies out depends on the initial data. There are, however, two trajectories that lead to the equilibrium point P_{IV}.

To end this section, we consider a simple example from chemical kinetics that deals with synthesis and dissociation. Let $mA+nB \rightleftarrows C$ be a chemical reaction, where m and n are natural numbers. Let $[A]$, $[B]$, and $[C]$ be the concentrations of the respective elements, which vary with time. The dynamical chemical equation is $[C]' = k[A]^m[B]^n - k'[C]$ where k and $k' > 0$ are the reaction speeds. The chemical equation translates into

$$\frac{1}{m}[A]' = \frac{1}{n}[B]' = -[C]'.$$

We put

$$\frac{1}{m}[A] + [C] = x, \quad \frac{1}{n}[B] + [C] = y, \quad [C] = z;$$

the system then becomes a system in normal form

$$x' = 0$$

$$y' = 0$$

$$z' = k[m(x - z)^m \cdot n(y - z)^n] - k'z,$$

with the conditions $x = x_0$, $y = y_0$, $0 < z < \min[x_0,y_0]$. The equilibrium positions are the solutions of the algebraic equation

$$P(z) = k[m(x_0 - z)^m n(y_0 - z)^n] - k'z = 0$$

that satisfy the aforementioned conditions. Since at $\alpha = \min[x_0,y_0]$ we have

$$P(\alpha) < 0, \quad P'(\alpha) < 0, \quad P''(\alpha) > 0,\dots,P^{2s}(\alpha) > 0, \quad P^{2s+1}(\alpha) < 0,\dots,$$

it follows that only the smallest of the roots can satisfy the condition $z_1 \le \min[x_0,y_0]$. On the other hand, $P(0) \ge 0$; thus, since $\lim_{z\to-\infty} P(z) = \infty$, it also follows that $z_1 \ge 0$.

We shall now verify that this solution is stable. The associated linear system is

$$X' = 0$$

$$Y' = 0$$

$$Z' = \{-kmn(x_0-z_1)^{m-1}(y_0-z_1)^{n-1}[m(y_0-z_1)+n(x_0-z_1)] - k'\}Z.$$

Since the coefficient of Z is negative, we have stability for the linear system. The rest of the proof is left for an exercise.

We finish by calculating the solutions explicitly for the case $m = n = 1$. We have

$$z_1 = \frac{1}{2}\left\{x_0 + y_0 + \frac{k'}{k} - \left[(x_0-y_0)^2 + \frac{k'^2}{k^2} + 2(x_0+y_0)\frac{k'}{k}\right]^{1/2}\right\}$$

$$z_2 = \frac{1}{2}\left\{x_0 + y_0 + \frac{k'}{k} + \left[(x_0-y_0)^2 + \frac{k'^2}{k^2} + 2(x_0+y_0)\frac{k'}{k}\right]^{1/2}\right\}$$

$$z(t) = \frac{(z_2 - z_0)z_1 e^{k(z_2-z_1)t} - z_2(z_1 - z_0)}{(z_2 + z_0)e^{k(z_2-z_1)t} - (z_1 - z_0)}.$$

In synthesis reactions ($k \gg k'$), it is of practical interest to know, for every $p \in \,]0,1[$, the time t_p for which $z(t_p) = pz_1$. In the preceding case we have

$$t_p = \frac{1}{k(z_2-z_1)} \ell g \left(\frac{z_2 - pz_1}{z_1(1-p)} \cdot \frac{z_1 - z_0}{z_2 - z_0}\right) .$$

If, in particular, $k' \sim 0$ and $z_0 = 0$, we have (if we suppose that $y_0 \geq x_0$)

$$z(t) \sim \frac{y_0 x_0 [e^{k(y_0-x_0)t} - 1]}{y_0 e^{k(y_0-x_0)t} - x_0}$$

$$t_p \sim \frac{1}{k(y_0-x_0)} \ell g \left[\frac{y_0 - px_0}{y_0(1-p)}\right].$$

The case when $x_0 = y_0$ is left as an exercise.

3.2. Problems in Automatic Control Theory

In this section we discuss the subject of feedback in automatic control. Roughly speaking, this means determining what must be done to control automatically an industrial process, a phase of economic systems, a nuclear reaction, the motion of a missile, etc. Although this idea has been known since antiquity, it was only in the last century that it was realized that it could be subjected to a rigorous mathematical analysis. This was first shown by the English physicist Maxwell in 1868, and then independently by the Russian engineer Vyshnegradskii around 1876. Both were concerned with small errors, that is, minor deviations from the desired result, and both approximated the real system of differential equations with a linear one. They observed that an improper use of the feedback could increase the error rather than reduce it. The rational use of feedback therefore leads to a stability analysis of the system under observation: does the error tend to zero or not?

We illustrate the situation with a concrete example. Consider the problem of displacing a solid in a viscous fluid by means of a force F that can either assist or impede the motion of the solid. The equation of motion is

$$y'' + \beta y' = F \tag{3.1}$$

where β is a positive constant that represents the coefficient of viscosity of the fluid. Suppose that we wish to control the velocity y' of the solid, that is, that we want $y' = a$ where a is a given constant. Our ability to control the movement of the body consists in our ability to choose F appropriately. Since we are interested only in the velocity, we may put $z = y'$ and get a first order equation

$$z' + \beta z = F.$$

We begin by considering what the engineers call "open circuit" control as opposed to so-called "closed circuit" control or feedback. In open circuit control, one has no information about the deviation from the desired result when one adjusts the control force F; there is no connection between the "output" z and the "input" F. Therefore, since there is no information about whether the solid is moving too fast or too slow, the best we can do is to take F constant. In this case, the solution for the velocity is

$$z(x) = z_0 e^{-\beta x} + \frac{F}{\beta}(1 - e^{-\beta x})$$

where z_0 is the initial velocity of the solid. Thus, as time $x \to \infty$, z tends asymptotically to F/β independently of the initial velocity z_0. Since we want the terminal velocity to be a, we must choose $F = \beta a$. This is the solution to our problem. It presents one disadvantage: if β is small, the velocity tends to a too slowly. Later on we shall note a further inconvenience in open circuit control.

We now compare the result with the feedback control

$$F = -\delta(z - a) + \beta a$$

where δ is a positive constant. The control force now depends linearly on the difference between the real and desired velocities; it is a matter of linear control. The differential equation of the controlled system becomes

$$z' + \beta z = -\delta(z - a) + \beta a$$

or

$$z' + (\beta + \delta)z = (\beta + \delta)a.$$

The feedback therefore has the effect of changing the coefficients of viscosity. The solution now is

$$z(x) = z_0 e^{-(\beta+\delta)x} + a(1 - e^{-(\beta+\delta)x}).$$

Again z tends to a independently of z_0, but more rapidly. Observe that if there is a small error in the control, that is, if $y' - a$ is small, the control force to be used is almost the same as before.

Although this alone demonstrates the convenience of using feedback control, a more important advantage is the major operational stability one gets with feedback. Suppose that we do not know the exact value of β or that we want a system that operates on an interval of values of β. For the sake of simplification, we suppose that the system has been designed for a coefficient of viscosity, β, but that the solid is in reality moving in a liquid with viscosity coefficient $\frac{1}{2}\beta$. (For example, the fluid is warmer.) The equation of motion for open circuit control is, under these circumstances,

$$z' + \frac{1}{2}\beta z = \beta a,$$

whereas under feedback control it is

$$z' + \frac{1}{2}\beta z = -\delta(z - a) + \beta a.$$

Under open circuit control, the velocity z tends to $2a$, so that there is an error or 100%, but under feedback control, z tends to

$$\frac{\beta + \delta}{\frac{1}{2}\beta + \delta} a,$$

so for δ sufficiently big the control is unaffected by the change of velocity. For example, if $\delta = 10\beta$, then y tends to $22a/21$, which involves an error of less than 5%. This is the other advantage of closed circuit over open circuit control; the feedback renders the system relatively insensitive to surrounding changes.

There is an additional advantage of feedback over open circuit control; in a certain sense, feedback control allows a system to adapt itself to its surroundings. We shall now modify the linear feedback control of the example just considered by means of a nonlinear feedback in such a way that the system adapts itself to changes in viscosity. One can say that the nonlinearity is essential for this type of adaptation. It is interesting to observe that we shall arrive at the desired result by using Liapunov functions, that is, by applying the results of Sec. 2.2.

We suppose that we have no information on β except that it is constant. We also allow the possibility that β may be negative, in which case the system under control is unstable. The controlled system then has the form

$$z' + \beta z = (b - \beta_0)(z - a) + ba$$

$$b' = (z - a)g(z).$$

The linear control represented by the right side of the first equation was chosen by thinking about what would happen if we knew β and then replacing β by an "adjustable" parameter b. The second differential equation determines the "adjustment" of b, and adjustment that must depend exclusively on the observation of the error $z - a$. The function g must be chosen, as must the positive constant β_0. Thus, the control does not presuppose any knowledge of the constant β. We want the velocity z to tend to a for every initial velocity z_0 and every initial value of the unknown function b. The function g, which determines the design of the mechanism to adjust b, can be chosen by using Liapunov's method. Since we want y to tend to a, and since we can expect that b tends to the unknown constant β, we may suppose that the auxiliary function

$$V(z,b) = \frac{1}{2}(z - a)^2 + \frac{1}{2}(b - \beta)^2$$

might serve our purposes. In fact, after an easy calculation, we get, for $F(z,b) = (-\beta z + (b - \beta_0)(z-a) + ba,\ g(z)(z-a))$:

$$\begin{aligned}(\operatorname{grad} V(z,b)|F(z,b)) &= (z - a)z' + (b - \beta)b' \\ &= -\beta_0(z-a)^2 + (b - \beta)(g + z(z - a)).\end{aligned}$$

We therefore take $g(z) = -z(z - a)$ and obtain

$$(\operatorname{grad} V(z,b)|F(z,b)) = -\beta_0(z - a)^2. \tag{3.2}$$

This means that $V(z(x),b(x))$ is always decreasing for every solution $z(x)$, $b(x)$ of the controlled system. Once we observe that the constant functions $z(x) \equiv a$, $b(x) \equiv \beta$ are solutions of the controlled system, we see that for every solution $z(x)$, $b(x)$, we have $\lim_{x\to\infty} z(x) = a$ and $\lim_{x\to\infty} b(x) = \beta$. The choice of β_0 is open, but (3.2) assures us that the bigger β_0 is, the faster y tends to a. Thus, independently of the value of β, this control system always reduces the error to zero, and the feedback adjusts itself to adapt to the surroundings.

4. THE METHOD OF RUNGE AND KUTTA

Although it is not within our scope to study specific numerical problems connected with differential equations, it seems interesting to give an example to show how the proboems of numerical integration are related to stability for ordinary differential equations. In this section, we introduce the well known method of Runge and Kutta, followed by a brief discussion of its applicability. We mention that in numerical analysis, the terms "stable" and "unstable" mean something completely different (related to the convergence of the approximating series); see the references at the end of this chapter.

4.1. The Fourth Order Runge-Kutta Algorithm

We shall now examine the most common Runge-Kutta method, that of the fourth order. Since we are primarily interested in the basic ideas and not the technical details, we shall only treat the case of a single equation. For other problems and general methods, see the texts on the subject, particularly those given in the bibliography.

To simplify the notation, we begin at a point $(0,y_0)$ and fix a step x for the approximations. If the equation is $y' = f(x,y)$, we put

$$\begin{cases} f_1 = f(0,y_0) & y_1 = y_0 + f_1 \cdot \frac{x}{2} \\ f_2 = f\left(\frac{x}{2}, y_1\right) & y_2 = y_0 + f_2 \cdot \frac{x}{2} \\ f_3 = f\left(\frac{x}{2}, y_2\right) & y_3 = y_0 + f_3 \cdot x \\ f_4 = f(x,y_3) & \end{cases} \tag{4.1}$$

and lastly $f_0 = \frac{1}{6}(f_1 + 2f_2 + 2f_3 + f_4)$. We next put

$$\tilde{y}(x) = y_0 + f_0 \cdot x. \tag{4.2}$$

(4.2) gives the approximation to the solution at the point x. We have the following fundamental theorem.

Theorem 4.1. Consider the equation

$$y' = f(x,y) \tag{4.3}$$

where f is of class C^3. If $\tilde{y}(x)$ is the value defined by (4.2) corresponding to the initial point $(0,y_0)$, then

$$\lim_{x\to 0} \frac{|\tilde{y}(x) - y(x)|}{|x|^4} = 0 \tag{4.4}$$

where $y(x)$ is the solution of (4.3) relative to the same initial data.

Proof: The theorem is equivalent to the claim that

$$\tilde{y}(x) = \sum_{i=0}^{4} \frac{1}{i!}\left(\frac{df}{dx_0}\right)^i (0,y_0)\cdot x^i + o(x^4). \tag{4.5}$$

We put

$$\Delta f = \left(\frac{\partial f}{\partial t}(0,y_0) + \frac{\partial f}{\partial y}(0,y_0)\cdot f(0,y_0)\right) = \left(\frac{\partial}{\partial t} + \frac{\partial}{\partial y}\cdot f\right)_{(0,y_0)} f,$$

with Δ^k having usual meaning; for example

$$\Delta^2 f = \left(\frac{\partial^2 f}{\partial t^2}(0,y_0) + 2\frac{\partial^2 f}{\partial t\partial y}(0,y_0)f(0,y_0) + \frac{\partial^2 f}{\partial y^2}(0,y_0)f^2(0,y_0)\right).$$

We furthermore put $f_y = \frac{\partial f}{\partial y}(0,y_0)$. Then the coefficients of (4.5) are given by

$$\left\{\begin{aligned} f(0,y_0) &= \Delta^0 f \\ \frac{df}{dx}(0,y_0) &= \Delta f \\ \frac{d^2f}{dx^2}(0,y_0) &= \Delta^2 f + f_y\Delta f \\ \frac{d^3f}{dx^3}(0,y_0) &= \Delta^3 f + f_y\Delta^2 f + 3\Delta f_y\cdot\Delta f + f_y^2\Delta f. \end{aligned}\right. \tag{4.6}$$

We next proceed to the calculation of the f_i' that appear in (4.1). We have

$$\begin{aligned} f_1 &= \Delta^0 f \\ f_2 &= \Delta^0 f + \Delta^1 f\cdot\frac{x}{2} + \frac{1}{2}\Delta^2 f\,\frac{x^2}{4} + \frac{1}{6}\Delta^3 f\,\frac{x^3}{8} + o(x^3) \\ f_3 &= \Delta^0 f + \Delta f\cdot\frac{x}{2} + \left(f_y\Delta f + \frac{1}{2}\Delta^2 f\right)\frac{x^2}{4} \\ &\quad + \left(\frac{1}{2}f_y\Delta^2 f + \Delta f_y\Delta f + \frac{1}{6}\Delta^3 f\right)\frac{x^3}{8} + o(x^3) \\ f_4 &= \Delta^0 f + \Delta f\cdot x + \frac{1}{2}(\Delta^2 f + f_y\Delta f)x^2 \\ &\quad + \left(\frac{1}{6}\Delta^3 f + \frac{1}{8}f_y\Delta^2 f + \frac{1}{2}\Delta f_y\Delta f + \frac{1}{4}f_y^2\Delta f\right)x^3 + o(x^3). \end{aligned}$$

We get (4.5) after taking the weighted mean (4.2) and comparing it with (4.6).

Note that the Cavalieri-Simpson integration formula

$$\int_0^x f(t)\,dt = \frac{x}{6}\left[f(0) + 4f\left(\frac{x}{2}\right) + f(x)\right] + o(x^4). \tag{4.7}$$

follows as a special case of Theorem 4.1 when f is C^3.

We have the following theorem on the convergence of the solutions.

Theorem 4.2. Given Eq. (4.3), let $f(x,y)$ be of class C^3 in a rectangle $R = \{(x,y) \mid |x - x_0| \le a, |y - y_0| \le b\}$ of R^2. Let $M = \max_R |f(x,y)|$. Let $y_k(x)$ be the solution obtained by finding $\tilde{y}$ using the Runge-Kutta method with step $1/k$ relative to the initial datum (x_0,y_0), and extending it linearly between the points at which it is calculated by (4.2). Let (x_0,y_0) be the center of the rectangle R; if $\delta = \min\{a,b/M\}$, then the sequence $y_k(x)$ converges uniformly in $]x_0-\delta,x_0+\delta[$ to $y(x)$, the solution of (4.3) with the same initial datum.

The technique of the proof is similar to that given in Section 2.1 of Chapter III and is left as an exercise for the reader. (Note that the solution is unique.)

4.2. Practical Use of the Runge-Kutta Method

From a theoretical point of view, the Runge-Kutta method can be applied whenever the hypothesis of Theorem 4.2 are satisfied. But in practice it is necessary to be able to give an estimate of the errors, in order to choose a suitable step for the algorithm. First of all, let us see what happens with the error in every single step. In order to grasp easily the idea, we start with a concrete example of numerical integration, Simpson's method. Consider

$$\int_0^2 x^4 dx = \frac{32}{5} .$$

Using Simpson's rule with step 2, we get $I_1 = 20/3$, with error $E_1 = 4/15$; using Simpson's rule twice with step 1, we get $I_2 = 77/12$, with error $E_2 = 1/60 = \frac{1}{16} E_1$. This relation between E_1 and E_2 is not incidental, because, in view of our method, the error is

$$E(h) = \frac{1}{90} f^N(x_0 + \nu h) (\frac{h}{2})^5 ,$$

where $0 < \nu < 1$. If the derivative is constant, we obtain

$$E(h/2) = \frac{1}{32} E(h).$$

Since we must perform two iterations, the total error with step one half, happens to be

$$E_2 = \frac{1}{16} E_1.$$

Under this assumption (constant fourth derivative), from the knowledge of I_1 and I_2 we know $E_1 - E_2$, hence

$$15E_2 = E_1 - E_2, \quad E_2 = \frac{1}{15}(E_1 - E_2).$$

In the general case the fourth derivative is not constant, thus the estimate

$$E_2 = \frac{1}{15}(E_1 - E_2) \tag{4.8}$$

is only a conjecture. But statistically it proves to be a good, reasonable estimate.

A similar argument holds for differential equations when we solve them using the Runge-Kutta method. The estimate (4.8) is used currently, in order to determine the length of the step. Unfortunately in this case errors of single steps are not cumulated, but they lead to an error on the initial value of the next step. Clearly if the equation is asymptotically stable in the sense described in the second section of this chapter, then the effect of this error tends to vanish, while, on the contrary, if there is instability, the errors grow. Also, statistically, the estimate (4.8) on the single step error tends to be an overestimate for stable equations, while it tends to be an underestimate for unstable equations.

Of course, the explosive behavior of the error can be avoided by choosing steps of short length (recall that there is no reason of keeping constant steps). Our previous remark, furthermore, suggests to require smaller errors in the first steps, especially for unstable equations. One more question arises: why *fourth order* method? In principle we can give methods for any order, but all even degrees allow us to reduce automatically the number of estimates by one.

Any problem has its optimal even degree; in the average, fourth order method has the advantage of not requiring long computer programs, and of being flexible enough to deal with most equations commonly used.

Exercise 1. Write up a fourth order Runge-Kutta algorithm for your computer.

Exercise 2. Compare the results obtained for the equations

$$y' = y, \quad y' = -y$$

starting from $y(0) = 1$, using the Runge-Kutta method on the interval $[0,1]$. Compare with the exact results and check (4.8) on the single steps with the actual errors.

Exercise 3. What is the Runge-Kutta method of order 2?

Exercise 4. Determine the Runge-Kutta algorithm of sixth order.

5. BIBLIOGRAPHICAL NOTES

For further results on Liapunov functions, see Burton [5], Chow and Yorke [7], Yorke [36], [37], Lakshmikantham and Leela [20] and Rouche, Habets and Laloy [41]. [20] contains a great number of "inverse theorems," that is, results on the relation between a given type of stability and the existence of a given type of Liapunov function. The type of Liapunov function introduced in the text is new and has been treated by Vidossich [35].

For asymptotic equivalence and the asymptotic behavior of solutions, see Brauer [4], Cesari [6] and Coppel [10]. In this kind of research, frequent use has been made of Alekseev's formula on the change of variables for ordinary equations (Exercise 5 of Sec. 3.2, Chapter I); see Brauer [4] and his bibliography.

For developments regarding Olech's method, cf. Olech [24], [25] and Hartman and Olech [17].

For invariant sets, see LaSalle [21] and his bibliography.

For the stability of equations of order n, see Reissig, Sansone, and Conti [29] and Yoshizawa [38].

For ad hoc methods to approach the study of particular equations, see Bernfeld and Lakshmikantham [3], Halanay [15], Lefschetz [22], Bernfeld and Salvadori [39], Bernfeld, Nagrini and Salvadori [40].

For the most recent points of view, see Artstein [1], whose work is in the same spirit of the theory of G-convergence treated in Chapter III, Marocco [23], where the topological degree is used, and, finally, Coppel [11].

For application of stability, see Gavalas [12], Hahn [14], Hale and LaSalle [16], and Popov [28]. In particular, the contents of Sec. 3.2 have been taken from Hale and LaSalle [16].

For a general introduction to numerical analysis, see Ortega [26], while for a more complete treatment see Collatz [8], John [18], Ortega and Rheinboldt [27], and Todd [34]. For the numerical approximation of the solutions to ordinary differential equations, see Babuska, Prager, and Vitásek [2], Collatz [9], Gear [13], Shampine and Gordon [30], Stetter [31], Stroud [33], and the works cited in their bibliographies.

[1] Z. Artstein, Uniform asymptotic stability via the limiting equation, *J. Diff. Eq., 27*(1978), 172-189.

[2] I. Babuska, M. Prager and V. Vitasek, *Numerical Processes in Differential Equations,* Interscience, New York, 1966.

[3] S. Bernfeld and V. Lakshmikantham, Differential inequalities and the Okamura function, *Ann. Mat. Pura Appl., 96*(1973), 89-105.

[4] F. Brauer, *Some Stability and Perturbation Problems for Differential and Integral Equations,* IMPA, Rio de Janeiro, 1976.

[5] T. A. Burton, Differential inequalities for Liapunov functions, *Nonlinear Anal., Theory, Methods, Appl., 1*(1977), 331-338.

[6] L. Cesari, *Asymptotic Behavior and Stability Problems in Ordinary Differential Equations,* Springer-Verlag, Berlin, 1963.

[7] S. N. Chow and J. A. Yorke, Lyapunov Theory and perturbation of stable and asymptotically stable systems, *J. Diff. Eq., 15*(1974), 308-321.

[8] L. Collatz, *Functional Analysis and Numerical Mathematics,* Academic Press, New York, 1966.

[9] L. Collatz, *The Numerical Treatment of Differential Equations,* Springer-Verlag, Berlin, 1966.

[10] W. A. Coppel, *Stability and Asymptotic Behavior of Differential Equations,* Heath, Boston, 1965.

[11] W. A. Coppel, *Dichotomies in Stability Theory,* Springer-Verlag, Lecture Notes in Math., vol. 629, 1978.

[12] G. R. Gavalas, *Nonlinear Differential Equations of Chemically Reacting Systems,* Springer-Verlag, Berlin, 1968.

[13] C. W. Gear, *Numerical Initial Value Problems in Ordinary Differential Equations,* Prentice-Hall, Englewood Cliffs, New Jersey, 1971.

[14] W. Hahn, *Stability of Motion,* Springer-Verlag, Berlin, 1967.

[15] A. Halanay, *Differential Equations: Stability, Oscillations, Time Lag,* Academic Press, New York, 1966.

[16] J. K. Hale and J. P. LaSalle, Differential equations: linearity versus nonlinearity, *SIAM Rev., 5*(1963), 249-272.

[17] P. Hartman and C. Olech, On global asymptotic stability of solutions of differential equations, *Trans. Amer. Math. Soc., 104*(1962), 154-178.

[18] F. John, *Lectures on Advanced Numerical Analysis,* Nelson, London, 1966.

[19] V. Jurdjevic, and J. P. Quinn, Controllability and Stability, *J. Diff. Eq., 28*(1978), 281-289.

[20] V. Lakshmikantham and S. Leela, *Differential and Integral Inequalities, vol. I,* Academic Press, New York, 1969.

[21] J. P. LaSalle, *The Stability of Dynamical Systems,* SIAM, Philadelphia, 1976.

[22] S. Lefschetz, *Differential Equations: Geometric Theory,* Interscience, New York, 1957.

[23] P. Marocco, A study of asymptotic behavior and stability of the solutions of Volterra integral equations using the topological degree, *J. Diff. Eq. 43*(1982), 235-248.

[24] C. Olech, On the global stability of an autonomous system on the plane, *Contrib. Diff. Eq., vol. I,* (1963), 389-400.

[25] C. Olech, Global phase-portrait of a plane autonomous system, *Ann. Inst. Fourier 14, 1*(1964), 87-98.

[26] J. M. Ortega, *Numerical Analysis: A Second Course,* Academic Press, New York, 1972.

[27] J. M. Ortega and W. C. Rheinboldt, *Iterative Solution of Nonlinear Equations in Several Variables,* Academic Press, New York, 1970.

[28] V. M. Popov, *Hyperstability of Control Systems,* Springer-Verlag, Berlin, 1973.

[29] R. Reissig, G. Sansone, and R. Conti, *Non-linear Differential Equations of Higher Order,* Nordhoff, Leyden, 1974.

[30] L. F. Shampine and M. K. Gordon, *Computer Solution of Ordinary Differential Equations (The Initial Value Problem),* Freeman, San Francisco, 1975.

[31] H. J. Stetter, *Analysis of Discretization Methods for Ordinary Differential Equations,* Springer-Verlag, Berlin, 1973.

[32] J. J. Stoker, On the stability of mechanical systems, *Comm. Pure Appl. Math., 8*(1955), 133-142.

[33] A. H. Stroud, *Numerical Quadrature and Solution of Ordinary Differential Equations,* Springer-Verlag, Berlin, 1974.

[34] J. Todd, *Survey of Numerical Analysis,* McGraw-Hill, New York, 1962.

[35] G. Vidossich, Two remarks on the stability of ordinary differential equations, *Nonlinear Anal., TMA,* 4(1980), 967-974.

[36] J. A. Yorke, A theorem on Liapunov functions using $\ddot{V}$, *Math. System T., 4*(1969), 40-45.

[37] J. A. Yorke, Differential inequalities and non-Lipschitz scalar functions, *Math. System T., 4*(1969), 140-153.

[38] T. Yoshizawa, *Stability Theory and the Existence of Periodic Solutions and Almost Periodic Solutions,* Springer-Verlag, Berlin, New York, 1975.

[39] S. R. Bernfeld and L. Salvadori, Generalized Hopf bifurcation and h-asymptotic stability, *Nonlin. Anal. TMA, 4*(1980), 1091-1107.

[40] S. R. Bernfeld, P. Nagrini, and L. Salvadori, Quasi-invariant manifolds, stability and generalized Hopf bufurcation, *Ann. Mat. Pure Appl. 130*(1982), 105-119.

[41] N. Rouche, P. Habets and M. Laloy, *Stability Theory by Liapunov's direct method,* Springer-Verlag, New York 1977.

Index

Adjoint equation, 100-101
A priori bounds, 146
Alekseev's formula, 38
Altman's theorem, 216
Analytic function of an operator, 103-107
Analyticity of solutions, 25-29
Arzèla-Ascoli theorem, 55, 133-138, 142, 152, 158, 165, 185, 196, 220, 223, 259, 327
Arzèla's theorem, 138
Ascoli's theorem, 133-138
Asymptotic equivalence, 346
Asymptotic stability, 333-346

Banach algebra, 88
Banach-Caccioppoli theorem, 45
Banach space, 47-51
Bolzano-Weierstrass theorem, 44, 131, 136
Borsuk's theorem, 211
Boundary point, 42
Brauer's theorem, 358
Brouwer's theorem, 213-217
Brush of Peano, 161

Cafiero's theorem, 179-182
Canonical form, 88-92
Cantor diagonalization, 139
Cauchy problem, 4, 19
Cauchy sequence, 45
Characteristic equation, 82

Closed set, 42
Closure, 42
Compactness, 42, 135
  G-, 196
Comparison theorem, 150
Complete space, 45
Connected set, 42
Continuity, 43
Continuous dependence, 12, 16, 29-40, 171, 172, 197
Continuum, 163
Contraction, 45, 53
Control theory, 370-373
Convergence
  in norm, 94
  G-, 125-131, 187-189, 198-202
  strong, 94
Convex set, 49
Critical point, 356